Analysis and Synthesis for Networked Multi-Rate Systems

This book presents novel state estimation methods for several classes of networked multi-rate systems including state estimation methods for networked multi-rate systems with various complex networked-induced phenomena and communication protocols. The systems investigated include stochastic nonlinear systems, time-delay systems, linear repetitive processes, and artificial neural networks. The techniques used are mainly the Lyapunov stability theory, the optimal estimation theory, the lifting technique, and certain convex optimization methods.

Features

- Gives a systematic investigation of the state estimation of multi-rate systems.
- Discusses results on state estimation problems under network-induced complexities.
- Studies different kinds of multi-rate systems including multi-rate nonlinear systems, multi-rate neural networks, and multi-rate linear repetitive processes.
- Explores network-enhanced complexities and communication protocols.
- Includes case studies showing the applicability of developed estimation algorithms including practical examples like DC servo systems and continuous stirred tank reactor systems.

Analysis and Synthesis for Networked Multi-Rate Systems is aimed at graduate students and researchers in signal processing, control systems, and electrical engineering.

Analysis and Synthesis for Networked Multi-Rate Systems

Yuxuan Shen, Zidong Wang and Hongli Dong

CRC Press
Taylor & Francis Group
Boca Raton London New York

CRC Press is an imprint of the
Taylor & Francis Group, an **informa** business

First edition published 2024
by CRC Press
2385 NW Executive Center Drive, Suite 320, Boca Raton FL 33431

and by CRC Press
4 Park Square, Milton Park, Abingdon, Oxon, OX14 4RN

CRC Press is an imprint of Taylor & Francis Group, LLC

ISBN: 978-1-032-55562-1 (hbk)
ISBN: 978-1-032-61949-1 (pbk)
ISBN: 978-1-032-61950-7 (ebk)

DOI: 10.1201/9781032619507

Typeset in CMR10
by KnowledgeWorks Global Ltd.

Publisher's note: This book has been prepared from camera-ready copy provided by the authors.

This book is dedicated to the Dream Dynasty consisting of a group of bright people who are devoted to filtering the noise from their lives ...

Contents

Preface

During the past two decades, networked systems have been found to have extensive applications in many areas. As such, it is of great significance to study the state estimation problems for networked systems to help us monitor the system state and improve the reliability of networked systems. In practical networked systems, the multi-rate sampling strategy is often applied because of the diverse physical features of the system plant and multiple sensors. In fact, using different rates to sample various signals for distinct purposes has clear engineering insight. On the other hand, the communication networks in networked systems are equipped with communication protocols whose purpose is to prevent data collisions during the information transmission. Both the multi-rate sampling and the communication protocols pose great impacts on the state estimation problem of networked systems and invalidate the traditional estimation methods. Consequently, it is urgent to develop novel state estimation methodologies that guarantee the desired state estimation performance with the existence of multi-rate sampling and communication protocols.

In this book, we present the novel state estimation methods for several classes of networked multi-rate systems. The content of this book can be divided into two parts. In the first part, we show the state estimation methods for networked multi-rate systems with various complex network-induced phenomena. The state estimation methods for networked multi-rate systems under the communication protocols are presented in the second part. The systems we investigated include stochastic non-linear systems, time-delay systems, linear repetitive processes, and artificial neural networks. The complex network-induced phenomena considered are time-delays, integral measurements, sensor resolutions, and dynamical bias. The communication protocols included are the Round-Robin protocol, the weighted Try-Once-Discard protocol, the p-persistent carrier sense multiple access protocol, and the dynamic event-triggered protocol. The techniques used are mainly the Lyapunov stability theory, the optimal estimation theory, the lifting technique, and certain convex optimization methods. This book provides valuable reference materials for researchers who wish to explore the state estimation for multi-rate systems.

The concise frame and description of this book are given as follows. Chapter 1 provides the recent advances on the state estimation problems for multi-rate systems. Chapter 2 studies the non-fragile H_∞ filtering problem for discrete multi-rate time-delayed systems over sensor networks characterized by Gilbert-Elliott models. Chapter 3 is concerned with the H_∞ filtering problem for multi-rate artificial neural networks with integral measurements. Chapter 4 investigates the recursive state estimation problem for multi-rate time-varying systems with multiplicative noises subject to sensor resolutions. Chapter 5 deals with the minimum-variance joint state and fault estimation problem for multi-rate time-varying systems with dynamic bias. Chapter 6 discusses the H_∞ filtering problem for multi-rate multi-sensor systems with randomly occurring sensor saturations under the p-persistent carrier sense multiple access protocol. Chapter 7 considers the l_2-l_∞ state estimation problem for delayed artificial neural networks under high-rate communication channels with Round-Robin protocol. Chapter 8 addresses the recursive state estimation problem for multi-rate multi-sensor systems with distributed time-delays under the Round-Robin protocol. Chapter 9 copes with the fusion estimation problem for multi-rate linear repetitive processes under weighted Try-Once-Discard protocol. In Chapter 10, the outlier-resistant recursive filtering problem is studied for multi-sensor multi-rate systems under the weighted Try-Once-Discard protocol, and the exponential boundedness of the filtering error dynamics is analyzed in the mean square sense. In Chapter 11, the dynamic event-based recursive filtering problem is studied for multi-rate systems with integral measurements over sensor networks. Chapter 12 gives the conclusion and some possible future research directions. This book is a research volume whose intended audience is graduate and postgraduate students as well as researchers.

Yuxuan Shen
Daqing, China

Zidong Wang
London, UK

Hongli Dong
Daqing, China

Authors

Yuxuan Shen received a PhD in control science and engineering from the Donghua University, Shanghai, China in 2020.

In 2018, he was a research assistant at the Texas A&M University at Qatar, Doha, Qatar. From 2018–2019, he was a visiting scholar with the Department of Computer Science, Brunel University London, London, UK. From 2020–2021, he worked as a lecturer in the Artificial Intelligence Energy Research Institute, Northeast Petroleum University, Daqing, China where he was promoted to associate professor in 2021. Dr. Shen has published over 30 papers in refereed international journals. His research interests include optimal filtering, multi-rate systems, communication protocols, as well as fault detection. He is a very active reviewer for several international journals. He was an outstanding reviewer for *Neurocomputing* in 2017 and for *Asian Journal of Control* in 2020.

Zidong Wang is professor of dynamical systems and computing at Brunel University London, London, UK. He received a BSc in mathematics in 1986 from Suzhou University, Suzhou, an MSc in applied mathematics in 1990, and a PhD in electrical and computer engineering in 1994, both from Nanjing University of Science and Technology, Nanjing.

He was appointed lecturer in 1990 and associate professor in 1994 at Nanjing University of Science and Technology. From 1997–1998, he was an Alexander von Humboldt Research Fellow with the Control Engineering Laboratory, Ruhr-University Bochum, Germany. From 1999–2001, he was a lecturer with the Department of Mathematics, University of Kaiserslautern, Germany. From 2001–2002, he was a University Senior Research Fellow with the School of Mathematical and Information Sciences, Coventry University, UK. In 2002, he joined the Department of Computer Science, Brunel University London, UK, as a lecturer, and was then promoted to reader in 2003 and to chair professor in 2007.

Prof. Wang's research interests include dynamical systems, signal processing, bioinformatics, control theory and applications. He has published more than 600 papers in refereed international journals. He was awarded the Humboldt Research Fellowship in 1996 from Alexander von Humboldt Foundation, the JSPS Research Fellowship in 1998 from Japan Society for the Promotion of Science, and the William Mong Visiting Research Fellowship in 2002 from the University of Hong Kong. He was a recipient of the State Natural Science Award from the State Council of China in 2014 and the

Outstanding Science and Technology Development Awards (in 2005 and twice in 1997) from the National Education Committee of China.

Prof. Wang is currently serving or has served as editor-in-chief for *International Journal of Systems Science* and *Neurocomputing*, executive editor for *Systems Science and Control Engineering*, subject editor for *Journal of The Franklin Institute*, associate editor for *IEEE Transactions on Automatic Control*, *IEEE Transactions on Control Systems Technology*, *IEEE Transactions on Systems, Man, and Cybernetics–Systems*, *Asian Journal of Control*, *Science China Information Sciences*, *IEEE/CAA Journal of Automatica Sinica*, *Control Theory and Technology*, action editor for *Neural Networks*, editorial board member for *Information Fusion*, *IET Control Theory & Applications*, *Complexity*, *International Journal of Systems Science*, *Neurocomputing*, *International Journal of General Systems*, *Studies in Autonomic, Data-Driven and Industrial Computing*, and Conference Editorial Board member for the IEEE Control Systems Society. He served as associate editor for *IEEE Transactions on Neural Networks*, *IEEE Transactions on Systems, Man, and Cybernetics–Part C*, *IEEE Transactions on Signal Processing*, *Circuits, Systems & Signal Processing*, and editorial board member for *International Journal of Computer Mathematics*.

Prof. Wang is a member of the Academia Europaea (section of Physics and Engineering Sciences), fellow of the IEEE (for contributions to networked control and complex networks), fellow of the Chinese Association of Automation, member of the IEEE Press Editorial Board, member of the EPSRC Peer Review College of the UK, fellow of the Royal Statistical Society, member of the program committee for many international conferences, and a very active reviewer for many international journals. He was nominated as an appreciated reviewer for *IEEE Transactions on Signal Processing* in 2006–2008 and 2011 and *IEEE Transactions on Intelligent Transportation Systems* in 2008, outstanding reviewer for *IEEE Transactions on Automatic Control* in 2004 and *Automatica* in 2000.

Hongli Dong received a PhD in control science and engineering from the Harbin Institute of Technology, Harbin, China in 2012.

She was a research assistant with the Department of Applied Mathematics, City University of Hong Kong from 2009–2010 and with the Department of Mechanical Engineering, University of Hong Kong from 2010–2011. From 2011–2012, she was a visiting scholar with the Department of Information Systems and Computing, Brunel University London, UK. From 2012–2014, she was an Alexander von Humboldt Research Fellow with the University of Duisburg-Essen, Duisburg, Germany. She is currently a professor with the Artificial Intelligence Energy Research Institute, Northeast Petroleum University, Daqing, China, where she is also the director of the Heilongjiang Provincial Key Laboratory of Networking and Intelligent Control. Her current research interests include robust control and networked control systems.

Dr. Dong is a very active reviewer for many international journals.

Acknowledgments

The authors would like to express their deep appreciation to those who have been directly involved in various aspects of the research leading to this book. Special thanks go to Prof. Bo Shen from Donghua University, Shanghai, China, Prof. Xiaohui Liu from Brunel University London, London, UK, Prof. Tingwen Huang from Texas A&M University at Qatar, Doha, Qatar, Prof. Fuad E. Alsaadi from King Abdulaziz University, Jeddah, Saudi Arabia, Prof. Fawaz E. Alsaadi from King Abdulaziz University, Jeddah, Saudi Arabia, Prof. Abdullah M. Dobaie from King Abdulaziz University, Jeddah, Saudi Arabia, Prof. Qing-Long Han from Swinburne University of Technology, Melbourne, Australia, Prof. Hongjian Liu from Anhui Polytechnic University, Wuhu, China. Last, but not least, the authors are especially grateful to their families for their encouragement and never-ending support when it was most required.

The writing of this book was supported in part by the National Natural Science Foundation of China under Grants 61933007, U21A2019, and 62103095, the Natural Science Foundation of Heilongjiang Province of China under Grant LH2021F005, the Heilongjiang Postdoctoral Sustentation Fund of China under Grant LBH-Z20119, the Hainan Province Science and Technology Special Fund of China under Grant ZDYF2022SHFZ105, the Royal Society of the UK, and the Alexander von Humboldt Foundation of Germany. The support of these organizations is gratefully acknowledged.

Symbols

$\mathbb{R}^n$	The n-dimensional Euclidean space
$\mathbb{R}^{n\times m}$	The set of $n \times m$ real matrices
$\mathbb{Z}^-$	The set of non-positive integers
$\mathbb{N}$	The set of non-negative integers
$\mathbb{N}^+$	The set of positive integers
$\\|X\\|$	The norm of matrix X defined by $\\|X\\| = \sqrt{\text{trace}(X^T X)}$
X^T	The transpose of the matrix X
X^{-1}	The inverse of the matrix X
$\text{tr}(X)$	The trace of the matrix X
I	An identity matrix with compatible dimension
0	A zero matrix with compatible dimension
$\mathbb{E}\{a\}$	The expectation of the stochastic variable a
$\delta(x)$	A binary function that equals 1 if $x = 0$ and equals 0 otherwise
$\lfloor\cdot\rfloor$	The floor function
$\lceil\cdot\rceil$	The ceiling function
$\text{mod}(x, y)$	The unique non-negative remainder on division of x by y
$\text{diag}\{\ldots\}$	The block-diagonal matrix
$\text{col}\{\ldots\}$	A column vector composed of elements
$\text{col}_n\{X\}$	$\text{col}\{X, \ldots, X\}$ with n blocks
$\text{diag}_n\{X\}$	$\text{diag}\{X, \ldots, X\}$ with n blocks
$X > Y$	$X - Y$ is positive definite where X and Y are real symmetric matrices

$X \geq Y$	$X - Y$ is positive semi-definite where X and Y are real symmetric matrices
$l_2([0,\infty);\mathbb{R}^n)$	The space of square summable n-dimensional vector-valued functions
$\mathrm{col}_{j=m}^{n}\{A(j)\}$	$\mathrm{col}\{A(m), A(m+1), \cdots, A(n)\}$
$\mathrm{diag}_{j=m}^{n}\{A(j)\}$	$\mathrm{diag}\{A(m), A(m+1), \cdots, A(n)\}$
$\sum_{i,j}$	$\sum_{i=1}^{m}\sum_{j=0}^{q_i-1}$
$\sum_{ij,ln}$	$\sum_{i=1}^{m}\sum_{l=1}^{m}\sum_{j=0}^{q_i-1}\sum_{n=0}^{q_l-1}$
$\sum_{ij\neq ln}$	$\sum_{i=1}^{m}\sum_{l=1}^{m}\sum_{j=0}^{q_i-1}\sum_{n=0}^{q_l-1}$ (excluding the case for $i = l$ and $j = n$)
$\mathrm{Prob}(\cdot)$	The occurrence probability of the event "·"

1

Introduction

With the rapid development of wireless communication technology and digital technology, networked systems have been widely applied in practical applications such as process monitoring, power grids, industrial control systems, and traffic systems [31, 46, 65, 111, 139, 152, 174]. In networked systems, the signals generated by system components, including the sensors, underlying plant, and controllers, are first sampled and then transmitted through communication networks. Compared to traditional systems, networked systems provide the advantages of low cost, easy maintenance, and high reliability. Nevertheless, the introduction of the communication network largely increases the complexity of networked systems and hence brings major challenges to the state estimation of networked systems. Consequently, the problem of state estimation for networked systems has been attracting ever-increasing research interest [9, 20, 56, 102, 195, 196].

Due to its large scale, it is often the case that the networked system contains numerous spatially distributed system components (including multiple sensors). As mentioned before, continuous signals from system components are first sampled into discrete signals with fixed sampling periods before being transmitted. Traditionally, to simplify the design of the state estimation algorithms, it is assumed that a unified sampling period is chosen for different signals [27, 92, 124, 125, 151]. Unfortunately, such an assumption is not realistic in real practice. Since different system components own different physical characteristics, it is quite difficult (if not impossible) to unify the sampling periods for different components. Furthermore, setting different sampling periods for different system components (i.e., multi-rate sampling) according to the importance of their signals is preferable in engineering practice [7, 116, 131, 132]. As such, the state estimation problems for networked multi-rate systems (MRSs), which reflect the real situation in practical engineering, have stirred the attentions from researchers.

In the state estimation problems for MRSs, due to the existence of the multiple time sequences, it is quite difficult to directly design state estimators for the considered MRSs. In other words, the state estimation methods developed for single-rate systems cannot be directly applied to MRSs. Therefore, much research effort has been devoted to solving state estimation problems for MRSs [2, 23, 26, 95, 127, 130, 138, 140, 150, 199]. In literature, a widely accepted way to deal with the state estimation problems for MRSs is that we first transmit the MRSs into equivalent single-rate systems, and then design state

DOI: 10.1201/9781032619507-1

estimators for the equivalent single-rate systems by using the renowned state estimation methods such as the H_∞ estimation method, the Kalman filtering method, the set-membership filtering method, etc. Once the estimate of the state of the transmitted single-rate systems is obtained, the estimate of the state of the MRSs can be easily obtained by some simple matrix operations.

On the other hand, due to the limited communication resources of the communication networks, some network-induced phenomena arise inevitably in networked systems. Such network-induced challenges include, but are not limited to, the missing measurements, the communication delays, the channel fading, and the packet disorders. If not properly handled, these network-induced phenomena would deteriorate the estimation performance in the state estimation of networked systems. Therefore, it is of great importance to find effective solutions to handle the above mentioned network-induced problems. Up to now, plenty of results have been obtained for state estimation problems for networked systems under network-induced phenomena, see [66, 110, 133, 169] for missing measurements, see [19, 82, 143, 173] for communication delays, see [51, 54, 100, 186] for channel fading, and see [25, 28, 97] for packet disorders.

Generally speaking, in the existing literature, there are mainly two ways to deal with the network-induced challenges in the state estimation problems. The first one is to passively design state estimation algorithms that are robust to the network-induced phenomena. That is, with the existence of the network-induced phenomena, designing state estimation algorithms such that the desired performance requirement is achieved; see for examples [18, 73, 87, 142]. The other way is to actively introduce communication protocols into the communication networks to alleviate/aviod the network-induced phenomena, and then develop state estimation algorithms whilst considering the influence of the introduced communication protocols. Some of the widely used communication protocols are the static event-triggered protocol [148, 167, 168, 208], the dynamic event-triggered protocol [11, 119, 190, 203], the self-triggered protocol [175, 176], the Round-Robin (RR) protocol [210], the weighted Try-Once-Discard (WTOD) protocol [74, 202], and the random access protocol [35, 80, 201, 209].

In this chapter, we aim to provide a systematic review of the existing results on the state estimation problems for networked MRSs. In Section 1.1, a general state-space model of the MRSs is presented and various methods that transform the MRSs into single-rate systems are surveyed. In Section 1.2, research results are discussed on the state estimation problems for MRSs where the methods in Section 1.1 are used to transmit the MRSs. Section 1.3 gives the whole structure of this book.

1.1 Multi-Rate Systems

In this section, we first introduce a general state-space model of the discrete-time MRSs. Then, some effective methods that transform the MRSs into single-rate systems are presented. Moreover, the corresponding estimators are designed for the transformed single-rate systems.

1.1.1 State-Space Model of the Multi-Rate Systems

Consider a discrete-time multi-sensor MRS described as follows:

$$\begin{cases} x(s_{k+1}) = A(s_k)x(s_k) + B(s_k)w(s_k), \\ y_i(t_k^i) = C_i(t_k^i)x(t_k^i) + D_i(t_k^i)v_i(t_k^i) \end{cases} \tag{1.1}$$

where $x(s_k) \in \mathbb{R}^{n_x}$ and $y_i(t_k^i) \in \mathbb{R}^{n_y}$ ($i \in [1, N]$) are the system state and the measurement output from the i-th sensor, respectively. $w(s_k)$ and $v_i(t_k^i)$ are the process noise and the measurement noise on the i-th sensor, respectively. $A(s_k)$, $B(s_k)$, $C_i(t_k^i)$, and $D_i(t_k^i)$ are known matrices with compatible dimensions.

The state update period of the system (1.1) is $h \triangleq s_{k+1} - s_k$, and the sampling period of the i-th sensor is $b_i h \triangleq t_{k+1}^i - t_k^i$ where $b_i \geq 1$ ($i \in [1, N]$) are allowed to be different. That is, in the MRS (1.1), the state update period of the system is allowed to be different with the sampling period of the sensor, while the sampling periods of different sensors are also allowed to be different. The MRS (1.1) can be seen as a combination of N multi-rate subsystems where the i-th subsystem consists of the state equation and the measurement equation of the i-th sensor.

1.1.2 Transformation from Multi-Rate Systems to Single-Rate Systems

After presenting the state-space model of the MRSs, in this subsection, we are going to introduce some effective methods that convert the MRSs into single-rate ones.

1.1.2.1 Using the Lifting Technique

The lifting technique proposed in [113] is one of the popular methods that are used to transform the MRSs. The main idea of the lifting technique is to obtain equivalent single-rate systems by increasing the state update period. Taking the i-th multi-rate subsystem of (1.1) as an example, the states $x(t_{k-1}^i + h), \ldots, x(t_k^i - h), x(t_k^i)$ in the interval $(t_{k-1}^i, t_k^i]$ are first augmented into a vector $\bar{x}_i(t_k^i)$. Then, a new state equation with the state update period $b_i h$ is obtained by the aid of the original state equation. Accordingly, the MRSs are transformed into single-rate systems. In the following, the detailed process is presented.

Denote $\bar{x}_i(t_k^i) \triangleq \mathrm{col}\{x(t_{k-1}^i + h), \dots, x(t_k^i - h), x(t_k^i)\}$. By recurring to (1.1), it is derived that

$$x(t_k^i + mh) = \tilde{A}_m^i(t_k^i)x(t_k^i) + \tilde{B}_m^i(t_k^i)\tilde{w}_i(t_k^i)$$

where

$$\begin{aligned}
\tilde{A}_m^i(t_k^i) &\triangleq \prod_{l=1}^{m} A(t_k^i + (m-l)h), \ \tilde{B}_m^i(t_k^i) \triangleq \tilde{F}_m^i(t_k^i)\breve{B}_i(t_k^i),\\
\tilde{F}_m^i(t_k^i) &\triangleq \begin{bmatrix} \tilde{A}_{m-1}^i(t_k^i) & \tilde{A}_{m-2}^i(t_k^i) & \cdots & A_1^i(t_k^i) & 0 & \cdots & 0 \end{bmatrix},\\
\breve{B}_i(t_k^i) &\triangleq \mathrm{diag}\{B(t_k^i), B(t_k^i + h), \cdots, B(t_{k+1}^i - h)\},\\
\tilde{w}_i(t_k^i) &\triangleq \mathrm{col}\{w(t_k^i), w(t_k^i + h), \dots, w(t_{k+1}^i - h)\}.
\end{aligned}$$

Then, it is obtained that

$$\begin{cases} \bar{x}_i(t_{k+1}^i) = \bar{A}_i(t_k^i)\bar{x}_i(t_k^i) + \bar{B}_i(t_k^i)\tilde{w}_i(t_k^i), \\ y_i(t_k^i) = \bar{C}_i(t_k^i)\bar{x}_i(t_k^i) + D_i(t_k^i)v_i(t_k^i) \end{cases} \tag{1.2}$$

where

$$\begin{aligned}
\bar{A}_i(t_k^i) &\triangleq \begin{bmatrix} 0 & \cdots & 0 & \tilde{A}_1^i(t_k^i) \\ 0 & \cdots & 0 & \tilde{A}_2^i(t_k^i) \\ \vdots & \vdots & \vdots & \vdots \\ 0 & \cdots & 0 & \tilde{A}_{b_i}^i(t_k^i) \end{bmatrix},\\
\bar{B}_i(t_k^i) &\triangleq \mathrm{col}\{\tilde{B}_1^i(t_k^i), \tilde{B}_2^i(t_k^i), \cdots, \tilde{B}_{b_i}^i(t_k^i)\},\\
\bar{C}_i(t_k^i) &\triangleq \begin{bmatrix} 0 & \cdots & 0 & C_i(t_k^i) \end{bmatrix}.
\end{aligned}$$

Following the similar process, the MRS (1.1) is transformed into N single-rate systems with uniform periods $b_i h$ ($i = 1, 2, \dots, N$). For the i-th single-rate system, the i-th local estimator is designed as

$$\hat{x}_i(t_{k+1}^i) = \bar{A}_i(t_k^i)\hat{x}_i(t_k^i) + K_i(t_k^i)(y_i(t_k^i) - \bar{C}_i(t_k^i)\hat{x}_i(t_k^i))$$

where $\hat{x}_i(t_k^i)$ is the estimate of $\bar{x}_i(t_k^i)$ and $K_i(t_k^i)$ is the estimator gain to be designed.

From the above process, we can see that the lifting technique is able to transform the linear MRSs to equivalent linear single-rate systems. Nevertheless, due to the augmentation of the states, the designed estimation algorithm will have a high computational cost. On the other hand, when applying the lifting technique to the non-linear MRSs, the transformation will be complicated due to the iteration of the non-linear function.

1.1.2.2 Iterating the State Equation

By simply iterating the state equation in (1.1), we have the following new state equation with a state update period b_ih:

$$x(t_{k+1}^i) = \tilde{A}_{b_i}^i(t_k^i)x(t_k^i) + \tilde{B}_{b_i}^i(t_k^i)\tilde{w}_i(t_k^i).$$

Accordingly, we have the following single-rate system:

$$\begin{cases} x(t_{k+1}^i) = \tilde{A}_{b_i}^i(t_k^i)x(t_k^i) + \tilde{B}_{b_i}^i(t_k^i)\tilde{w}_i(t_k^i), \\ y_i(t_k^i) = C_i(t_k^i)x_i(t_k^i) + D_i(t_k^i)v_i(t_k^i). \end{cases} \tag{1.3}$$

For the single-rate system (1.3), the i-th local estimator is designed as

$$\hat{x}_i(t_{k+1}^i) = \tilde{A}_{b_i}^i(t_k^i)\hat{x}_i(t_k^i) + K_i(t_k^i)(y_i(t_k^i) - C_i(t_k^i)\hat{x}_i(t_k^i))$$

where $\hat{x}_i(t_k^i)$ is the estimate of $x(t_k^i)$ and $K_i(t_k^i)$ is the estimator gain to be designed.

By iterating the state equation, the i-th multi-rate subsystem of (1.1) is transformed into the single-rate system (1.3). Compared to the single-rate system (1.2) obtained by using the lifting technique, the computational cost of the estimation algorithm designed for the single-rate system (1.3) is low. Nevertheless, the designed estimation algorithm only estimates the states at the measurement sampling instants (i.e., t_k^i $(k = 0, 1, 2, \ldots)$) and other states are not estimated.

1.1.2.3 Compensating Measurements with Zero

In the above two approaches, the MRSs are transformed into single-rate systems by increasing the state update period of the state equation. Intuitively, we can also complete the transformation by decreasing the sampling period of the measurement equation. A method to achieve this goal is compensating the measurements at non-sampling instants with zero.

By introducing a variable

$$\lambda_i(s_k) = \begin{cases} 1, & \text{if } \mathrm{mod}(s_k, b_ih) = 0; \\ 0, & \text{otherwise} \end{cases}$$

with $\mathrm{mod}(x, y)$ being the unique non-negative remainder on division of x by y, the compensated measurement is formulated as

$$y_i(s_k) = \lambda_i(s_k)C_i(s_k)x(s_k) + \lambda_i(s_k)D_i(s_k)v_i(s_k).$$

Then, a single-rate system is derived as follows:

$$\begin{cases} x(s_{k+1}) = A(s_k)x(s_k) + B(s_k)w(s_k), \\ y_i(s_k) = \lambda_i(s_k)C_i(s_k)x(s_k) + \lambda_i(s_k)D_i(s_k)v_i(s_k). \end{cases} \tag{1.4}$$

For the single-rate system (1.4), the i-th local estimator is designed as

$$\hat{x}_i(s_{k+1}) = A(s_k)\hat{x}_i(s_k) + K_i(s_k)(y_i(s_k) - \lambda_i(s_k)C_i(s_k)\hat{x}_i(s_k))$$

where $\hat{x}_i(s_k)$ is the estimate of $x(s_k)$ and $K_i(s_k)$ is the estimator gain to be designed.

The advantages of such an approach are that the computational cost is low, and the designed estimation algorithm can estimate the state at all state update instants. Unfortunately, the estimation accuracy may be low since the estimate is obtained by prediction at non-measurement-sampling instants.

1.1.2.4 Compensating Measurements with Zero-Order Holder

Suppose that a zero-order holder is used to compensate the measurements at the non-sampling instants. After the compensation, the measurement available is

$$\bar{y}_i(s_k) \triangleq y_i(t_m^i), \quad t_m^i \leq s_k < t_{m+1}^i.$$

Inspired by the input delay approach which has been used to transform the discrete-time control input into the delayed control input [43], we define a variable $\rho_{s_k}^i$ as follows:

$$\rho_{s_k}^i \triangleq s_k - t_m^i, \quad t_m^i \leq s_k < t_{m+1}^i.$$

Noting that $t_m^i = s_k - (s_k - t_m^i) = s_k - \rho_{s_k}^i$ holds for $t_m^i \leq s_k < t_{m+1}^i$, the measurement $\bar{y}_i(s_k)$ is reformulated as

$$\begin{aligned}\bar{y}_i(s_k) &= y_i(s_k - \rho_{s_k}^i) \\ &= C_i(s_k - \rho_{s_k}^i)x(s_k - \rho_{s_k}^i) \\ &\quad + D_i(s_k - \rho_{s_k}^i)v_i(s_k - \rho_{s_k}^i).\end{aligned}$$

Therefore, the i-th multi-rate subsystem of (1.1) is transformed into a single-rate system

$$\begin{cases} x(s_{k+1}) = A(s_k)x(s_k) + B(s_k)w(s_k), \\ y_i(s_k) = C_i(s_k - \rho_{s_k}^i)x(s_k - \rho_{s_k}^i) \\ \qquad\qquad + D_i(s_k - \rho_{s_k}^i)v_i(s_k - \rho_{s_k}^i). \end{cases} \tag{1.5}$$

For the single-rate system (1.5), the i-th local estimator is designed as

$$\hat{x}_i(s_{k+1}) = A(s_k)\hat{x}_i(s_k) + K_i(s_k)(\bar{y}_i(s_k) - C_i(s_k - \rho_{s_k}^i)\hat{x}(s_k - \rho_{s_k}^i))$$

where $\hat{x}_i(s_k)$ is the estimate of $x(s_k)$ and $K_i(s_k)$ is the estimator gain to be designed.

Remark 1.1 *In this section, for simplification of the formulas, we only consider the case that the system state update period and the sampling period are different. Note that, the transformation methods provided in this section can be easily extended to the case where the system state update period, the sampling period and the estimate update period are mutually different.*

1.2 Estimation Problems for Multi-Rate Systems

Due to the practical engineering significance of the multi-rate sampling, the state estimation problem for MRSs has become an active research topic. Up to now, a number of estimation approaches have been applied to the MRS state estimation problems. According to the performance indices used, the applied estimation approaches can be categorized into the H_∞ estimation approach [68, 69, 84, 85, 105], the Kalman filtering approach and its variants [60, 166, 182], the set-membership filtering approach [17, 99, 205], the moving horizon estimation approach [183], etc. In the following, the available results are reviewed on state estimation problems for MRSs.

1.2.1 H_∞ Estimation Approach

In the past few decades, the H_∞ estimation approach has received particular research interest since it is able to attenuate the influence of the energy-bounded external disturbances on the estimation performance. The core idea of the H_∞ estimation is to design an estimator such that a predefined disturbance attenuation level (i.e. the H_∞ performance index) is achieved. Based on the methods like the linear matrix inequality (LMI), the Hamilton-Jacobi inequality as well as the Riccati equation, the H_∞ state estimation problems have been concerned for various systems [57, 83].

In the MRS state estimation problems, the H_∞ estimation approach has been widely used and a large body of literature has been available [41, 93, 103, 137, 141, 161]. In [137], the H_2 and H_∞ filtering problems have been studied for multi-rate linear time-invariant systems with a state update period h and a sampling period nh. First, a standard filter with a periodic filter gain and an update period h has been proposed to cater for the multi-rate sampling. Then, the state update period, the sampling period, and the estimate update period have been uniformed by iterating the system state equation and the filter equation. Finally, the H_2 and H_∞ filters have been designed by solving certain LMIs where non-convex constraints have been introduced due to the multi-rate sampling. Subsequently, the authors of [137] have extended the results to multi-rate linear time-invariant systems with packet dropouts in [93] where the state update period, the sampling period as well as the estimate update period are different. A sufficient condition has been provided on the existence of a stable filter. In [141], the H_∞ estimation problem has been concerned for multi-sensor multi-rate linear time-invariant systems where the estimate update period is an integer multiple of the sampling period/state update period. The measurements of sensors have been transmitted to a fusion center in a competitive way. By fusing the measurements received in one estimate update period, the estimator in the fusion centre has estimated the system state. A sufficient condition has been obtained on the mean-square stability as well as the H_∞ performance and the H_∞ estimator has been characterized.

For time-varying MRSs, the quantized finite-horizon H_∞ filtering problem has been studied in [103] under the stochastic communication protocol. The lifting technique has been first applied to handle the MRSs, and then the desired H_∞ filter has been designed by solving certain Riccati differential equations. Similar to [103], the variance-constrained finite-horizon H_∞ state estimation problem has been considered in [161] for time-varying MRSs. In [41], the sequential fusion H_∞ filtering problem has been investigated for time-varying MRSs where p asynchronous sensors are used to measure the system and send the measurements to a fusion center asynchronously. In the fusion center, the estimator updates the estimate once a measurement is received. With such a multi-rate sampling strategy, a sequential fusion H_∞ filter has been designed based on the Krein-space approach.

The fault detection problem is a long-standing research topic that has been widely studied [70, 76, 188, 204, 207]. In the MRSs, due to reasons like sensor aging, random sensor failure, or harsh environment, the fault may occur which largely degrades the system performance. Therefore, it is of great importance to design a fault detection algorithm to detect the fault. Recently, the H_∞ fault detection problems for MRSs have received initial research interest. In [200], the H_∞ fault detection problem has been investigated for MRSs with asynchronous state update rate and sampling rate. By iterating the state equation, the equivalent single-rate systems have been derived, and then an observer-based fault detection filter has been designed. With the help of the LMI method, sufficient conditions have been developed such that the H_∞ norm from the noises and faults to the fault estimation error is less that a given attenuation level. In [197], the intermittent fault detection problem has been concerned for a class of non-uniformly sampled MRSs. The non-uniform sampling interval is governed by a Markov process with partly unknown and uncertain transition probabilities. Due to the existence of the Markov process, the MRSs has been transformed into single-rate systems with Markovian jumping parameters. Sampling-interval-dependent fault detection filters have been designed such that the residual estimation error satisfies the H_- and H_∞ performances simultaneously.

1.2.2 Kalman Filtering Approach and Its Variants

The classic Kalman filtering approach is one of the most celebrated filtering approaches. For linear systems with Gaussian noises whose statistics are exactly known, the classic Kalman filtering is an optimal filtering approach under which the filtering error covariance is minimized at each time instant. Nevertheless, when applied to non-linear systems or systems with non-Gaussian noises, the classic Kalman filtering is no longer applicable. Therefore, some modified Kalman filtering approaches have been developed that includes the extended Kalman approach [86,159], the unscented Kalman approach [91,144], the cubature Kalman approach [16, 79] and so on.

For state estimation problems concerning the MRSs, the classic Kalman filtering approach is another popular approach that has been widely studied [29, 48, 67, 78, 109, 178, 179]. In [178], the fusion estimation problem has been studied for multi-sensor systems by using the traditional Kalman filtering approach, where the sampling periods of the sensors are asynchronous. By using the lifting technique, the multi-sensor MRS has been transformed into synchronous single-rate single-sensor systems. Local Kalman filters has been first designed to obtain the local estimates which have been then fused to a global estimate with the help of the fusion method in [12]. In [179], the problem in [178] has been reconsidered and a novel fusion estimation algorithm has been proposed which fuses the local estimates in a recursive form and is optimal in the linear minimum variance sense. In [67], the result in [178] has been extended to multi-sensor systems with arbitrary number of sensors and arbitrary sampling rates. Both the centralized asynchronous fusion estimation algorithm and the distributed asynchronous fusion estimation algorithm have been developed based on the Kalman filtering. It has been verified that the estimation performances of both fusion estimation algorithms are equivalent under the full-rate communication assumption, while the centralized one outperforms the distributed one when the communications are constrained. In [181], the federated Kalman filtering problem has been investigated for asynchronous multi-sensor MRSs with random missing measurements.

The Kalman fusion estimation algorithm has been developed in [193] for MRSs over the sensor networks. The system state update period, the measurement transmission period, and the estimate update period have been allowed to be different. The local Kalman filters have been first designed based on the single-rate systems derived by the lifting technique. Then, the local estimates have been fused with the weighted by matrix fusion method. Due to the asynchronism of the local estimates, at a certain time instant, only the available estimates have been fused. In [191], the hierarchical fusion estimation problem has been coped with for MRSs over sensor networks. The underlying sensor network consists of N sensor clusters and each sensor cluster contains multiple sensors and a cluster head. First, the estimator in the cluster head has generated a local estimate based on the fused measurement which has been obtained by fusing the available measurements from the sensors in this cluster with the sequential fusion method. Then, the cluster head has communicated with other cluster heads to generate the fused estimate. In the estimate fusion process, only the available local estimates have been used also due to the asynchronism of the local estimates. In [122], the Kalman fusion estimation problem has been also investigated for MRSs over sensor networks where the state update period and the sampling period are different. Different from [193], the estimates at the non-sampling instants have been obtained through prediction based on the state equation. Therefore, the estimates from all local estimators have been available in the estimation fusion process at each time instant. The distributed fusion estimation problem has been considered in [72] for multi-rate linear systems where the estimation process and the fusion process are similar to those in [193].

The fault detection problems for MRSs based on the Kalman filtering approach have received little research attentions. In [90], the Kalman filtering problem has been investigated for a class of non-uniformly sampled MRSs. After transforming the MRSs into single-rate systems with the lifting technique, a filter has been designed for the obtained single-rate systems and the stability as well as the convergence of the filtering error has been analyzed. Then, the fault detection and isolation problems have been concerned for the underlying non-uniformly sampled MRSs.

Other than the classic Kalman filtering approach, the modified Kalman filtering approaches have also been widely used in the state estimation problems for MRSs. In [50], the extended Kalman filtering problem has been discussed for non-linear systems where the availability of the *primary* and the *secondary* measurements are asynchronous. Several methods that effective for the asynchronous measurements have been discussed and modified in the extended Kalman filtering framework. In [180], the state estimation problem has been dealt with for non-linear multi-sensor MRSs by using the modified sigma point Kalman filtering approach. A modified unscented Kalman filter has been proposed in [38] for the multi-rate INS/GPS integrated navigation systems. The state estimation problem has been investigated in [55] for non-linear systems with a normal measurement and an infrequent integral measurement. The systems have been first reformulated to equivalent variable dimension systems. Then, a variable dimension unscented Kalman filter has been designed. In [118], a joint-unscented Kalman filtering algorithm has been proposed for a continuous stirred-tank reactor system with asynchronous sensors.

1.2.3 Other Estimation Approaches

In the state estimation problems for MRSs, the moving horizon estimation approach, the particle filtering approach, the l_2-l_∞ filtering approach and the set-membership filtering approach have been taken into consideration either [3, 4, 37, 96, 108, 211]. In the moving horizon estimation, the estimate is generated by solving a predefined optimization problem based on the measurements in a moving time interval with fixed length. Due to its efficiency in handling non-linear systems, the moving horizon estimation has received ever-increasing research interest in the past decade [52, 96, 211]. In [96], the state estimation problem has been solved for mobile robot systems with asynchronous sensors according to the moving horizon estimation. The sampling rates of the sensors have been uniformed by compensating the measurements from the slow-rate sensor with a prediction value. Then, a moving horizon estimation algorithm has been developed by solving a regularized least-squares problem. The moving horizon estimation problem has been concerned in [211] for linear systems with multi-rate measurements and correlated noises. By introducing a switching variable, the multi-rate measurements have been combined in a new measurement model. Based on the obtained single-rate systems, the desired moving horizon estimator has been designed.

The particle filtering approach, roots on the sampling-based approximation techniques, is an effective filtering approach for non-linear systems or systems with non-Gaussian noises. In [58], both the particle filtering and extended Kalman filtering have been applied to an intensified chemical process subject to asynchronous measurements. The l_2-l_∞ filtering problem has been studied in [185] for MRSs. The lifting technique has been used to handle the MRSs and the l_2-l_∞ filter has been designed according to the solution of LMIs with a non-convex constraint. The set-membership filtering problem has been concerned in [104] for MRSs over sensor networks. A set of local filters have been designed such that the filtering errors have been constrained in a given ellipsoid. In [120], the zonotopes-based distributed set-membership filtering problem has been investigated for MRSs where the state estimates belong to the computed sets.

1.2.4 Handling the Network-Induced Challenges

In networked systems, due to the limited resources of the communication networks, the network-induced phenomena inevitably occur [61, 126]. Up to now, considerable research attentions have been paid on the state estimation problems for networked MRSs subject to network-induced phenomena and plenty of results have been available. In the following, some typical network-induced phenomena are introduced and the corresponding state estimation problems for networked MRSs are summarized.

The packet dropout is one of the most frequently occurred network-induced phenomena which may be caused by many reasons such as intermittent sensor failures, network congestion and so on. In [47], the linear-minimum-mean-square-error observer design problem has been considered for multi-sensor MRSs with multiple packet dropouts. The observer has been designed by minimizing the estimation error covariances. In [206], the almost surely state estimation problem has been studied for MRSs subject to both Markovian packet dropouts and random packet dropouts characterized by the bernoulli distributed random variables. An estimator has been designed such that the estimation error is almost surely exponentially stable.

The signal quantization and saturation are two ubiquitous network-induced phenomena due to the inherent nature of the digital transmission and the physical constraints of the hardware, respectively [15, 123]. In [198], the variance-constrained H_∞ state estimation problem has been studied for MRSs subject to measurement quantization. The quantization effect has been characterized by a logarithmic quantizer and transformed to sector-bounded uncertainties. The desired H_∞ filter has been designed by resorting to the stochastic analysis approach and the Lyapunov theory. In [81], the multi-objective filtering problem has been concerned for MRSs with random sensor

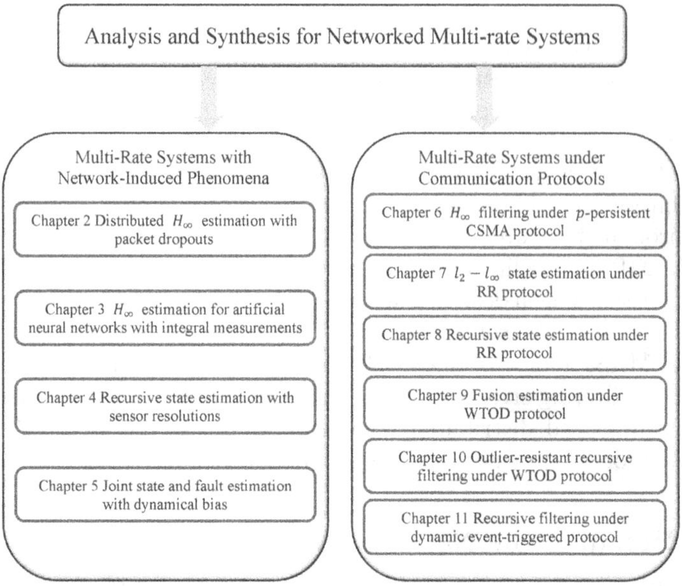

FIGURE 1.1: The structure of this book.

saturations. The saturation function has been rewritten as a combination of a linear term and a non-linear function satisfying the sector condition.

1.3 Outline of This Book

This book is divided into twelve chapters, where Chapter 2–Chapter 5 mainly focus on the state estimation problems for MRSs with network-induced phenomena and Chapter 6–Chapter 11 are concerned with the state estimation problems for MRSs under the communication protocols. The organizational structure of this book is presented in Fig. 1.1. To be specific, the outline of this book is given as follows:

- Chapter 1 provides a timely and systematic review with respect to the research progress in state estimation problems for networked MRSs, and lists the outline of the book.
- In Chapter 2, the non-fragile H_∞ filtering problem is investigated for a class of discrete multi-rate time-delayed systems over sensor networks. The probabilistic packet dropout occurs during the information transmissions among the sensor nodes in the sensor network characterized by the Gilbert-Elliott model. In order to take the multi-rate sampling into account, the

state update period of the system and the sampling period of the sensors are allowed to be different. The variation of the filter gain is considered to reflect the physical errors with the filter implementation. By using the Lyapunov-Krasovskii functional approach, a sufficient condition is derived that ensures the exponential mean-square stability and the H_∞ performance requirement of the filtering error dynamics. Then, the filter gains are characterized in terms of the solution to a set of matrix inequalities. Finally, a simulation example is provided to demonstrate the effectiveness of the proposed filtering scheme.

- In Chapter 3, the H_∞ filtering problem is concerned for a class of multi-rate artificial neural networks (ANNs) with integral measurements. A novel method, rather than the widely used lifting technique, is proposed to transform the multi-rate ANNs to single-rate switched ones. First, with the help of the Lyapunov-Krasovskii functional and the switched system approach, sufficient conditions are derived under which the existence of the desired filter is ensured. Then, the characterization of the filter gains is realized by solving certain LMIs. Finally, two illustrative examples are given that confirm the usefulness of the developed H_∞ filtering scheme and reveal the influence of the multi-rate sampling on the filtering performance.

- In Chapter 4, the recursive state estimation problem is discussed for a class of MRSs with multiplicative noises where the measurement outputs are collected from sensors with certain resolutions. Due to the existence of the sensor resolution, the actual measurement output of the sensor might deviate from its true value and such a deviation, if not adequately taken into account, would lead to serious degradation of the estimation performance, and we are therefore motivated to develop an effective state estimation algorithm that is insensitive to the sensor-resolution-induced measurement distortions. A state estimator is designed such that an upper bound on the estimation error covariance is minimized. Moreover, a simulation example with application background on moving target tracking problem is presented to verify the validity of the developed recursive state estimation algorithm.

- In Chapter 5, the joint state and fault estimation problem is dealt with for a class of MRSs with dynamical bias. To reflect real practice, the multi-rate sampling is considered which allows the sensor sampling rate and the state update rate to be different. The sensor is subject to the sensor fault that changes according to a dynamic equation. Instead of applying the traditional lifting technique, we introduce a time-varying delay into the measurement equation so as to transform the MRSs into single-rate ones. The recursion of the estimation error covariance is first derived, and appropriate estimator gains are then characterized that minimizes the estimation error covariance. A simulation example on the DC servo system is given to confirm the usefulness of the developed recursive state and fault estimation algorithm.

- In Chapter 6, the H_∞ filtering problem is concerned for a class of discrete networked multi-rate multi-sensor systems with randomly occurring sensor saturations under the p-persistent carrier sense multiple access (CSMA) protocol. A set of mutually independent Bernoulli distributed white sequences is introduced to characterize the random occurrence of the sensor saturations. The p-persistent carrier sense multiple access protocol is employed to decide which sensor is allowed to transmit its measurement to the filter at a certain time instant. By using the lifting technique, the MRS is converted to a single-rate one for convenient analysis. Sufficient conditions are established on the existence of the desired H_∞ filters and the corresponding filter gains are then characterized by resorting to the feasibility of certain matrix inequalities. Finally, a numerical example is given to illustrate the effectiveness of the proposed filtering scheme.

- In Chapter 7, the l_2-l_∞ state estimation problem is concerned for a class of delayed ANNs under high-rate communication channels with RR protocol. To estimate the state of the ANNs, numerous sensors are deployed to measure the ANNs. The sensors communicate with the remote state estimator through a shared high-rate communication channel. In the high-rate communication channel, the RR protocol is utilized to schedule the transmission sequence of the numerous sensors. First, sufficient conditions are given which guarantee the existence of the desired l_2-l_∞ state estimator. Then, the estimator gains are obtained by solving two sets of matrix inequalities. Finally, numerical examples are provided to verify the effectiveness of the developed l_2-l_∞ state estimation scheme.

- In Chapter 8, the problem of recursive state estimation is investigated for a class of multi-rate multi-sensor systems with distributed time-delays under RR protocol. The state update period of the system and the sampling period of the sensors are allowed to be different so as to reflect the engineering practice. An iterative method is presented to transform the MRS into a single-rate one, thereby facilitating the system analysis. The RR protocol is introduced to determine the transmission sequence of sensors with aim to alleviate undesirable data collisions. Under the RR protocol scheduling, only one sensor can get the access to transmit its measurement at each sampling time instant. A recursive state estimation scheme is developed such that an upper bound on the estimation error covariance is guaranteed and then locally minimized through adequately designing the estimator parameter. Finally, simulation examples are provided to show the effectiveness of the proposed estimator design scheme.

- In Chapter 9, the fusion estimation problem is studied for a class of discrete time-varying multi-rate linear repetitive processes (LRPs) under WTOD protocol. The LRPs are measured by multiple sensors that are allowed to have different sampling periods, and the state update period of the LRPs is also allowed to be different from the sampling periods of the asynchronous sensors. To facilitate the estimator design, the lifting technique is applied to transform the multi-rate LRPs to single-rate ones. Moreover, due to limited communication capability, the WTOD protocol is adopted to schedule the asynchronous sensors. A set of local estimators is designed such that the upper bounds on the local estimation error covariances are guaranteed, and such upper bounds are then minimized by appropriately designing the estimator gains. Furthermore, the estimates from the local estimators are fused by recurring to the sequential covariance intersection fusion method. Finally, a simulation example is given to demonstrate the effectiveness of the proposed fusion estimation scheme.

- In Chapter 10, a new outlier-resistant recursive filtering problem is concerned for a class of multi-sensor multi-rate networked systems under the WTOD protocol. The sensors are sampled with a period that is different from the state update period of the system. In order to lighten the communication burden and alleviate the network congestions, the WTOD protocol is implemented in the sensor-to-filter channel to schedule the order of the data transmission of the sensors. In case of the measurement outliers, a saturation function is employed in the filter structure to constrain the innovations contaminated by the measurement outliers, thereby maintaining a satisfactory filtering performance. By resorting to the solution to a matrix difference equation, an upper bound is first obtained on the covariance of the filtering error, and the gain matrix of the filter is then characterized to minimize the derived upper bound. Furthermore, the exponential boundedness of the filtering error dynamics is analyzed in the mean square sense. Finally, the usefulness of the proposed outlier-resistant recursive filtering scheme is verified by simulation examples.

- In Chapter 11, the dynamic event-based recursive filtering problem is investigated for MRSs over sensor networks. The state update rate of the plant and the sampling rate of the sensors are allowed to be different in order to reflect the multi-rate sampling strategy. Moreover, the phenomenon of integral measurements is considered to cater for the real engineering practice. To reduce unnecessary data transmissions, the dynamic event-based mechanism is implemented in the communication channels among sensor nodes. An upper bound on the filtering error covariance is first derived by solving a matrix Riccati equation, and then minimized at each sampling instant by choosing appropriate filter gains. Comprehensive simulations are conducted on a numerical example and a practical example to show the effectiveness and superiority of the proposed dynamic event-based recursive filtering scheme.

- Chapter 12 draws some conclusions on the book, and points out some potential research directions related to the work done in this book.

2

Non-Fragile H_∞ Filtering for Multi-Rate Time-Delayed Systems over Sensor Networks

The sensor networks have received an ever-increasing research interest due mainly to their extensive applications in many areas such as environment monitoring, target tracking, power grids, and industrial control systems. Different from the traditional networked systems, a sensor network consists of a group of sensor nodes which are distributed spatially, where the sensor node can receive information not only from the plant but also from its neighboring sensor nodes according to the given topology. The complicated coupling among the sensor nodes brings substantial difficulties in the filter design problem and, accordingly, the distributed filtering problem for sensor networks has recently become a research focus leading to a rich body of literature.

The filtering problem with time-delays has long been an interesting research topic in the area of signal processing that has received persistent attention from many researchers. Several types of time-delays (e.g. discrete, distributed, and stochastic time-delays) have been introduced into the filter design problem and their impacts on the filter performance have been widely investigated. For example, the robust fault detection problem has been studied in [34] for Takagi-Sugeno fuzzy systems with random mixed time-delays, and the event-triggered state estimation problem has been investigated in [88] for neural networks with mixed time-delays. It is worth noting that, in most available results regarding filtering problems with time-delays, the concerned system has been assumed to be sampled with a single rate. The corresponding filtering problem for multi-rate time-delay systems has, unfortunately, received very little research attention so far despite its significant engineering importance. It remains an open problem as how the multi-rate sampling impacts the filter performance when coupled with time-delays, and this constitutes one motivation of this chapter.

Probabilistic packet dropouts are arguably one of the most frequently encountered network-induced phenomena that have been extensively studied in the context of networked control systems. In most available literature, a common approach to characterizing the packet dropouts is to use the Bernoulli

DOI: 10.1201/9781032619507-2

distributed random variables. Due to circumstance changes such as stochastic electromagnetic interference or sudden environmental variations, the packet dropouts are likely to occur with different probabilities in different periods, and such a feature might not be captured by the traditional Bernoulli-type description.

The Gilbert-Elliott model [36, 49] has been widely used in characterizing burst-error in wireless communication channel [117], where the condition of the wireless communication channel is characterized by a Markov chain with two modes: "Good" and "Bad". In the "Good" mode, the channel is free of packet dropouts while, in the "Bad" mode, the channel may be subjected to packet dropouts with a given probability. Compared with the traditional Bernoulli-type packet dropout model, the Gilbert-Elliott model is clearly more realistic since the packet dropout probability is not required to be a constant over different time periods. As such, it would be interesting to look into the H_∞ filtering problem over sensor networks with packet dropouts characterized by the Gilbert-Elliott model.

In practical engineering, it is not uncommon that the filter gains cannot be precisely implemented due to a variety of reasons such as finite precision, rounding errors, analogue-to-digital conversion, and finite word length in computation. In other words, the filter gains might experience undesired fluctuations that would greatly deteriorate the filtering performance, and this gives rise to the non-fragile filter design issue whose idea is to design a filter that can maintain the satisfactory performance in case of gain fluctuations, see e.g. [59, 133]. For example, in [59], the filter gain variations have been taken into account and a non-fragile state estimator has been designed such that the estimation error dynamics is exponentially stable. The non-fragile filtering problem has been investigated in [133] for singular Markovian jump systems with missing measurements and time-varying delays. However, when it comes to the multi-rate sensor networks, the relevant results have really scattered, and it makes practical sense to investigate into the non-fragile distributed filtering problem for MRSs over sensor networks.

In this chapter, we aim to study the non-fragile H_∞ filtering problem for discrete multi-rate time-delayed systems over sensor networks. The main contributions of this chapter are highlighted as follows: 1) the influence from multi-rate samplings and time-delays on the filtering performance is, for the first time, investigated in the framework of sensor networks; 2) a Gilbert-Elliott-model-based communication scheme is proposed to reflect the probabilistic packet dropouts occurring in the communication channels among the sensor nodes; and 3) a set of non-fragile filters is designed to mitigate the performance deterioration induced by filter gain fluctuations during the implementation.

2.1 Problem Formulation

Consider a sensor network whose topology is described by a directed graph $\mathscr{G} = (\mathcal{V}, \mathscr{E}, \mathcal{R})$ of order N with the set of nodes $\mathcal{V} = \{1, 2, \ldots, N\}$, set of edges $\mathscr{E} \subseteq \mathcal{V} \times \mathcal{V}$ and weighted adjacency matrix $\mathcal{R} = [a_{ij}]$ with non-negative adjacency elements a_{ij}. An edge of $\mathscr{G}$ is denoted by (i, j). The adjacency element associated with the edge of the graph is defined as $a_{ij} > 0 \Leftrightarrow (i, j) \in \mathscr{E}$, which means that the i-th sensor node can receive information from the j-th sensor node. Throughout this chapter, we assume that $a_{ii} = 0$ for $i \in \mathcal{V}$. For sensor node $i \in \mathcal{V}$, the neighboring sensor nodes are denoted by the set $\mathcal{V}_i \triangleq \{j \in \mathcal{V} : (i, j) \in \mathscr{E}\}$.

Consider a plant described by the following discrete time-delayed systems

$$\begin{cases} x(T_{k+1}) = Ax(T_k) + Bx(T_k - dh) + Ew(T_k) \\ z(T_k) = Mx(T_k) \\ x(T_k) = \phi(T_k), \quad \forall\, k \in \mathbb{Z}^- \end{cases} \tag{2.1}$$

where $x(T_k) \in \mathbb{R}^{n_x}$ is the state vector, $\phi(T_k) \in \mathbb{R}^{n_x}$ are the initial values of the state, $z(T_k) \in \mathbb{R}^{n_z}$ is the signal to be estimated, $w(T_k)$ is the process noise belonging to $l_2([0, \infty), \mathbb{R}^{n_w})$, and A, B, E, and M are known matrices with appropriate dimensions. $h \triangleq T_{k+1} - T_k$ is the state update period of system (2.1) and d is a known positive integer. Moreover, it is assumed that $w(T_k) = 0$ for $k \in \mathbb{Z}^-$.

The measurement output $y_i(t_k) \in \mathbb{R}^{n_y}$ from the i-th ($i = 1, 2, \ldots, N$) sensor node is modeled as

$$y_i(t_k) = C_i x(t_k) + D_i v_i(t_k) \tag{2.2}$$

where $v_i(t_k)$ is the measurement noise on the i-th sensor node belonging to $l_2([0, \infty), \mathbb{R}^{n_v})$. C_i and D_i are known matrices with appropriate dimensions. The sampling period of the sensors is integer multiple of h, i.e. $t_{k+1} - t_k \triangleq bh$ where $b \in \mathbb{N}^+$ is a known positive integer.

In this chapter, the packet dropouts in communication channels among sensor nodes are characterized by the Gilbert-Elliott model. According to the Gilbert-Elliott model, the actual information $\bar{y}_{ji}(t_k)$ received by the filter on sensor node i from sensor node j can be described as

$$\bar{y}_{ji}(t_k) = \delta(\theta_{ji}(t_k), 0) y_j(t_k) + \delta(\theta_{ji}(t_k), 1) \beta_{ji}(t_k) y_j(t_k) \tag{2.3}$$

where $\delta(\cdot, \cdot)$ is the Kronecker delta function, $\theta_{ji}(t_k)$ ($j \in \mathcal{V}_i, i \in \mathcal{V}$) are discrete homogeneous Markov chains taking values in the finite state space $\mathcal{I} = \{0, 1\}$, and $\beta_{ji}(t_k)$ ($j \in \mathcal{V}_i, i \in \mathcal{V}$) are mutually independent Bernoulli distributed white sequences taking values on 0 or 1 with

$$\begin{cases} \text{Prob}\{\beta_{ji}(t_k) = 1\} = \mu_{ji}, \\ \text{Prob}\{\beta_{ji}(t_k) = 0\} = 1 - \mu_{ji} \end{cases}$$

where $\mu_{ji} \in [0, 1]$ ($j \in \mathcal{V}_i, i \in \mathcal{V}$) are known constants.

The transition probability matrix of the Markov chains $\theta_{ji}(t_k)$ ($j \in \mathcal{V}_i, i \in \mathcal{V}$) is given by $\Lambda \triangleq [p_{rt}]$ with

$$p_{rt} \triangleq \Pr\{\theta_{ji}(t_{k+1}) = t | \theta_{ji}(t_k) = r\}, \ \forall r, t \in \mathcal{I} \tag{2.4}$$

where $p_{rt} \geq 0$ ($r, t \in \mathcal{I}$) is the transition rate from mode r to mode t and $\sum_{t=0}^{1} p_{rt} = 1$ ($\forall r \in \mathcal{I}$). In addition, it is assumed that $\theta_{ji}(t_k) = 1$ and $\beta_{ji}(t_k) = 0$ for all $j \notin \mathcal{V}_i, i \in \mathcal{V}$.

Throughout this chapter, the random variables $\theta_{ji}(t_k)$ and $\beta_{ji}(t_k)$ ($j \in \mathcal{V}_i, i \in \mathcal{V}$) are assumed to be mutually independent.

Remark 2.1 *In* (2.3), $\theta_{ji}(t_k) = 0$ ($j \in \mathcal{V}_i, i \in \mathcal{V}$) *means that the communication channel from sensor node j to sensor node i is in "Good" mode and has no packet dropout at time instant t_k, while $\theta_{ji}(t_k) = 1$ ($j \in \mathcal{V}_i, i \in \mathcal{V}$) means that the communication channel from sensor node j to sensor node i is in "Bad" mode and may subject to packet dropout with a given probability at time instant t_k. Note that, in the case where $\mu_{ji} = 0$ for all $i, j \in \mathcal{V}$, the original Gilbert-Elliott model will be reduced to the simplified one which has been considered in [146].*

Remark 2.2 *In wireless communication channel, the packet dropouts are mainly caused by weak channel conditions such as network congestions and harsh external environment, and the probability of the packet dropout depends on the condition of the channel. In the real world, the channel condition changes over time with a pace much slower than that of the packet transmission, and therefore the packet dropouts are likely to happen in burst. The Gilbert-Elliott model proposed in [36, 49] is suitable for characterizing such a bursty nature and has proven to be capable of modeling the bursty packet dropouts with sufficient accuracy in [171]. So far, some initial research results have been available on filtering/control problems for networked systems with packet dropouts characterized by simplified Gilbert-Elliott model, see e.g. [146]. However, as shown in [184], the simplified Gilbert-Elliott model has less accuracy than the original one when modeling bursty packet dropouts. As such, the packet dropouts are characterized by the original Gilbert-Elliott model in this chapter.*

Noting that the state update period of the system and the sampling period of the sensors are allowed to be different, the system under consideration is essentially a MRS. In what follows, in order to facilitate the system analysis and synthesis, the MRS will be technically converted into a single-rate one. For notation simplicity, in the following, dh is simply denoted by d.

By iterating (2.1) recursively, the following equation can be obtained for $j = 1, 2, \ldots, b$:

$$x(t_k + j) = \mathfrak{F}_j^0 x(t_k) + \sum_{i=1}^{j} \mathfrak{F}_j^i x(t_k - id) + \sum_{l=1}^{j} \sum_{i=0}^{j-l} \mathfrak{F}_{j-l}^i E w(t_k + l - 1 - id)$$

where $\mathfrak{F}_j^0 \triangleq A^j$, $\mathfrak{F}_j^j \triangleq B^j$, and $\mathfrak{F}_j^i \triangleq \mathfrak{F}_{j-1}^{i-1}B + \mathfrak{F}_{j-1}^i A$.

By denoting

$$\begin{aligned}
\bar{x}(t_k) &\triangleq \text{col}\{x(t_{k-1}+1), x(t_{k-1}+2), \cdots, x(t_k)\},\\
\bar{z}(t_k) &\triangleq \text{col}\{z(t_{k-1}+1), z(t_{k-1}+2), \cdots, z(t_k)\},\\
\bar{w}(t_k) &\triangleq \text{col}\{w(t_k), w(t_k+1), \cdots, w(t_k+b-1)\},\\
\bar{M} &\triangleq \text{diag}_b\{M\}, \quad \mathfrak{A} \triangleq \text{col}\{\mathfrak{F}_1^0, \mathfrak{F}_2^0, \cdots, \mathfrak{F}_b^0\},\\
\bar{G}_\tau &\triangleq \begin{bmatrix} \mathfrak{G}_{b-1}^\tau & \mathfrak{G}_{b-2}^\tau & \cdots & \mathfrak{G}_\tau^\tau & \underbrace{0_{bn_x\times n_w} \quad \cdots \quad 0_{bn_x\times n_w}}_{b-1} \end{bmatrix},\\
\bar{A} &\triangleq \begin{bmatrix} \underbrace{0_{n_x\times bn_x} \quad \cdots \quad 0_{n_x\times bn_x}}_{b-1} & \mathfrak{A} \end{bmatrix},\\
\bar{B}_\tau &\triangleq \begin{bmatrix} \underbrace{0_{n_x\times bn_x} \quad \cdots \quad 0_{n_x\times bn_x}}_{b-1} & \mathfrak{B}_\tau \end{bmatrix},\\
\mathfrak{B}_\tau &\triangleq \text{col}\{\underbrace{0_{n_x\times n_x}, \cdots, 0_{n_x\times n_x}}_{\tau-1}, \mathfrak{F}_\tau^\tau, \mathfrak{F}_{\tau+1}^\tau, \cdots, \mathfrak{F}_b^\tau\}\\
\bar{C}_i &\triangleq \begin{bmatrix} \underbrace{0_{n_y\times n_x} \quad \cdots \quad 0_{n_y\times n_x}}_{b-1} & C_i \end{bmatrix},\\
\mathfrak{G}_i^\tau &\triangleq \text{col}\{\underbrace{0_{n_x\times n_w}, \cdots, 0_{n_x\times n_w}}_{b-i+\tau-1}, \mathfrak{F}_\tau^\tau E, \mathfrak{F}_{\tau+1}^\tau E, \cdots, \mathfrak{F}_i^\tau E\},
\end{aligned}$$

we obtain that

$$\begin{aligned}
\bar{x}(t_{k+1}) &= \bar{A}\bar{x}(t_k) + \sum_{\tau=1}^{b} \bar{B}_\tau \bar{x}(t_k - \tau d) + \sum_{\tau=0}^{b-1} \bar{G}_\tau \bar{w}(t_k - \tau d), \qquad (2.5)\\
y_i(t_k) &= \bar{C}_i \bar{x}(t_k) + D_i v_i(t_k),\\
\bar{z}(t_k) &= \bar{M}\bar{x}(t_k).
\end{aligned}$$

From $\bar{x}(t_k - \tau d) = \text{col}\{x(t_{k-1}+1-\tau d), x(t_{k-1}+2-\tau d), \cdots, x(t_k - \tau d)\}$, it is easily known that there exists an element $x(t_{k-1}+i-\tau d)$ in $\bar{x}(t_k - \tau d)$ such that $x(t_{k-1}+i-\tau d) = x(t_{k-1} - n_\tau b)$ with

$$n_\tau \triangleq \frac{\tau d - i}{b}$$

being a non-negative integer. Then, we rewrite $\bar{x}(t_k - \tau d)$ as

$$\bar{x}(t_k - \tau d) = \psi_{1\tau}\bar{x}(t_k - (n_\tau + 1)b) + \psi_{2\tau}\bar{x}(t_k - n_\tau b) \qquad (2.6)$$

where

$$\psi_{1\tau} \triangleq \begin{bmatrix} 0 & I_{\tau d - n_\tau b} \\ 0 & 0 \end{bmatrix}, \quad \psi_{2\tau} \triangleq \begin{bmatrix} 0 & 0 \\ I_{b-(\tau d - n_\tau b)} & 0 \end{bmatrix}.$$

Accordingly, (2.5) can be further reformulated as

$$\begin{aligned}\bar{x}(t_{k+1}) = &\bar{A}\bar{x}(t_k) + \sum_{\tau=1}^{b} \bar{B}_\tau \psi_{1\tau}\bar{x}(t_k - (n_\tau + 1)b) \\ &+ \sum_{\tau=1}^{b} \bar{B}_\tau \psi_{2\tau}\bar{x}(t_k - n_\tau b) + \sum_{\tau=0}^{b-1} \bar{G}_\tau \bar{w}(t_k - \tau d). \end{aligned} \tag{2.7}$$

For convenience of later analysis, the Markov chains $\theta_{ji}(t_k)$ $(j \in \mathcal{V}_i, i \in \mathcal{V})$ in (2.3) are mapped to one Markov chain according to the following lemma.

Lemma 2.1 *[213] The Markov chains $\theta_{ij}(t_k)$ $(i \in \mathcal{V}_j, j \in \mathcal{V})$ and constants $\theta_{ij}(t_k)$ $(i \notin \mathcal{V}_j, j \in \mathcal{V})$ can be mapped to a new sequence $r(t_k) \in \bar{\mathcal{I}} \subset \{1, 2, \ldots, 2^{N^2}\}$ by the one-to-one mapping $\Theta(\cdot)$ defined by*

$$r(t_k) \triangleq \Theta(\theta_{ij}(t_k)) \triangleq 1 + \sum_{j=1}^{N}\sum_{i=1}^{N} \theta_{ij}(t_k) 2^{l_{ij}}$$

where $l_{ij} \triangleq (j-1)N + i - 1$. Moreover, if $r(t_k)$ is given, the value of $\theta_{ij}(t_k)$ can be calculated by the following equation

$$\theta_{ij}(t_k) = \varphi_{ij}(r(t_k)) \triangleq \mathrm{mod}\left(\left\lfloor \frac{r(t_k) - 1}{2^{l_{ij}}} \right\rfloor, 2\right)$$

where $\lfloor a \rfloor$ represents the largest integer that is no larger than a.

Obviously, $r(t_k)$ is a Markov chain with the state space $\bar{\mathcal{I}}$ and has the transition probability matrix $\bar{\Lambda} = [\bar{p}_{mn}]$ with

$$\begin{aligned}\bar{p}_{mn} &\triangleq \Pr\{r(t_{k+1}) = n | r(t_k) = m\} \\ &= \prod_{j=1}^{N}\prod_{i=1}^{N} \Pr\{\theta_{ij}(t_{k+1}) = \varphi_{ij}(n) | \theta_{ij}(t_k) = \varphi_{ij}(m)\}\end{aligned}$$

where $\bar{p}_{mn} \geq 0$ $(m, n \in \bar{\mathcal{I}})$ is the transition rate from m to n and $\sum_{n \in \bar{\mathcal{I}}} \bar{p}_{mn} = 1$ $(\forall m \in \bar{\mathcal{I}})$.

According to Lemma 2.1, (2.3) can be rewritten as

$$\bar{y}_{ji}(t_k) = \delta(\varphi_{ji}(r(t_k)), 0) y_j(t_k) + \delta(\varphi_{ji}(r(t_k)), 1) \beta_{ji}(t_k) y_j(t_k). \tag{2.8}$$

For presentation convenience, in the following, a function $f(r(t_k))$ ($f(\cdot)$ represents an arbitrary function) is denoted by f^r and a matrix $A(r(t_k))$ (A represents an arbitrary matrix) is denoted by A_r when $r(t_k) = r$ $(r \in \bar{\mathcal{I}})$.

In the sensor network, the filter on sensor node i receives information not only from sensor node i but also from its neighboring sensor nodes. Based on this fact, the non-fragile filter on sensor node i is constructed as follows:

$$\begin{cases} \hat{x}_i(t_{k+1}) = \bar{A}\hat{x}_i(t_k) + \sum_{\tau=1}^{b} \bar{B}_\tau \psi_{1\tau}\hat{x}_i(t_k - (n_\tau + 1)b) \\ \qquad + \sum_{\tau=1}^{b} \bar{B}_\tau \psi_{2\tau}\hat{x}_i(t_k - n_\tau b) \\ \qquad + (O_{i,r} + \Delta O_{i,r})\big(y_i(t_k) - \bar{C}_i\hat{x}_i(t_k)\big) \\ \qquad + \sum_{j\in\mathcal{V}_i} a_{ij}\big(H_{ij}(\varphi_{ji}^r) + \Delta H_{ij}(\varphi_{ji}^r)\big) \\ \qquad\qquad \times \big(\bar{y}_{ji}(t_k) - (\delta(\varphi_{ji}^r, 0) + \delta(\varphi_{ji}^r, 1)\mu_{ji})\bar{C}_j\hat{x}_j(t_k)\big) \\ \hat{z}_i(t_k) = \bar{M}\hat{x}_i(t_k) \end{cases} \tag{2.9}$$

where $\hat{x}_i(t_k)$ and $\hat{z}_i(t_k)$ are the estimates for $\bar{x}(t_k)$ and $\bar{z}(t_k)$ from the filter on sensor node i, respectively. $O_{i,r}$ and $H_{ij}(\varphi_{ji}^r)$ are the filter gains to be designed. $\Delta O_{i,r}$ and $\Delta H_{ij}(\varphi_{ji}^r)$ are the gain fluctuations. As described in [94], it is assumed that

$$\Delta H_{ij}(\varphi_{ji}^r) = H_{ij}(\varphi_{ji}^r)\Upsilon_1\Sigma_1(t_k)\Upsilon_2, \;\; \Delta O_{i,r} = O_{i,r}\Upsilon_3\Sigma_2(t_k)\Upsilon_4$$

where Υ_1, Υ_2, Υ_3, and Υ_4 are known matrices with appropriate dimensions, and $\Sigma_1(t_k)$ and $\Sigma_2(t_k)$ are unknown matrices satisfying $\Sigma_1^T(t_k)\Sigma_1(t_k) \le I$ and $\Sigma_2^T(t_k)\Sigma_2(t_k) \le I$, respectively.

By denoting

$$\begin{aligned} K_{ij}(\varphi_{ji}^r) &\triangleq H_{ij}(\varphi_{ji}^r) + \Delta H_{ij}(\varphi_{ji}^r), \;\; \mathcal{O}_{i,r} \triangleq O_{i,r} + \Delta O_{i,r}, \\ \mathcal{O}_r &\triangleq \mathrm{diag}\{\mathcal{O}_{1,r}, \mathcal{O}_{2,r}, \cdots, \mathcal{O}_{N,r}\}, \\ e_i(t_k) &\triangleq \bar{x}(t_k) - \hat{x}_i(t_k), \;\; \tilde{z}_i(t_k) \triangleq \bar{z}(t_k) - \hat{z}_i(t_k), \\ e(t_k) &\triangleq \mathrm{col}\{e_1(t_k), e_2(t_k), \cdots, e_N(t_k)\}, \\ \tilde{z}(t_k) &\triangleq \mathrm{col}\{\tilde{z}_1(t_k), \tilde{z}_2(t_k), \cdots, \tilde{z}_N(t_k)\}, \;\; \bar{\beta}_{ji}(t_k) \triangleq \beta_{ji}(t_k) - \mu_{ji}, \\ \tilde{x}(t_k) &\triangleq \mathrm{col}_N\{\bar{x}(t_k)\}, \;\; \bar{\Gamma}_i(t_k) \triangleq \mathrm{diag}\{\bar{\beta}_{1i}(t_k)I, \bar{\beta}_{2i}(t_k)I, \cdots, \bar{\beta}_{Ni}(t_k)I\}, \\ \Gamma_i &\triangleq \mathrm{diag}\{\mu_{1i}I, \mu_{2i}I, \cdots, \mu_{Ni}I\}, \\ \bar{v}(t_k) &\triangleq \mathrm{col}\{v_1(t_k), v_2(t_k), \cdots, v_N(t_k)\}, \\ \Phi_{i,r}^1 &\triangleq \mathrm{diag}\{\delta(\varphi_{1i}^r, 1)I, \delta(\varphi_{2i}^r, 1)I, \cdots, \delta(\varphi_{Ni}^r, 1)I\}, \\ \Phi_{i,r}^0 &\triangleq \mathrm{diag}\{\delta(\varphi_{1i}^r, 0)I, \delta(\varphi_{2i}^r, 0)I, \cdots, \delta(\varphi_{Ni}^r, 0)I\}, \\ \tilde{A} &\triangleq \mathrm{diag}_N\{\bar{A}\}, \;\; \tilde{B}_\tau \triangleq \mathrm{diag}_N\{\bar{B}_\tau\}, \\ \tilde{G}_\tau &\triangleq \mathrm{col}_N\{\bar{G}_\tau\}, \;\; \tilde{M} \triangleq \mathrm{diag}_N\{\bar{M}\}, \\ \bar{\psi}_{1\tau} &\triangleq \mathrm{diag}_N\{\psi_{1\tau}\}, \;\; \bar{\psi}_{2\tau} \triangleq \mathrm{diag}_N\{\psi_{2\tau}\}, \\ \bar{D} &\triangleq \mathrm{diag}\{D_1, D_2, \cdots, D_N\}, \;\; \tilde{C} \triangleq \mathrm{diag}\{\bar{C}_1, \bar{C}_2, \cdots, \bar{C}_N\}, \end{aligned}$$

$$Z_i \triangleq \text{diag}\{\underbrace{0,\cdots,0}_{i-1}, I, \underbrace{0,\cdots,0}_{N-i}\},$$

it follows that

$$\begin{cases} e(t_{k+1}) = \tilde{A}e(t_k) - \mathcal{O}_r\tilde{C}e(t_k) - \mathcal{O}_r\bar{D}\bar{v}(t_k) \\ \qquad + \sum_{\tau=1}^{b} \tilde{B}_\tau\bar{\psi}_{1\tau}e(t_k - (n_\tau+1)b) + \sum_{\tau=1}^{b} \tilde{B}_\tau\bar{\psi}_{2\tau}e(t_k - n_\tau b) \\ \qquad + \sum_{\tau=0}^{b-1} \tilde{G}_\tau\bar{w}(t_k - \tau d) - \sum_{i=1}^{N} Z_iK_r\Phi^0_{i,r}\tilde{C}e(t_k) \\ \qquad - \sum_{i=1}^{N} Z_iK_r\Phi^1_{i,r}\Gamma_i\tilde{C}e(t_k) - \sum_{i=1}^{N} Z_iK_r\Phi^0_{i,r}\bar{D}\bar{v}(t_k) \\ \qquad - \sum_{i=1}^{N} Z_iK_r\Phi^1_{i,r}\Gamma_i\bar{D}\bar{v}(t_k) - \sum_{i=1}^{N} Z_iK_r\Phi^1_{i,r}\bar{\Gamma}_i(t_k)\tilde{C}\tilde{x}(t_k) \\ \qquad - \sum_{i=1}^{N} Z_iK_r\Phi^1_{i,r}\bar{\Gamma}_i(t_k)\bar{D}\bar{v}(t_k) \\ \tilde{z}(t_k) = \tilde{M}e(t_k) \end{cases} \tag{2.10}$$

where $K_r \triangleq H_r + \Delta H_r$ with $H_r \triangleq \left[\mathscr{H}_{ij}(\varphi^r_{ji})\right]_{N\times N}$, $\Delta H_r \triangleq \left[\Delta\mathscr{H}_{ij}(\varphi^r_{ji})\right]_{N\times N}$ and

$$\mathscr{H}_{ij}(\varphi^r_{ji}) \triangleq \begin{cases} a_{ij}H_{ij}(\varphi^r_{ji}), & i=1,2,\cdots,N,\ j\in\mathcal{V}_i, \\ 0, & i=1,2,\cdots,N,\ j\notin\mathcal{V}_i, \end{cases}$$

$$\Delta\mathscr{H}_{ij}(\varphi^r_{ji}) \triangleq \begin{cases} a_{ij}\Delta H_{ij}(\varphi^r_{ji}), & i=1,2,\cdots,N,\ j\in\mathcal{V}_i, \\ 0, & i=1,2,\cdots,N,\ j\notin\mathcal{V}_i. \end{cases}$$

It is not difficult to see that H_r and ΔH_r are both sparse matrices satisfying

$$H_r \in \Re_{bn_x\times n_y}, \quad \Delta H_r \in \Re_{bn_x\times n_y}$$

where $\Re_{bn_x\times n_y} \triangleq \{\bar{U} = [U_{ij}] \in \mathbb{R}^{Nbn_x\times Nn_y} | U_{ij} \in \mathbb{R}^{bn_x\times n_y}, U_{ij} = 0 \text{ if } j\notin\mathcal{V}_i\}$.

Finally, by setting $\eta(t_k) \triangleq \text{col}\{\tilde{x}(t_k), e(t_k)\}$ and denoting

$$\begin{aligned} \mathcal{A}_r &\triangleq \text{diag}\{\tilde{A}, \tilde{A} - \mathcal{O}_r\tilde{C}\},\ \mathcal{B}_\tau \triangleq \text{diag}\{\tilde{B}_\tau, \tilde{B}_\tau\}, \\ \mathcal{G}_\tau &\triangleq \text{col}\{\tilde{G}_\tau, \tilde{G}_\tau\},\ \tilde{\psi}_{1\tau} \triangleq \text{diag}\{\bar{\psi}_{1\tau}, \bar{\psi}_{1\tau}\}, \\ \tilde{\psi}_{2\tau} &\triangleq \text{diag}\{\bar{\psi}_{2\tau}, \bar{\psi}_{2\tau}\},\ \vec{C} \triangleq \text{diag}\{\tilde{C}, \tilde{C}\}, \\ \mathcal{Z}_{i,r} &\triangleq \text{diag}\{Z_iK_r, Z_iK_r\},\ \tilde{\Gamma}_{i,r} \triangleq \text{diag}\{0, \Phi^0_{i,r} + \Phi^1_{i,r}\Gamma_i\}, \\ \bar{\mathcal{F}}_{ij,r} &\triangleq \text{diag}\{0, \Phi^1_{i,r}F_j\},\ \bar{\mathcal{O}}_r \triangleq \text{diag}\{\mathcal{O}_r, \mathcal{O}_r\}, \\ \mathcal{D} &\triangleq \text{col}\{0, \bar{D}\},\ \mathcal{M} \triangleq \begin{bmatrix} 0 & \tilde{M} \end{bmatrix}, \\ \mathcal{F}_{ij,r} &\triangleq \begin{bmatrix} 0 & 0 \\ \Phi^1_{i,r}F_j & 0 \end{bmatrix},\ F_j \triangleq \text{diag}\{\underbrace{0,\cdots,0}_{j-1}, I, \underbrace{0,\cdots,0}_{N-j}\}, \end{aligned}$$

we arrive at the following augmented system

$$
\begin{cases}
\eta(t_{k+1}) = \mathcal{A}_r\eta(t_k) + \sum_{\tau=1}^{b} \mathcal{B}_\tau\tilde{\psi}_{2\tau}\eta(t_k - n_\tau b) \\
\quad + \sum_{\tau=1}^{b} \mathcal{B}_\tau\tilde{\psi}_{1\tau}\eta(t_k - (n_\tau + 1)b) \\
\quad + \sum_{\tau=0}^{b-1} \mathcal{G}_\tau\bar{w}(t_k - \tau d) - \sum_{i=1}^{N} \mathcal{Z}_{i,r}\tilde{\Gamma}_{i,r}\vec{\mathcal{C}}\eta(t_k) \\
\quad - \sum_{i=1}^{N} \mathcal{Z}_{i,r}\tilde{\Gamma}_{i,r}\mathcal{D}\bar{v}(t_k) - \bar{\mathcal{O}}_r\mathcal{D}\bar{v}(t_k) \\
\quad - \sum_{i=1}^{N}\sum_{j=1}^{N} \bar{\beta}_{ji}(t_k)\mathcal{Z}_{i,r}\mathcal{F}_{ij,r}\vec{\mathcal{C}}\eta(t_k) \\
\quad - \sum_{i=1}^{N}\sum_{j=1}^{N} \bar{\beta}_{ji}(t_k)\mathcal{Z}_{i,r}\bar{\mathcal{F}}_{ij,r}\mathcal{D}\bar{v}(t_k) \\
\tilde{z}(t_k) = \mathcal{M}\eta(t_k).
\end{cases}
\tag{2.11}
$$

Definition 2.1 *[34] The augmented system* (2.11) *with* $\bar{w}(t_k) = 0$ *and* $\bar{v}(t_k) = 0$ *is said to be exponentially mean-square stable if there exist constants* $\lambda > 0$ *and* $0 < \hbar < 1$ *such that*

$$
\mathbb{E}\{\|\eta(t_k)\|^2\} \le \lambda\hbar^{t_k} \sup_{i\in\mathbb{Z}^-} \mathbb{E}\{\|\eta(t_i)\|^2\}.
$$

The main purpose of this chapter is to design a set of non-fragile filters in the form of (2.9) such that the following requirements are satisfied simultaneously:

1. the zero-solution of the augmented system (2.11) with $\bar{w}(t_k) = 0$ and $\bar{v}(t_k) = 0$ is exponentially mean-square stable;
2. under zero initial condition, for a given disturbance attenuation level $\gamma > 0$ and all non-zero $\bar{w}(t_k)$ and $\bar{v}(t_k)$, the filtering error $\tilde{z}(t_k)$ satisfies the following H_∞ performance constraint:

$$
\sum_{k=0}^{\infty} \mathbb{E}\{\|\tilde{z}(t_k)\|^2\} < \gamma^2 \sum_{k=0}^{\infty} \|\bar{w}(t_k)\|^2 + \gamma^2 \sum_{k=0}^{\infty} \|\bar{v}(t_k)\|^2. \tag{2.12}
$$

2.2 Non-Fragile H_∞ Filter Design

In this section, we will investigate the non-fragile H_∞ filter design problem for discrete multi-rate time-delayed systems over sensor networks. First, let us consider the exponential mean-square stability and the H_∞ performance of the augmented system (2.11).

The following lemmas will be used in deriving the main results.

Lemma 2.2 *[8] For constant matrices S_1, S_2, and S_3 which satisfy $S_1 = S_1^T$ and $0 < S_2 = S_2^T$, we have $S_1 + S_3^T S_2^{-1} S_3 < 0$ if and only if*

$$\begin{bmatrix} S_1 & S_3^T \\ S_3 & -S_2 \end{bmatrix} < 0 \quad \text{or} \quad \begin{bmatrix} -S_2 & S_3 \\ S_3^T & S_1 \end{bmatrix} < 0.$$

Lemma 2.3 *[8] Let $\Psi = \Psi^T$, X and Y be real matrices with appropriate dimensions. For F satisfying $F^T F \leq I$, we have*

$$\Psi + XFY + Y^T F^T X^T < 0$$

if and only if there exists a positive scalar α such that

$$\Psi + \alpha^{-1} X X^T + \alpha Y^T Y < 0.$$

Lemma 2.4 *[134] Let $P_i \in \mathbb{R}^{bn_x \times bn_x}$ be an invertible matrix. For $X_i \in \mathbb{R}^{bn_x \times Nn_y}$, if $Y_i = P_i X_i$, then we have $X_i \in \bar{\Re}_{bn_x \times n_y} \Longleftrightarrow Y_i \in \bar{\Re}_{bn_x \times n_y}$ where $\bar{\Re}_{bn_x \times n_y} = \{\bar{U}_i = \begin{bmatrix} U_{i1} & U_{i2} & \cdots & U_{iN} \end{bmatrix} \in \mathbb{R}^{bn_x \times Nn_y} | U_{ij} \in \mathbb{R}^{bn_x \times n_y}, U_{ij} = 0 \text{ if } j \notin \mathcal{V}_i\}$.*

In the following theorem, a sufficient condition is provided under which the exponential mean-square stability as well as the H_∞ performance requirement are guaranteed.

Theorem 2.1 *Let the disturbance attenuation level $\gamma > 0$ and filter gains K_r, $\mathcal{O}_r$ ($r \in \bar{\mathcal{I}}$) be given. The augmented system (2.11) is exponentially mean-square stable (with $\bar{w}(t_k) = 0$ and $\bar{v}(t_k) = 0$) and the H_∞ performance constraint (2.12) is satisfied (for all non-zero $\bar{w}(t_k)$ and $\bar{v}(t_k)$) if there exist a set of positive definite matrices $\mathcal{P}_r$ ($r \in \bar{\mathcal{I}}$), positive definite matrices Q and R such that the following LMIs hold:*

$$\bar{\Pi}_r \triangleq \begin{bmatrix} \bar{\Pi}_{1_r} & \Pi_{2_r} \\ * & \bar{\Pi}_{4_r} \end{bmatrix} < 0 \tag{2.13}$$

where

$$
\begin{aligned}
\bar{\Pi}_{1_r} &\triangleq \begin{bmatrix} \bar{\Pi}_{11_r} & \Pi_{12_r} & \Pi_{13_r} \\ * & \bar{\mathcal{B}}_1^T \bar{\mathcal{P}}_r \bar{\mathcal{B}}_1 - \bar{Q} & \bar{\mathcal{B}}_1^T \bar{\mathcal{P}}_r \bar{\mathcal{B}}_2 \\ * & * & \bar{\mathcal{B}}_2^T \bar{\mathcal{P}}_r \bar{\mathcal{B}}_2 - \bar{R} \end{bmatrix}, \\
\Pi_{2_r} &\triangleq \begin{bmatrix} \Pi_{14_r} & \Pi_{15_r} \\ -\Pi_{24_r} & \bar{\mathcal{B}}_1^T \bar{\mathcal{P}}_r \bar{\mathcal{G}} \\ -\Pi_{34_r} & \bar{\mathcal{B}}_2^T \bar{\mathcal{P}}_r \bar{\mathcal{G}} \end{bmatrix}, \ \bar{\Pi}_{4_r} \triangleq \begin{bmatrix} \bar{\Pi}_{44_r} & -\Pi_{45_r} \\ * & \bar{\mathcal{G}}^T \bar{\mathcal{P}}_r \bar{\mathcal{G}} - \frac{\gamma^2}{b} I \end{bmatrix}, \\
\bar{\Pi}_{11_r} &\triangleq \Pi_{11_r} + \mathcal{M}^T \mathcal{M}, \ \bar{\Pi}_{44_r} \triangleq \Pi_{44_r} - \gamma^2 I, \\
\Pi_{12_r} &\triangleq \mathcal{A}_r^T \bar{\mathcal{P}}_r \bar{\mathcal{B}}_1 - \sum_{i=1}^{N} \vec{C}^T \tilde{\Gamma}_{i,r}^T \mathcal{Z}_{i,r}^T \bar{\mathcal{P}}_r \bar{\mathcal{B}}_1, \\
\Pi_{11_r} &\triangleq bQ + bR + \mathcal{A}_r^T \bar{\mathcal{P}}_r \mathcal{A}_r - \mathcal{P}_r - 2 \sum_{i=1}^{N} \mathcal{A}_r^T \bar{\mathcal{P}}_r \mathcal{Z}_{i,r} \tilde{\Gamma}_{i,r} \vec{C} \\
&\quad + \left(\sum_{i=1}^{N} \mathcal{Z}_{i,r} \tilde{\Gamma}_{i,r} \vec{C} \right)^T \bar{\mathcal{P}}_r \left(\sum_{i=1}^{N} \mathcal{Z}_{i,r} \tilde{\Gamma}_{i,r} \vec{C} \right) \\
&\quad + \sum_{i=1}^{N} \sum_{j=1}^{N} \ell_{ji}^2 \vec{C}^T \mathcal{F}_{ij,r}^T \mathcal{Z}_{i,r}^T \bar{\mathcal{P}}_r \mathcal{Z}_{i,r} \mathcal{F}_{ij,r} \vec{C}, \\
\Pi_{14_r} &\triangleq - \sum_{i=1}^{N} \mathcal{A}_r^T \bar{\mathcal{P}}_r \mathcal{Z}_{i,r} \tilde{\Gamma}_{i,r} \mathcal{D} + \sum_{i=1}^{N} \vec{C}^T \tilde{\Gamma}_{i,r}^T \mathcal{Z}_{i,r}^T \bar{\mathcal{P}}_r \bar{\mathcal{O}}_r \mathcal{D} \\
&\quad + \left(\sum_{i=1}^{N} \mathcal{Z}_{i,r} \tilde{\Gamma}_{i,r} \vec{C} \right)^T \bar{\mathcal{P}}_r \left(\sum_{i=1}^{N} \mathcal{Z}_{i,r} \tilde{\Gamma}_{i,r} \mathcal{D} \right) \\
&\quad - \mathcal{A}_r^T \bar{\mathcal{P}}_r \bar{\mathcal{O}}_r \mathcal{D} + \sum_{i=1}^{N} \sum_{j=1}^{N} \ell_{ji}^2 \vec{C}^T \mathcal{F}_{ij,r}^T \mathcal{Z}_{i,r}^T \bar{\mathcal{P}}_r \mathcal{Z}_{i,r} \bar{\mathcal{F}}_{ij,r} \mathcal{D}, \\
\Pi_{44_r} &\triangleq \left(\sum_{i=1}^{N} \mathcal{Z}_{i,r} \tilde{\Gamma}_{i,r} \mathcal{D} \right)^T \bar{\mathcal{P}}_r \left(\sum_{i=1}^{N} \mathcal{Z}_{i,r} \tilde{\Gamma}_{i,r} \mathcal{D} \right) + \mathcal{D}^T \bar{\mathcal{O}}_r^T \bar{\mathcal{P}}_r \bar{\mathcal{O}}_r \mathcal{D} \\
&\quad + 2 \sum_{i=1}^{N} \mathcal{D}^T \bar{\mathcal{O}}_r^T \bar{\mathcal{P}}_r \mathcal{Z}_{i,r} \tilde{\Gamma}_{i,r} \mathcal{D} \\
&\quad + \sum_{i=1}^{N} \sum_{j=1}^{N} \ell_{ji}^2 \mathcal{D}^T \bar{\mathcal{F}}_{ij,r}^T \mathcal{Z}_{i,r}^T \bar{\mathcal{P}}_r \mathcal{Z}_{i,r} \bar{\mathcal{F}}_{ij,r} \mathcal{D}, \\
\Pi_{13_r} &\triangleq \mathcal{A}_r^T \bar{\mathcal{P}}_r \bar{\mathcal{B}}_2 - \sum_{i=1}^{N} \vec{C}^T \tilde{\Gamma}_{i,r}^T \mathcal{Z}_{i,r}^T \bar{\mathcal{P}}_r \bar{\mathcal{B}}_2, \\
\Pi_{15_r} &\triangleq \mathcal{A}_r^T \bar{\mathcal{P}}_r \bar{\mathcal{G}} - \sum_{i=1}^{N} \vec{C}^T \tilde{\Gamma}_{i,r}^T \mathcal{Z}_{i,r}^T \bar{\mathcal{P}}_r \bar{\mathcal{G}},
\end{aligned}
$$

$$\Pi_{24_r} \triangleq \bar{\mathcal{B}}_1^T \bar{\mathcal{P}}_r \bar{\mathcal{O}}_r \mathcal{D} + \sum_{i=1}^{N} \bar{\mathcal{B}}_1^T \bar{\mathcal{P}}_r \mathcal{Z}_{i,r} \tilde{\Gamma}_{i,r} \mathcal{D},$$

$$\Pi_{34_r} \triangleq \bar{\mathcal{B}}_2^T \bar{\mathcal{P}}_r \bar{\mathcal{O}}_r \mathcal{D} + \sum_{i=1}^{N} \bar{\mathcal{B}}_2^T \bar{\mathcal{P}}_r \mathcal{Z}_{i,r} \tilde{\Gamma}_{i,r} \mathcal{D},$$

$$\Pi_{45_r} \triangleq \mathcal{D}^T \bar{\mathcal{O}}_r^T \bar{\mathcal{P}}_r \bar{\mathcal{G}} + \sum_{i=1}^{N} \mathcal{D}^T \tilde{\Gamma}_{i,r}^T \mathcal{Z}_{i,r}^T \bar{\mathcal{P}}_r \bar{\mathcal{G}},$$

$$\bar{\mathcal{B}}_1 \triangleq \begin{bmatrix} \mathcal{B}_1 \tilde{\psi}_{11} & \mathcal{B}_2 \tilde{\psi}_{12} & \cdots & \mathcal{B}_b \tilde{\psi}_{1b} \end{bmatrix}, \ \bar{\mathcal{P}}_r \triangleq \sum_{t \in \bar{\mathcal{I}}} \bar{p}_{rt} \mathcal{P}_t,$$

$$\bar{\mathcal{B}}_2 \triangleq \begin{bmatrix} \mathcal{B}_1 \tilde{\psi}_{21} & \mathcal{B}_2 \tilde{\psi}_{22} & \cdots & \mathcal{B}_b \tilde{\psi}_{2b} \end{bmatrix}, \ \bar{\mathcal{G}} \triangleq \begin{bmatrix} \mathcal{G}_0 & \mathcal{G}_1 & \cdots & \mathcal{G}_{b-1} \end{bmatrix},$$

$$\ell_{ji}^2 \triangleq \mu_{ji}(1-\mu_{ji}), \ \bar{Q} \triangleq \mathrm{diag}_b\{Q\}, \ \bar{R} \triangleq \mathrm{diag}_b\{R\}.$$

Proof *Choose the following Lyapunov functional*

$$V(\eta(t_k), r(t_k)) \triangleq V_1(\eta(t_k), r(t_k)) + V_2(\eta(t_k), r(t_k)) + V_3(\eta(t_k), r(t_k)) \quad (2.14)$$

where

$$V_1(\eta(t_k), r(t_k)) \triangleq \eta^T(t_k) \mathcal{P}_{r(t_k)} \eta(t_k),$$

$$V_2(\eta(t_k), r(t_k)) \triangleq \sum_{\tau=1}^{b} \sum_{\chi=t_{k-n_\tau-1}}^{t_{k-1}} \eta^T(\chi) Q \eta(\chi),$$

$$V_3(\eta(t_k), r(t_k)) \triangleq \sum_{\tau=1}^{b} \sum_{\chi=t_{k-n_\tau}}^{t_{k-1}} \eta^T(\chi) R \eta(\chi).$$

By denoting

$$\eta_{1b}(t_k) \triangleq \mathrm{col}\{\eta(t_{k-n_1-1}), \eta(t_{k-n_2-1}), \cdots, \eta(t_{k-n_b-1})\},$$

$$\eta_{2b}(t_k) \triangleq \mathrm{col}\{\eta(t_{k-n_1}), \eta(t_{k-n_2}), \cdots, \eta(t_{k-n_b})\},$$

$$\bar{w}_b(t_k) \triangleq \mathrm{col}\{\bar{w}(t_k), \bar{w}(t_k - d), \cdots, \bar{w}(t_k - (b-1)d)\},$$

it follows from (2.11) *that*

$$\begin{aligned}
&\mathbb{E}\{V_1(\eta(t_{k+1}), r(t_{k+1}))|\eta(t_k), r(t_k) = r\} - V_1(\eta(t_k), r) \\
&= \mathbb{E}\{\eta^T(t_k)(\mathcal{A}_r^T \bar{\mathcal{P}}_r \mathcal{A}_r - \mathcal{P}_r)\eta(t_k) + 2\eta^T(t_k)\mathcal{A}_r^T \bar{\mathcal{P}}_r \bar{\mathcal{B}}_1 \eta_{1b}(t_k) \\
&\quad + 2\eta^T(t_k)\mathcal{A}_r^T \bar{\mathcal{P}}_r \bar{\mathcal{B}}_2 \eta_{2b}(t_k) + 2\eta^T(t_k)\mathcal{A}_r^T \bar{\mathcal{P}}_r \bar{\mathcal{G}} \bar{w}_b(t_k) \\
&\quad - 2\eta^T(t_k)\mathcal{A}_r^T \bar{\mathcal{P}}_r \bar{\mathcal{O}}_r \mathcal{D} \bar{v}(t_k) + \eta_{1b}^T(t_k) \bar{\mathcal{B}}_1^T \bar{\mathcal{P}}_r \bar{\mathcal{B}}_1 \eta_{1b}(t_k) \\
&\quad + 2\eta_{1b}^T(t_k) \bar{\mathcal{B}}_1^T \bar{\mathcal{P}}_r \bar{\mathcal{B}}_2 \eta_{2b}(t_k) + 2\eta_{1b}^T(t_k) \bar{\mathcal{B}}_1^T \bar{\mathcal{P}}_r \bar{\mathcal{G}} \bar{w}_b(t_k) \\
&\quad - 2\eta_{1b}^T(t_k) \bar{\mathcal{B}}_1^T \bar{\mathcal{P}}_r \bar{\mathcal{O}}_r \mathcal{D} \bar{v}(t_k) + \eta_{2b}^T(t_k) \bar{\mathcal{B}}_2^T \bar{\mathcal{P}}_r \bar{\mathcal{B}}_2 \eta_{2b}(t_k) \\
&\quad + 2\eta_{2b}^T(t_k) \bar{\mathcal{B}}_2^T \bar{\mathcal{P}}_r \bar{\mathcal{G}} \bar{w}_b(t_k) - 2\eta_{2b}^T(t_k) \bar{\mathcal{B}}_2^T \bar{\mathcal{P}}_r \bar{\mathcal{O}}_r \mathcal{D} \bar{v}(t_k)
\end{aligned}$$

$$
\begin{aligned}
&+ \bar{w}_b^T(t_k)\bar{\mathcal{G}}^T\bar{\mathcal{P}}_r\bar{\mathcal{G}}\bar{w}_b(t_k) - 2\bar{w}_b^T(t_k)\bar{\mathcal{G}}^T\bar{\mathcal{P}}_r\bar{\mathcal{O}}_r\mathcal{D}\bar{v}(t_k) \\
&+ \bar{v}^T(t_k)\mathcal{D}^T\bar{\mathcal{O}}_r^T\bar{\mathcal{P}}_r\bar{\mathcal{O}}_r\mathcal{D}\bar{v}(t_k) \\
&- 2\eta^T(t_k)\mathcal{A}_r^T\bar{\mathcal{P}}_r\left(\sum_{i=1}^{N}\mathcal{Z}_{i,r}\tilde{\Gamma}_{i,r}\vec{C}\eta(t_k)\right) \\
&- 2\eta^T(t_k)\mathcal{A}_r^T\bar{\mathcal{P}}_r\left(\sum_{i=1}^{N}\mathcal{Z}_{i,r}\tilde{\Gamma}_{i,r}\mathcal{D}\bar{v}(t_k)\right) \\
&- 2\eta_{1b}^T(t_k)\bar{\mathcal{B}}_1^T\bar{\mathcal{P}}_r\left(\sum_{i=1}^{N}\mathcal{Z}_{i,r}\tilde{\Gamma}_{i,r}\vec{C}\eta(t_k)\right) \\
&- 2\eta_{1b}^T(t_k)\bar{\mathcal{B}}_1^T\bar{\mathcal{P}}_r\left(\sum_{i=1}^{N}\mathcal{Z}_{i,r}\tilde{\Gamma}_{i,r}\mathcal{D}\bar{v}(t_k)\right) \\
&- 2\eta_{2b}^T(t_k)\bar{\mathcal{B}}_2^T\bar{\mathcal{P}}_r\left(\sum_{i=1}^{N}\mathcal{Z}_{i,r}\tilde{\Gamma}_{i,r}\vec{C}\eta(t_k)\right) \\
&- 2\eta_{2b}^T(t_k)\bar{\mathcal{B}}_2^T\bar{\mathcal{P}}_r\left(\sum_{i=1}^{N}\mathcal{Z}_{i,r}\tilde{\Gamma}_{i,r}\mathcal{D}\bar{v}(t_k)\right) \\
&- 2\bar{w}_b^T(t_k)\bar{\mathcal{G}}^T\bar{\mathcal{P}}_r\left(\sum_{i=1}^{N}\mathcal{Z}_{i,r}\tilde{\Gamma}_{i,r}\vec{C}\eta(t_k)\right) \\
&- 2\bar{w}_b^T(t_k)\bar{\mathcal{G}}^T\bar{\mathcal{P}}_r\left(\sum_{i=1}^{N}\mathcal{Z}_{i,r}\tilde{\Gamma}_{i,r}\mathcal{D}\bar{v}(t_k)\right) \\
&+ 2\bar{v}^T(t_k)\mathcal{D}^T\bar{\mathcal{O}}_r^T\bar{\mathcal{P}}_r\left(\sum_{i=1}^{N}\mathcal{Z}_{i,r}\tilde{\Gamma}_{i,r}\vec{C}\eta(t_k)\right) \\
&+ 2\bar{v}^T(t_k)\mathcal{D}^T\bar{\mathcal{O}}_r^T\bar{\mathcal{P}}_r\left(\sum_{i=1}^{N}\mathcal{Z}_{i,r}\tilde{\Gamma}_{i,r}\mathcal{D}\bar{v}(t_k)\right) \\
&+ \sum_{i=1}^{N}\sum_{j=1}^{N}\eta^T(t_k)\vec{C}^T\tilde{\Gamma}_{i,r}^T\mathcal{Z}_{i,r}^T\bar{\mathcal{P}}_r\mathcal{Z}_{j,r}\tilde{\Gamma}_{j,r}\vec{C}\eta(t_k) \\
&+ 2\sum_{i=1}^{N}\sum_{j=1}^{N}\eta^T(t_k)\vec{C}^T\tilde{\Gamma}_{i,r}^T\mathcal{Z}_{i,r}^T\bar{\mathcal{P}}_r\mathcal{Z}_{j,r}\tilde{\Gamma}_{j,r}\mathcal{D}\bar{v}(t_k) \\
&+ \sum_{i=1}^{N}\sum_{j=1}^{N}\bar{v}^T(t_k)\mathcal{D}^T\tilde{\Gamma}_{i,r}^T\mathcal{Z}_{i,r}^T\bar{\mathcal{P}}_r\mathcal{Z}_{j,r}\tilde{\Gamma}_{j,r}\mathcal{D}\bar{v}(t_k) \\
&+ \sum_{i=1}^{N}\sum_{j=1}^{N}\ell_{ji}^2\eta^T(t_k)\vec{C}^T\mathcal{F}_{ij,r}^T\mathcal{Z}_{i,r}^T\bar{\mathcal{P}}_r\mathcal{Z}_{i,r}\mathcal{F}_{ij,r}\vec{C}\eta(t_k)
\end{aligned}
$$

$$+\sum_{i=1}^{N}\sum_{j=1}^{N}\ell_{ji}^2\bar{v}^T(t_k)\mathcal{D}^T\bar{\mathcal{F}}_{ij,r}^T\mathcal{Z}_{i,r}^T\bar{\mathcal{P}}_r\mathcal{Z}_{i,r}\bar{\mathcal{F}}_{ij,r}\mathcal{D}\bar{v}(t_k)$$
$$+2\sum_{i=1}^{N}\sum_{j=1}^{N}\ell_{ji}^2\eta^T(t_k)\vec{\mathcal{C}}^T\mathcal{F}_{ij,r}^T\mathcal{Z}_{i,r}^T\bar{\mathcal{P}}_r\mathcal{Z}_{i,r}\bar{\mathcal{F}}_{ij,r}\mathcal{D}\bar{v}(t_k)\}.$$

Moreover, it is not difficult to see that

$$\begin{aligned}
&\mathbb{E}\{V_2(\eta(t_{k+1}),r(t_{k+1}))|\eta(t_k),r(t_k)=r\}-V_2(\eta(t_k),r)\\
&=\mathbb{E}\{\sum_{\tau=1}^{b}\sum_{\chi=t_{k-n_\tau}}^{t_k}\eta^T(\chi)Q\eta(\chi)-\sum_{\tau=1}^{b}\sum_{\chi=t_{k-n_\tau-1}}^{t_{k-1}}\eta^T(\chi)Q\eta(\chi)\}\\
&=\sum_{\tau=1}^{b}\mathbb{E}\{\eta^T(t_k)Q\eta(t_k)\}-\sum_{\tau=1}^{b}\mathbb{E}\{\eta^T(t_{k-n_\tau-1})Q\eta(t_{k-n_\tau-1})\}\\
&=\sum_{\tau=1}^{b}\mathbb{E}\{\eta^T(t_k)Q\eta(t_k)\}-\mathbb{E}\{\eta_{1b}^T(t_k)\bar{Q}\eta_{1b}(t_k)\}
\end{aligned}$$

and

$$\begin{aligned}
&\mathbb{E}\{V_3(\eta(t_{k+1}),r(t_{k+1}))|\eta(t_k),r(t_k)=r\}-V_3(\eta(t_k),r)\\
&=\mathbb{E}\{\sum_{\tau=1}^{b}\sum_{\chi=t_{k-n_\tau+1}}^{t_k}\eta^T(\chi)R\eta(\chi)-\sum_{\tau=1}^{b}\sum_{\chi=t_{k-n_\tau}}^{t_{k-1}}\eta^T(\chi)R\eta(\chi)\}\\
&=\sum_{\tau=1}^{b}\mathbb{E}\{\eta^T(t_k)R\eta(t_k)\}-\sum_{\tau=1}^{b}\mathbb{E}\{\eta^T(t_{k-n_\tau})R\eta(t_{k-n_\tau})\}\\
&=\sum_{\tau=1}^{b}\mathbb{E}\{\eta^T(t_k)R\eta(t_k)\}-\mathbb{E}\{\eta_{2b}^T(t_k)\bar{R}\eta_{2b}(t_k)\}.
\end{aligned}$$

From the above equations, it can be easily obtained that

$$\begin{aligned}
\mathbb{E}\{\Delta V(t_k)\}=&\mathbb{E}\{V(\eta(t_{k+1}),r(t_{k+1}))|\eta(t_k),r(t_k)=r\}-V(\eta(t_k),r)\\
=&\mathbb{E}\{\xi^T(t_k)\Pi_r\xi(t_k)\}
\end{aligned}\tag{2.15}$$

where

$$\begin{aligned}
&\xi(t_k)\triangleq\begin{bmatrix}\eta^T(t_k) & \eta_{1b}^T(t_k) & \eta_{2b}^T(t_k) & \bar{v}^T(t_k) & \bar{w}_b^T(t_k)\end{bmatrix}^T,\\
&\Pi_r\triangleq\begin{bmatrix}\Pi_{1_r} & \Pi_{2_r}\\ * & \Pi_{4_r}\end{bmatrix},\ \Pi_{4_r}\triangleq\begin{bmatrix}\Pi_{44_r} & -\Pi_{45_r}\\ * & \bar{\mathcal{G}}^T\bar{\mathcal{P}}_r\bar{\mathcal{G}}\end{bmatrix},\\
&\Pi_{1_r}\triangleq\begin{bmatrix}\Pi_{11_r} & \Pi_{12_r} & \Pi_{13_r}\\ * & \bar{\mathcal{B}}_1^T\bar{\mathcal{P}}_r\bar{\mathcal{B}}_1-\bar{Q} & \bar{\mathcal{B}}_1^T\bar{\mathcal{P}}_r\bar{\mathcal{B}}_2\\ * & * & \bar{\mathcal{B}}_2^T\bar{\mathcal{P}}_r\bar{\mathcal{B}}_2-\bar{R}\end{bmatrix}.
\end{aligned}$$

Now, we are ready to show that the system (2.11) is exponentially mean-square stable and satisfies the H_∞ performance constraint. First, with $\bar{w}(t_k) = 0$ and $\bar{v}(t_k) = 0$, (2.15) results in

$$\mathbb{E}\{\Delta V(t_k)\} = \mathbb{E}\{\bar{\xi}^T(t_k)\Pi_{1_r}\bar{\xi}(t_k)\} \tag{2.16}$$

where $\bar{\xi}(t_k) \triangleq \text{col}\{\eta(t_k), \eta_{1b}(t_k), \eta_{2b}(t_k)\}$. From (2.13), we know that $\Pi_{1_r} < 0$. Then, there exists a sufficiently small scalar $\vartheta > 0$ such that

$$\mathbb{E}\{\Delta V(t_k)\} + \vartheta\mathbb{E}\{\|\eta(t_k)\|^2\} < 0.$$

By following the similar lines as the proof of Lemma 1 in [165], we know that the system (2.11) with $\bar{w}(t_k) = 0$ and $\bar{v}(t_k) = 0$ is exponentially mean-square stable.

In the following, we will consider the H_∞ performance constraint. First, the following index is introduced:

$$J(n) \triangleq \mathbb{E}\{\sum_{k=0}^{n} \|\tilde{z}(t_k)\|^2 - \gamma^2 \sum_{k=0}^{n} \|\bar{w}(t_k)\|^2 - \gamma^2 \sum_{k=0}^{n} \|\bar{v}(t_k)\|^2\}.$$

Under the zero initial condition, it is known from (2.13) that

$$\begin{aligned} J(n) =& \mathbb{E}\{\sum_{k=0}^{n} \|\tilde{z}(t_k)\|^2 - \frac{\gamma^2}{b} \sum_{k=0}^{n}\sum_{\tau=0}^{b-1} \|\bar{w}(t_k)\|^2 - \gamma^2 \sum_{k=0}^{n} \|\bar{v}(t_k)\|^2\} \\ \leq& \mathbb{E}\{\sum_{k=0}^{n} \|\tilde{z}(t_k)\|^2 - \frac{\gamma^2}{b} \sum_{k=0}^{n}\sum_{\tau=0}^{b-1} \|\bar{w}(t_k - \tau d)\|^2 - \gamma^2 \sum_{k=0}^{n} \|\bar{v}(t_k)\|^2\} \\ \leq& \sum_{k=0}^{n} \mathbb{E}\{\|\tilde{z}(t_k)\|^2 - \frac{\gamma^2}{b} \|\bar{w}_b(t_k)\|^2 - \gamma^2 \|\bar{v}(t_k)\|^2 + \Delta V(t_k)\} \\ =& \sum_{k=0}^{n} \mathbb{E}\{\xi^T(t_k)\bar{\Pi}_r\xi(t_k)\} < 0. \end{aligned}$$

Letting $n \to \infty$, it is obvious that

$$\sum_{k=0}^{\infty} \mathbb{E}\{\|\tilde{z}(t_k)\|^2\} < \gamma^2 \sum_{k=0}^{\infty} \|\bar{w}(t_k)\|^2 + \gamma^2 \sum_{k=0}^{\infty} \|\bar{v}(t_k)\|^2,$$

which means that the H_∞ performance constraint (2.12) is satisfied. The proof is complete.

In Theorem 2.1, the exponential mean-square stability and the H_∞ performance are analyzed. Based on the analysis result obtained in Theorem 2.1, in the following theorem, the filter design problem is solved for discrete multi-rate time-delayed systems over sensor networks.

For notation simplify, we denote

$$
\begin{aligned}
\mathcal{N}_{1r} &\triangleq \begin{bmatrix}0 & \aleph_{1r}^T\bar{\mathcal{P}}_r & 0\end{bmatrix}^T, \ \mathcal{N}_{2r} \triangleq \begin{bmatrix}\aleph_{2r} & 0 & 0\end{bmatrix},\\
\mathcal{N}_{3r} &\triangleq \begin{bmatrix}0 & \aleph_{3r}^T\bar{\mathcal{P}}_r & 0\end{bmatrix}^T, \ \mathcal{N}_{4} \triangleq \begin{bmatrix}\aleph_{4} & 0 & 0\end{bmatrix},\\
\mathcal{N}_{5r} &\triangleq \begin{bmatrix}0 & 0 & \aleph_{5r}^T\tilde{\mathcal{P}}_r\end{bmatrix}^T, \ \mathcal{N}_{6r} \triangleq \begin{bmatrix}\aleph_{6r} & 0 & 0\end{bmatrix},\\
\bar{\Upsilon}_i &\triangleq \mathrm{diag}_N\{\Upsilon_i\}, \ \tilde{\Upsilon}_i \triangleq \ \mathrm{diag}_2\{\bar{\Upsilon}_i\} \ (i=1,2,3,4),\\
\bar{\Sigma}_1(t_k) &\triangleq \mathrm{diag}_N\{\Sigma_1(t_k)\}, \ \tilde{\Sigma}_1(t_k) \triangleq \mathrm{diag}_2\{\bar{\Sigma}_1(t_k)\},\\
\hat{\Sigma}_1(t_k) &\triangleq \mathrm{diag}_N\{\tilde{\Sigma}_1(t_k)\}, \ \breve{\Sigma}_1(t_k) \triangleq \mathrm{diag}_N\{\hat{\Sigma}_1(t_k)\},\\
\bar{\Sigma}_2(t_k) &\triangleq \mathrm{diag}_N\{\Sigma_2(t_k)\}, \ \tilde{\Sigma}_2(t_k) \triangleq \mathrm{diag}_2\{\bar{\Sigma}_2(t_k)\},\\
\bar{Z}_i &\triangleq \mathrm{diag}_2\{Z_i\}, \ \bar{H}_r \triangleq \mathrm{diag}_2\{H_r\}, \ \bar{\mathcal{A}} \triangleq \mathrm{diag}_2\{\tilde{A}\},\\
\Delta\bar{H}_r &\triangleq \mathrm{diag}_2\{\Delta H_r\}, \ \tilde{\mathcal{P}}_r \triangleq \mathrm{diag}_{N^2}\{\bar{\mathcal{P}}_r\},\\
\tilde{\mathcal{O}}_r &\triangleq \mathrm{diag}_2\{\bar{O}_r\}, \ \Delta\tilde{\mathcal{O}}_r \triangleq \mathrm{diag}_2\{\Delta\bar{O}_r\},\\
\bar{O}_r &\triangleq \mathrm{diag}\{O_{1,r}, O_{2,r}, \cdots, O_{N,r}\},\\
\Delta\bar{O}_r &\triangleq \mathrm{diag}\{\Delta O_{1,r}, \Delta O_{2,r}, \cdots, \Delta O_{N,r}\},\\
\aleph_{1r} &\triangleq \begin{bmatrix}-\bar{Z}_1\bar{H}_r\tilde{\Upsilon}_1 & -\bar{Z}_2\bar{H}_r\tilde{\Upsilon}_1 & \cdots & -\bar{Z}_N\bar{H}_r\tilde{\Upsilon}_1\end{bmatrix},\\
\aleph_{2r} &\triangleq \begin{bmatrix}\aleph_{21r}^T & \aleph_{22r}^T & \cdots & \aleph_{2Nr}^T\end{bmatrix}^T, \ \aleph_{3r} \triangleq -\tilde{\mathcal{O}}_r\tilde{\Upsilon}_3,\\
\aleph_{2ir} &\triangleq \begin{bmatrix}\tilde{\Upsilon}_2\tilde{\Gamma}_{i,r}\vec{C} & 0 & 0 & \tilde{\Upsilon}_2\tilde{\Gamma}_{i,r}\mathcal{D} & 0\end{bmatrix}, \ \aleph_4 \triangleq \begin{bmatrix}\tilde{\Upsilon}_4\mathcal{C} & 0 & 0 & \tilde{\Upsilon}_4\mathcal{D} & 0\end{bmatrix},\\
\aleph_{5r} &\triangleq \mathrm{diag}\{\aleph_{51r}, \aleph_{52r}, \cdots, \aleph_{5Nr}\}, \ \aleph_{5ir} \triangleq \mathrm{diag}_N\{\bar{Z}_i\bar{H}_r\tilde{\Upsilon}_1\},\\
\aleph_{6r} &\triangleq \begin{bmatrix}\mathcal{S}_{3r}\vec{C} & 0 & 0 & \mathcal{S}_{4r}\mathcal{D} & 0\end{bmatrix}, \ \mathcal{S}_{3r} \triangleq \begin{bmatrix}\mathcal{S}_{31r}^T & \mathcal{S}_{32r}^T & \cdots & \mathcal{S}_{3Nr}^T\end{bmatrix}^T,\\
\mathcal{S}_{4r} &\triangleq \begin{bmatrix}\mathcal{S}_{41r}^T & \mathcal{S}_{42r}^T & \cdots & \mathcal{S}_{4Nr}^T\end{bmatrix}^T,\\
\mathcal{S}_{3ir} &\triangleq \begin{bmatrix}\ell_{1i}\mathcal{F}_{i1,r}^T\tilde{\Upsilon}_2^T & \ell_{2i}\mathcal{F}_{i2,r}^T\tilde{\Upsilon}_2^T & \cdots & \ell_{Ni}\mathcal{F}_{iN,r}^T\tilde{\Upsilon}_2^T\end{bmatrix}^T,\\
\mathcal{S}_{4ir} &\triangleq \begin{bmatrix}\ell_{1i}\bar{\mathcal{F}}_{i1,r}^T\tilde{\Upsilon}_2^T & \ell_{2i}\bar{\mathcal{F}}_{i2,r}^T\tilde{\Upsilon}_2^T & \cdots & \ell_{Ni}\bar{\mathcal{F}}_{iN,r}^T\tilde{\Upsilon}_2^T\end{bmatrix}^T,\\
\Delta\bar{\mathcal{A}}_r &\triangleq \begin{bmatrix}0 & 0\\ 0 & -\Delta\bar{O}_r\tilde{C}\end{bmatrix}, \ \mathcal{C} \triangleq \begin{bmatrix}0 & 0\\ 0 & \tilde{C}\end{bmatrix},\\
\tilde{\Pi}_r &\triangleq \mathrm{diag}\{-\mathcal{P}_r + bQ + bR + \mathcal{M}^T\mathcal{M}, -\bar{Q}, -\bar{R}, -\gamma^2 I, -\frac{\gamma^2}{b}I\},\\
\Xi_{1r} &\triangleq \begin{bmatrix}\mathcal{A}_r - \sum_{i=1}^N \mathcal{Z}_{i,r}\tilde{\Gamma}_{i,r}\vec{C} & \bar{\mathcal{B}}_1 & \bar{\mathcal{B}}_2 & -\bar{\mathcal{O}}_r\mathcal{D} - \sum_{i=1}^N \mathcal{Z}_{i,r}\tilde{\Gamma}_{i,r}\mathcal{D} & \bar{\mathcal{G}}\end{bmatrix},\\
\bar{\Xi}_{1r} &\triangleq \Big[\ \bar{\mathcal{A}} - \tilde{\mathcal{O}}_r\mathcal{C} - \textstyle\sum_{i=1}^N \bar{Z}_i\bar{H}_r\tilde{\Gamma}_{i,r}\vec{C} \quad \bar{\mathcal{B}}_1 \quad \bar{\mathcal{B}}_2\\
&\qquad\qquad -\tilde{\mathcal{O}}_r\mathcal{D} - \textstyle\sum_{i=1}^N \bar{Z}_i\bar{H}_r\tilde{\Gamma}_{i,r}\mathcal{D} \quad \bar{\mathcal{G}}\ \Big],\\
\Delta\bar{\Xi}_{1r} &\triangleq \begin{bmatrix}-\sum_{i=1}^N \bar{Z}_i\Delta\bar{H}_r\tilde{\Gamma}_{i,r}\vec{C} & 0 & 0 & -\sum_{i=1}^N \bar{Z}_i\Delta\bar{H}_r\tilde{\Gamma}_{i,r}\mathcal{D} & 0\end{bmatrix},\\
\tilde{\Xi}_{1r} &\triangleq \Big[\ \bar{\mathcal{P}}_r\bar{\mathcal{A}} - \bar{X}_r\mathcal{C} - \textstyle\sum_{i=1}^N \bar{Z}_i\bar{Y}_r\tilde{\Gamma}_{i,r}\vec{C} \quad \bar{\mathcal{P}}_r\bar{\mathcal{B}}_1 \quad \bar{\mathcal{P}}_r\bar{\mathcal{B}}_2
\end{aligned}
$$

$$-\bar{X}_r\mathcal{D}-\sum_{i=1}^{N}\bar{Z}_i\bar{Y}_r\tilde{\Gamma}_{i,r}\mathcal{D} \quad \bar{\mathcal{P}}_r\bar{\mathcal{G}} \Big],$$

$$\bar{\mathcal{S}}_{1ir} \triangleq \left[\ell_{1i}\mathcal{F}_{i1,r}^T\bar{H}_r^T\bar{Z}_i^T \quad \ell_{2i}\mathcal{F}_{i2,r}^T\bar{H}_r^T\bar{Z}_i^T \quad \cdots \quad \ell_{Ni}\mathcal{F}_{iN,r}^T\bar{H}_r^T\bar{Z}_i^T\right]^T,$$

$$\bar{\mathcal{S}}_{2ir} \triangleq \left[\ell_{1i}\bar{\mathcal{F}}_{i1,r}^T\bar{H}_r^T\bar{Z}_i^T \quad \ell_{2i}\bar{\mathcal{F}}_{i2,r}^T\bar{H}_r^T\bar{Z}_i^T \quad \cdots \quad \ell_{Ni}\bar{\mathcal{F}}_{iN,r}^T\bar{H}_r^T\bar{Z}_i^T\right]^T,$$

$$\tilde{\mathcal{S}}_{1ir} \triangleq \left[\ell_{1i}\mathcal{F}_{i1,r}^T\bar{Y}_r^T\bar{Z}_i^T \quad \ell_{2i}\mathcal{F}_{i2,r}^T\bar{Y}_r^T\bar{Z}_i^T \quad \cdots \quad \ell_{Ni}\mathcal{F}_{iN,r}^T\bar{Y}_r^T\bar{Z}_i^T\right]^T,$$

$$\tilde{\mathcal{S}}_{2ir} \triangleq \left[\ell_{1i}\bar{\mathcal{F}}_{i1,r}^T\bar{Y}_r^T\bar{Z}_i^T \quad \ell_{2i}\bar{\mathcal{F}}_{i2,r}^T\bar{Y}_r^T\bar{Z}_i^T \quad \cdots \quad \ell_{Ni}\bar{\mathcal{F}}_{iN,r}^T\bar{Y}_r^T\bar{Z}_i^T\right]^T,$$

$$\Delta\bar{\mathcal{S}}_{1ir} \triangleq \left[\ell_{1i}\mathcal{F}_{i1,r}^T\Delta\bar{H}_r^T\bar{Z}_i^T \quad \ell_{2i}\mathcal{F}_{i2,r}^T\Delta\bar{H}_r^T\bar{Z}_i^T \quad \cdots \quad \ell_{Ni}\mathcal{F}_{iN,r}^T\Delta\bar{H}_r^T\bar{Z}_i^T\right]^T,$$

$$\Delta\bar{\mathcal{S}}_{2ir} \triangleq \left[\ell_{1i}\bar{\mathcal{F}}_{i1,r}^T\Delta\bar{H}_r^T\bar{Z}_i^T \quad \ell_{2i}\bar{\mathcal{F}}_{i2,r}^T\Delta\bar{H}_r^T\bar{Z}_i^T \quad \cdots \quad \ell_{Ni}\bar{\mathcal{F}}_{iN,r}^T\Delta\bar{H}_r^T\bar{Z}_i^T\right]^T.$$

Theorem 2.2 *Given the disturbance attenuation level $\gamma > 0$, the augmented system (2.11) is exponentially mean-square stable and satisfies the H_∞ performance constraint (2.12) if there exist a set of positive definite matrices $\mathcal{P}_r \triangleq \mathrm{diag}\{P_r, P_r\}$ with $P_r \triangleq \mathrm{diag}\{P_{1r}, P_{2r}, \cdots, P_{Nr}\}$ $(r \in \bar{\mathcal{I}})$, a set of matrices $X_r \triangleq \mathrm{diag}\{X_{1r}, X_{2r}, \cdots, X_{Nr}\}$ $(r \in \bar{\mathcal{I}})$, a set of matrices $Y_r \triangleq \begin{bmatrix}Y_{1r}^T & Y_{2r}^T & \cdots & Y_{Nr}^T\end{bmatrix}^T$ $(r \in \bar{\mathcal{I}})$ satisfying the constraints $Y_{ir} \in \Re_{bn_x \times n_y}$ $(i = 1, 2, \cdots, N)$, two positive definite matrices Q and R and positive constant scalars α_1, α_2, and α_3 such that*

$$\begin{bmatrix} \tilde{\Psi}_r & \bar{\mathcal{N}}_{1r} & \alpha_1\mathcal{N}_{2r}^T & \bar{\mathcal{N}}_{3r} & \alpha_2\mathcal{N}_4^T & \bar{\mathcal{N}}_{5r} & \alpha_3\mathcal{N}_{6r}^T \\ * & -\alpha_1 I & 0 & 0 & 0 & 0 & 0 \\ * & * & -\alpha_1 I & 0 & 0 & 0 & 0 \\ * & * & * & -\alpha_2 I & 0 & 0 & 0 \\ * & * & * & * & -\alpha_2 I & 0 & 0 \\ * & * & * & * & * & -\alpha_3 I & 0 \\ * & * & * & * & * & * & -\alpha_3 I \end{bmatrix} < 0 \tag{2.17}$$

where

$$\tilde{\Psi}_r \triangleq \begin{bmatrix} \tilde{\Pi}_r & \tilde{\Xi}_{1r}^T & \tilde{\Xi}_{2r}^T \\ * & -\bar{\mathcal{P}}_r & 0 \\ * & * & -\tilde{\mathcal{P}}_r \end{bmatrix}, \quad \bar{\mathcal{P}}_r \triangleq \sum_{t\in\bar{\mathcal{I}}} \bar{p}_{rt}\mathcal{P}_t,$$

$$\bar{X}_r \triangleq \mathrm{diag}\{X_r, X_r\}, \ \bar{Y}_r \triangleq \mathrm{diag}\{Y_r, Y_r\},$$

$$\tilde{\Xi}_{2r}^T \triangleq \begin{bmatrix}\tilde{\mathcal{S}}_{1r}\vec{\mathcal{C}} & 0 & 0 & \tilde{\mathcal{S}}_{2r}\mathcal{D} & 0\end{bmatrix}, \ \tilde{\mathcal{S}}_{1r} \triangleq \begin{bmatrix}\tilde{\mathcal{S}}_{11r}^T & \tilde{\mathcal{S}}_{12r}^T & \cdots & \tilde{\mathcal{S}}_{1Nr}^T\end{bmatrix}^T,$$

$$\tilde{\mathcal{S}}_{2r} \triangleq \begin{bmatrix}\tilde{\mathcal{S}}_{21r}^T & \tilde{\mathcal{S}}_{22r}^T & \cdots & \tilde{\mathcal{S}}_{2Nr}^T\end{bmatrix}^T, \ \bar{\mathcal{N}}_{1r} \triangleq \begin{bmatrix}0 & \bar{\aleph}_{1r}^T & 0\end{bmatrix}^T,$$

$$\bar{\mathcal{N}}_{3r} \triangleq \begin{bmatrix}0 & \bar{\aleph}_{3r}^T & 0\end{bmatrix}^T, \ \bar{\aleph}_{3r} \triangleq -\bar{X}_r\tilde{\Upsilon}_3, \ \bar{\mathcal{N}}_{5r} \triangleq \begin{bmatrix}0 & 0 & \bar{\aleph}_{5r}^T\end{bmatrix}^T,$$

$$\bar{\aleph}_{1r} \triangleq \begin{bmatrix}-\bar{Z}_1\bar{Y}_r\tilde{\Upsilon}_1 & -\bar{Z}_2\bar{Y}_r\tilde{\Upsilon}_1 & \cdots & -\bar{Z}_N\bar{Y}_r\tilde{\Upsilon}_1\end{bmatrix},$$

$$\bar{\aleph}_{5r} \triangleq \mathrm{diag}\{\bar{\aleph}_{51r}, \bar{\aleph}_{52r}, \cdots, \bar{\aleph}_{5Nr}\}, \ \bar{\aleph}_{5ir} \triangleq \mathrm{diag}_N\{\bar{Z}_i\bar{Y}_r\tilde{\Upsilon}_1\}.$$

Furthermore, if the inequality (2.17) is feasible, the gains of filters can be obtained by

$$H_r = \begin{bmatrix}H_{1r}^T & H_{2r}^T & \cdots & H_{Nr}^T\end{bmatrix}^T, \ O_{i,r} = \bar{P}_{ir}^{-1}X_{ir}$$

where $H_{ir} \triangleq \bar{P}_{ir}^{-1}Y_{ir}$ *and* $\bar{P}_{ir} \triangleq \sum_{t\in\mathcal{I}} \bar{p}_{rt}P_{it}$ *for* $i = 1, 2, \ldots, N$.

Proof *It is easily known that*

$$\bar{\mathcal{O}}_r = \tilde{\mathcal{O}}_r + \Delta\tilde{\mathcal{O}}_r,\ \mathcal{Z}_{i,r} = \bar{Z}_i\bar{H}_r + \bar{Z}_i\Delta\bar{H}_r,\ \mathcal{A}_r = \bar{\mathcal{A}} - \tilde{\mathcal{O}}_r\mathcal{C} + \Delta\bar{\mathcal{A}}_r.$$

The matrix $\bar{\Pi}_r$ *in* (2.13) *can be rewritten as*

$$\bar{\Pi}_r = \tilde{\Pi}_r + \Xi_{1r}^T\bar{\mathcal{P}}_r\Xi_{1r} + \Xi_{2r}^T\tilde{\mathcal{P}}_r\Xi_{2r}$$

where

$$\begin{aligned}
\Xi_{2r} &\triangleq \begin{bmatrix}\mathcal{S}_{1r}\vec{C} & 0 & 0 & \mathcal{S}_{2r}\mathcal{D} & 0\end{bmatrix},\ \mathcal{S}_{1r} \triangleq \begin{bmatrix}\mathcal{S}_{11r}^T & \mathcal{S}_{12r}^T & \cdots & \mathcal{S}_{1Nr}^T\end{bmatrix}^T,\\
\mathcal{S}_{2r} &\triangleq \begin{bmatrix}\mathcal{S}_{21r}^T & \mathcal{S}_{22r}^T & \cdots & \mathcal{S}_{2Nr}^T\end{bmatrix}^T,\\
\mathcal{S}_{1ir} &\triangleq \begin{bmatrix}\ell_{1i}\mathcal{F}_{i1,r}^T\mathcal{Z}_{i,r}^T & \ell_{2i}\mathcal{F}_{i2,r}^T\mathcal{Z}_{i,r}^T & \cdots & \ell_{Ni}\mathcal{F}_{iN,r}^T\mathcal{Z}_{i,r}^T\end{bmatrix}^T,\\
\mathcal{S}_{2ir} &\triangleq \begin{bmatrix}\ell_{1i}\bar{\mathcal{F}}_{i1,r}^T\mathcal{Z}_{i,r}^T & \ell_{2i}\bar{\mathcal{F}}_{i2,r}^T\mathcal{Z}_{i,r}^T & \cdots & \ell_{Ni}\bar{\mathcal{F}}_{iN,r}^T\mathcal{Z}_{i,r}^T\end{bmatrix}^T.
\end{aligned}$$

Based on Lemma 2.2, $\bar{\Pi}_r < 0$ *is equivalent to*

$$\Psi_r = \begin{bmatrix}\tilde{\Pi}_r & \Xi_{1r}^T\bar{\mathcal{P}}_r & \Xi_{2r}^T\tilde{\mathcal{P}}_r\\ * & -\bar{\mathcal{P}}_r & 0\\ * & * & -\tilde{\mathcal{P}}_r\end{bmatrix} < 0.$$

Moreover, Ψ_r *can be written as*

$$\Psi_r = \bar{\Psi}_r + \Delta\bar{\Psi}_{1r} + \Delta\bar{\Psi}_{2r} + \Delta\bar{\Psi}_{3r}$$

where

$$\begin{aligned}
\bar{\Psi}_r &\triangleq \begin{bmatrix}\tilde{\Pi}_r & \bar{\Xi}_{1r}^T\bar{\mathcal{P}}_r & \bar{\Xi}_{2r}^T\tilde{\mathcal{P}}_r\\ * & -\bar{\mathcal{P}}_r & 0\\ * & * & -\tilde{\mathcal{P}}_r\end{bmatrix},\ \Delta\bar{\Psi}_{1r} \triangleq \begin{bmatrix}0 & \Delta\bar{\Xi}_{1r}^T\bar{\mathcal{P}}_r & 0\\ * & 0 & 0\\ * & * & 0\end{bmatrix},\\
\Delta\bar{\Psi}_{2r} &\triangleq \begin{bmatrix}0 & \Delta\tilde{\Xi}_{1r}^T\bar{\mathcal{P}}_r & 0\\ * & 0 & 0\\ * & * & 0\end{bmatrix},\ \Delta\bar{\Psi}_{3r} \triangleq \begin{bmatrix}0 & 0 & \Delta\bar{\Xi}_{2r}^T\tilde{\mathcal{P}}_r\\ * & 0 & 0\\ * & * & 0\end{bmatrix},\\
\Delta\tilde{\Xi}_{1r} &\triangleq \begin{bmatrix}\Delta\bar{\mathcal{A}}_r & 0 & 0 & -\Delta\tilde{\mathcal{O}}_r\mathcal{D} & 0\end{bmatrix},\ \bar{\Xi}_{2r} \triangleq \begin{bmatrix}\bar{\mathcal{S}}_{1r}\vec{C} & 0 & 0 & \bar{\mathcal{S}}_{2r}\mathcal{D} & 0\end{bmatrix},\\
\bar{\mathcal{S}}_{1r} &\triangleq \begin{bmatrix}\bar{\mathcal{S}}_{11r}^T & \bar{\mathcal{S}}_{12r}^T & \cdots & \bar{\mathcal{S}}_{1Nr}^T\end{bmatrix}^T,\ \bar{\mathcal{S}}_{2r} \triangleq \begin{bmatrix}\bar{\mathcal{S}}_{21r}^T & \bar{\mathcal{S}}_{22r}^T & \cdots & \bar{\mathcal{S}}_{2Nr}^T\end{bmatrix}^T,\\
\Delta\bar{\Xi}_{2r} &\triangleq \begin{bmatrix}\Delta\bar{\mathcal{S}}_{1r}\vec{C} & 0 & 0 & \Delta\bar{\mathcal{S}}_{2r}\mathcal{D} & 0\end{bmatrix},\\
\Delta\bar{\mathcal{S}}_{1r} &\triangleq \begin{bmatrix}\Delta\bar{\mathcal{S}}_{11r}^T & \Delta\bar{\mathcal{S}}_{12r}^T & \cdots & \Delta\bar{\mathcal{S}}_{1Nr}^T\end{bmatrix}^T,\\
\Delta\bar{\mathcal{S}}_{2r} &\triangleq \begin{bmatrix}\Delta\bar{\mathcal{S}}_{21r}^T & \Delta\bar{\mathcal{S}}_{22r}^T & \cdots & \Delta\bar{\mathcal{S}}_{2Nr}^T\end{bmatrix}^T.
\end{aligned}$$

From $\Delta H_{ij}(\varphi_{ji}^r) = H_{ij}(\varphi_{ji}^r)\Upsilon_1\Sigma_1(t_k)\Upsilon_2$ *and* $\Delta O_{i,r} = O_{i,r}\Upsilon_3\Sigma_2(t_k)\Upsilon_4$, *we know that* $\Delta\bar{H}_r = \bar{H}_r\tilde{\Upsilon}_1\tilde{\Sigma}_1(t_k)\tilde{\Upsilon}_2$, $\Delta\tilde{\mathcal{O}}_r = \tilde{\mathcal{O}}_r\tilde{\Upsilon}_3\tilde{\Sigma}_2(t_k)\tilde{\Upsilon}_4$ *and* $\Delta\bar{\mathcal{A}}_r = -\tilde{\mathcal{O}}_r\tilde{\Upsilon}_3\tilde{\Sigma}_2(t_k)\tilde{\Upsilon}_4\mathcal{C}$.

Noting that $\Delta\bar{\bar{\Xi}}_{1r} = \aleph_{1r}\hat{\Sigma}_1(t_k)\aleph_{2r}$*,* $\Delta\tilde{\bar{\Xi}}_{1r} = \aleph_{3r}\tilde{\Sigma}_2(t_k)\aleph_4$*, and* $\Delta\bar{\bar{\Xi}}_{2r} = \aleph_{5r}\breve{\Sigma}_1(t_k)\aleph_{6r}$*, we have*

$$\begin{aligned}\Psi_r =&\bar{\Psi}_r + \mathcal{N}_{1r}\hat{\Sigma}_1(t_k)\mathcal{N}_{2r} + \mathcal{N}_{2r}^T\hat{\Sigma}_1^T(t_k)\mathcal{N}_{1r}^T + \mathcal{N}_{3r}\tilde{\Sigma}_2(t_k)\mathcal{N}_4 + \mathcal{N}_4^T\tilde{\Sigma}_2^T(t_k)\mathcal{N}_{3r}^T\\ &+ \mathcal{N}_{5r}\breve{\Sigma}_1(t_k)\mathcal{N}_{6r} + \mathcal{N}_{6r}^T\breve{\Sigma}_1^T(t_k)\mathcal{N}_{5r}^T < 0. \end{aligned}\tag{2.18}$$

It is easily verified that $\hat{\Sigma}_1^T(t_k)\hat{\Sigma}_1(t_k) \leq I$*,* $\tilde{\Sigma}_2^T(t_k)\tilde{\Sigma}_2(t_k) \leq I$ *and* $\breve{\Sigma}_1^T(t_k)\breve{\Sigma}_1(t_k) \leq I$*. Thus, according to Lemma 2.3, it can be seen that* (2.18) *is equivalent to*

$$\begin{aligned}&\bar{\Psi}_r + \alpha_1^{-1}\mathcal{N}_{1r}\mathcal{N}_{1r}^T + \alpha_1\mathcal{N}_{2r}^T\mathcal{N}_{2r} + \alpha_2^{-1}\mathcal{N}_{3r}\mathcal{N}_{3r}^T\\ &\quad + \alpha_2\mathcal{N}_4^T\mathcal{N}_4 + \alpha_3^{-1}\mathcal{N}_{5r}\mathcal{N}_{5r}^T + \alpha_3\mathcal{N}_{6r}^T\mathcal{N}_{6r} < 0. \end{aligned}\tag{2.19}$$

By using Lemma 2.2 again, the inequality (2.19) *holds if*

$$\begin{bmatrix} \bar{\Psi}_r & \mathcal{N}_{1r} & \alpha_1\mathcal{N}_{2r}^T & \mathcal{N}_{3r} & \alpha_2\mathcal{N}_4^T & \mathcal{N}_{5r} & \alpha_3\mathcal{N}_{6r}^T\\ * & -\alpha_1 I & 0 & 0 & 0 & 0 & 0\\ * & * & -\alpha_1 I & 0 & 0 & 0 & 0\\ * & * & * & -\alpha_2 I & 0 & 0 & 0\\ * & * & * & * & -\alpha_2 I & 0 & 0\\ * & * & * & * & * & -\alpha_3 I & 0\\ * & * & * & * & * & * & -\alpha_3 I \end{bmatrix} < 0$$

holds. Letting $\bar{\mathcal{P}}_r\bar{Z}_i\bar{H}_r = \bar{Z}_i\bar{Y}_r$ *and* $\bar{\mathcal{P}}_r\tilde{\mathcal{O}}_r = \bar{X}_r$ $(i = 1, 2, \cdots, N,\ r \in \bar{\mathcal{I}})$*, we obtain the inequality* (2.17) *which ensures that the augmented system* (2.11) *is exponentially mean-square stable and satisfies the* H_∞ *performance constraint* (2.12)*.*

Now, it remains to calculate filter gains $O_{i,r}$ *and* $H_{ij}(\varphi_{ji}^r)$*. Letting*

$$H_{ir} = \begin{bmatrix} H_{i1r} & H_{i2r} & \cdots & H_{iNr}\end{bmatrix}$$

be the i*th row of the block matrix* H_r *and* Y_{ir} *be the* i*th row of the block matrix* Y_r*, we have*

$$P_{ir}H_{ir} = Y_{ir},\ P_{ir}O_{i,r} = X_{ir},\ (i = 1, 2, \cdots, N,\ r \in \bar{\mathcal{I}}).$$

According to Lemma 2.4, we know that $H_{ir} \in \bar{\Re}_{bn_x\times n_y}$ *and therefore* $H_r \in \Re_{bn_x\times n_y}$*, which completes the proof.*

Remark 2.3 *In this chapter, we aim to propose a non-fragile* H_∞ *filtering scheme for the discrete multi-rate time-delayed systems over sensor networks with packet dropouts. In Theorem 2.1, sufficient conditions are given such that the filtering error dynamics is exponentially mean-square stable and the* H_∞ *performance constraint is satisfied. In Theorem 2.2, the filter gains are characterized in terms of the solutions to a set of LMIs. Note that, in Theorem 2.2, the influences from the time-delays, the packet dropouts, the multi-rate sampling, and the topology of the sensor network to the filtering performance are all reflected in the matrix inequality* (2.17)*.*

Remark 2.4 *Until now, we have solved the non-fragile H_∞ filtering problem for the discrete multi-rate time-delayed systems over sensor networks. Comparing with the existing results, the main advantages of the main results are highlighted as follows: 1) the packet dropouts occurring in the communication channels among the sensor nodes are characterized by the Gilbert-Elliott model which is suitable to cater for engineering practice; 2) the influence of the time-delays is, for the first time, studied in the filtering problem for MRSs; and 3) the non-fragile H_∞ filtering problem is investigated in the framework of sensor networks.*

2.3 Some Special Cases

In this section, we aim to show that the proposed system model and H_∞ filtering algorithm can be specialized to the following two kinds of systems: 1) MRSs without time-delays and packet dropouts and 2) single-rate time-delayed systems without packet dropouts.

Case 2.1 *By setting $d = 0$ and $p_{00} = p_{10} = 1$, the proposed system is specialized to the MRS without time-delays and packet dropouts. The corresponding filter design problem is solved in the following corollary.*

Corollary 2.1 *Given the disturbance attenuation level $\gamma > 0$, the H_∞ filtering problem for system (2.1) with $d = 0$ and $p_{00} = p_{10} = 1$ is solved by filter (2.9) if there exist a set of positive definite matrices $\mathcal{P} \triangleq \text{diag}\{P, P\}$ with $P \triangleq \text{diag}\{P_1, P_2, \cdots, P_N\}$, a set of matrices $X \triangleq \text{diag}\{X_1, X_2, \cdots, X_N\}$, a set of matrices $Y \triangleq \text{col}\{Y_1, Y_2, \cdots, Y_N\}$ satisfying the constraints $Y_i \in \bar{\Re}_{bn_x \times n_y}$ $(i = 1, 2, \cdots, N)$, and positive scalars α_1 and α_2 such that*

$$\begin{bmatrix} \tilde{\Psi} & \bar{\mathcal{N}}_1 & \alpha_1\mathcal{N}_2^T & \bar{\mathcal{N}}_3 & \alpha_2\mathcal{N}_4^T \\ * & -\alpha_1 I & 0 & 0 & 0 \\ * & * & -\alpha_1 I & 0 & 0 \\ * & * & * & -\alpha_2 I & 0 \\ * & * & * & * & -\alpha_2 I \end{bmatrix} < 0 \tag{2.20}$$

where

$$\tilde{\Psi} \triangleq \begin{bmatrix} \tilde{\Pi} & \tilde{\Xi}_1^T \\ * & -\mathcal{P} \end{bmatrix}, \ \bar{\mathcal{N}}_1 \triangleq \begin{bmatrix} 0 & \bar{\aleph}_1^T \end{bmatrix}^T, \ \mathcal{N}_2 \triangleq \begin{bmatrix} \aleph_2 & 0 \end{bmatrix},$$

$$\bar{\mathcal{N}}_3 \triangleq \begin{bmatrix} 0 & \bar{\aleph}_3^T \end{bmatrix}^T, \ \mathcal{N}_4 \triangleq \begin{bmatrix} \aleph_4 & 0 \end{bmatrix}, \ \aleph_2 \triangleq \begin{bmatrix} \aleph_{21}^T & \aleph_{22}^T & \cdots & \aleph_{2N}^T \end{bmatrix}^T,$$

$$\aleph_{2i} \triangleq \begin{bmatrix} \tilde{\Upsilon}_2\tilde{\Gamma}\vec{\mathcal{C}} & \tilde{\Upsilon}_2\tilde{\Gamma}\mathcal{D} & 0 \end{bmatrix}, \ \aleph_4 \triangleq \begin{bmatrix} \tilde{\Upsilon}_4\mathcal{C} & \tilde{\Upsilon}_4\mathcal{D} & 0 \end{bmatrix},$$

$$\tilde{\Pi} \triangleq \text{diag}\{-\mathcal{P} + \mathcal{M}^T\mathcal{M}, -\gamma^2 I, -\frac{\gamma^2}{b}I\}, \ \tilde{\Gamma} \triangleq \text{diag}\{0, I\},$$

$$\Xi_1 \triangleq \begin{bmatrix} \mathcal{A} - \sum_{i=1}^{N} \mathcal{Z}_i \tilde{\Gamma} \vec{\mathcal{C}} & -\bar{\mathcal{O}}\mathcal{D} - \sum_{i=1}^{N} \mathcal{Z}_i \tilde{\Gamma} \mathcal{D} & \mathcal{G} \end{bmatrix},$$

$$\bar{\Xi}_1 \triangleq \begin{bmatrix} \bar{\mathcal{A}} - \tilde{\mathcal{O}}\mathcal{C} - \sum_{i=1}^{N} \bar{\mathcal{Z}}_i \bar{H} \tilde{\Gamma} \vec{\mathcal{C}} & -\tilde{\mathcal{O}}\mathcal{D} - \sum_{i=1}^{N} \bar{\mathcal{Z}}_i \bar{H} \tilde{\Gamma} \mathcal{D} & \bar{\mathcal{G}} \end{bmatrix},$$

$$\Delta\bar{\Xi}_1 \triangleq \begin{bmatrix} -\sum_{i=1}^{N} \bar{\mathcal{Z}}_i \Delta \bar{H} \tilde{\Gamma} \vec{\mathcal{C}} & -\sum_{i=1}^{N} \bar{\mathcal{Z}}_i \Delta \bar{H} \tilde{\Gamma} \mathcal{D} & 0 \end{bmatrix},$$

$$\tilde{\Xi}_1 \triangleq \begin{bmatrix} \mathcal{P}\bar{\mathcal{A}} - \bar{X}\mathcal{C} - \sum_{i=1}^{N} \bar{\mathcal{Z}}_i \bar{Y} \tilde{\Gamma} \vec{\mathcal{C}} & -\bar{X}\mathcal{D} - \sum_{i=1}^{N} \bar{\mathcal{Z}}_i \bar{Y} \tilde{\Gamma} \mathcal{D} & \mathcal{P}\bar{\mathcal{G}} \end{bmatrix}$$

and other notations are the same with the corresponding ones in section 2.2 (without the subscript r).

Furthermore, if the inequality (2.20) *is feasible, the filter gains can be obtained by*

$$H = \begin{bmatrix} H_1^T & H_2^T & \cdots & H_N^T \end{bmatrix}^T, \; H_i = P_i^{-1} Y_i, \; O_i = P_i^{-1} X_i.$$

Proof *This proof is completed immediately by setting* $d = 0$ *and* $p_{00} = p_{10} = 1$ *in Theorem 2.2.*

Case 2.2 *By setting* $b = 1$ *and* $p_{00} = p_{10} = 1$*, the proposed system is specialized to the single-rate time-delayed system without packet dropouts. The corresponding filter design problem is solved in the following corollary.*

Corollary 2.2 *Given the disturbance attenuation level* $\gamma > 0$*, the* H_∞ *filtering problem for system* (2.1) *with* $b = 1$ *and* $p_{00} = p_{10} = 1$ *is solved by filter* (2.9) *if there exist a set of positive definite matrices* $\mathcal{P} = \text{diag}\{P, P\}$ *with* $P = \text{diag}\{P_1, P_2, \cdots, P_N\}$*, a set of matrices* $X = \text{diag}\{X_1, X_2, \cdots, X_N\}$*, a set of matrices* $Y = \text{col}\{Y_1, Y_2, \cdots, Y_N\}$ *satisfying the constraints* $Y_i \in \bar{\Re}_{n_x \times n_y}$ $(i = 1, 2, \cdots, N)$*, positive definite matrix* Q*, and positive scalars* α_1 *and* α_2 *such that*

$$\begin{bmatrix} \tilde{\Psi} & \bar{\mathcal{N}}_1 & \alpha_1 \mathcal{N}_2^T & \bar{\mathcal{N}}_3 & \alpha_2 \mathcal{N}_4^T \\ * & -\alpha_1 I & 0 & 0 & 0 \\ * & * & -\alpha_1 I & 0 & 0 \\ * & * & * & -\alpha_2 I & 0 \\ * & * & * & * & -\alpha_2 I \end{bmatrix} < 0 \tag{2.21}$$

where

$$\tilde{\Psi} \triangleq \begin{bmatrix} \tilde{\Pi} & \tilde{\Xi}_1^T \\ * & -\mathcal{P} \end{bmatrix}, \; \bar{\mathcal{N}}_1 \triangleq \begin{bmatrix} 0 & \bar{\aleph}_1^T \end{bmatrix}^T, \; \mathcal{N}_2 \triangleq \begin{bmatrix} \aleph_2 & 0 \end{bmatrix},$$

$$\bar{\mathcal{N}}_3 \triangleq \begin{bmatrix} 0 & \bar{\aleph}_3^T \end{bmatrix}^T, \; \mathcal{N}_4 \triangleq \begin{bmatrix} \aleph_4 & 0 \end{bmatrix}, \; \aleph_2 \triangleq \begin{bmatrix} \aleph_{21}^T & \aleph_{22}^T & \cdots & \aleph_{2N}^T \end{bmatrix}^T,$$

$$\aleph_4 \triangleq \begin{bmatrix} \tilde{\Upsilon}_4 \mathcal{C} & 0 & \tilde{\Upsilon}_4 \mathcal{D} & 0 \end{bmatrix}, \; \aleph_{2i} \triangleq \begin{bmatrix} \tilde{\Upsilon}_2 \tilde{\Gamma} \vec{\mathcal{C}} & 0 & \tilde{\Upsilon}_2 \tilde{\Gamma} \mathcal{D} & 0 \end{bmatrix},$$

$$\tilde{\Pi} \triangleq \text{diag}\{-\mathcal{P} + \mathcal{M}^T \mathcal{M}, -\gamma^2 I, -\frac{\gamma^2}{b} I\}, \; \tilde{\Gamma} \triangleq \text{diag}\{0, I\},$$

$$\bar{\Xi}_1 \triangleq \begin{bmatrix} \bar{\mathcal{A}} - \tilde{\mathcal{O}}\mathcal{C} - \sum_{i=1}^{N} \bar{\mathcal{Z}}_i \bar{H} \tilde{\Gamma} \vec{\mathcal{C}} & \bar{\mathcal{B}}_1 & -\tilde{\mathcal{O}}\mathcal{D} - \sum_{i=1}^{N} \bar{\mathcal{Z}}_i \bar{H} \tilde{\Gamma} \mathcal{D} & \mathcal{G} \end{bmatrix},$$

$$\Delta\bar{\Xi}_1 \triangleq \begin{bmatrix} -\sum_{i=1}^{N} \bar{Z}_i \Delta \bar{H} \tilde{\Gamma} \vec{C} & 0 & -\sum_{i=1}^{N} \bar{Z}_i \Delta \bar{H} \tilde{\Gamma} \mathcal{D} & 0 \end{bmatrix},$$

$$\tilde{\Xi}_1 \triangleq \begin{bmatrix} \mathcal{P}\bar{\mathcal{A}} - \bar{X}\mathcal{C} - \sum_{i=1}^{N} \bar{Z}_i \bar{Y} \tilde{\Gamma} \vec{C} & \mathcal{P}\bar{\mathcal{B}}_1 & -\bar{X}\mathcal{D} - \sum_{i=1}^{N} \bar{Z}_i \bar{Y} \tilde{\Gamma} \mathcal{D} & \mathcal{P}\bar{\mathcal{G}} \end{bmatrix}$$

and other notations are the same with the corresponding ones in section 2.2 (without the subscript r).

Furthermore, if the inequality (2.21) *is feasible, the filter gains can be obtained by*

$$H = \begin{bmatrix} H_1^T & H_2^T & \cdots & H_N^T \end{bmatrix}^T, \; H_i = P_i^{-1} Y_i, \; O_i = P_i^{-1} X_i.$$

Proof *This proof is completed immediately by setting* $b = 1$ *and* $p_{00} = p_{10} = 1$ *in Theorem 2.2.*

Remark 2.5 *The system model proposed in this chapter closely reflects some real-world phenomena such as multi-rate sampling, time-delays, packet dropouts, and filter gain variations. Moreover, the proposed system model is quite general since it includes the systems with individual, or part of the above-mentioned phenomena as special cases. For example, the proposed system model is specialized to MRSs without time-delays when* $d = 0$, *to single-rate time-delayed systems with packet dropouts when* $b = 1$, *and to multi-rate time-delayed systems without packet dropouts when* $p_{00} = p_{10} = 1$. *Similarly, the non-fragile filter* (2.9) *is transformed to the traditional filter by setting* $\Upsilon_1 = \Upsilon_2 = \Upsilon_3 = \Upsilon_4 = 0$.

2.4 Simulation Examples

Consider a sensor network whose topology is described by a directed graph $\mathscr{G} = (\mathcal{V}, \mathscr{E}, \mathcal{R})$ with the set of nodes $\mathcal{V} = \{1, 2, 3\}$, the set of edges $\mathscr{E} = \{(1,2), (2,3), (3,1)\}$ and the weighted adjacency matrix $\mathcal{R} = [a_{ij}]_{3\times 3}$ with adjacency element $a_{ij} = 1$ for $(i, j) \in \mathscr{E}$ and $a_{ij} = 0$ otherwise.

Example 2.1

The discrete time-delayed system is described by (2.1) with the following matrix parameters:

$$A = \begin{bmatrix} 0.13 & 0.11 \\ 0.25 & -0.20 \end{bmatrix}, \quad B = \begin{bmatrix} -0.23 & 0.35 \\ -0.26 & 0.28 \end{bmatrix},$$

$$E = \begin{bmatrix} 0.11 \\ -0.17 \end{bmatrix}, \quad M = \begin{bmatrix} 0.55 & -0.15 \end{bmatrix}.$$

The measurements from sensor nodes are modeled by (2.2) with the following parameters:

$$C_1 = \begin{bmatrix}1.12 & -1.08\end{bmatrix}, \; C_2 = \begin{bmatrix}1.13 & -1.35\end{bmatrix}, \; C_3 = \begin{bmatrix}1.14 & -1.23\end{bmatrix},$$
$$D_1 = 0.53, \; D_2 = 0.51, \; D_3 = 0.56.$$

The Markov chains $\theta_{13}(t_k)$, $\theta_{21}(t_k)$, and $\theta_{32}(t_k)$ take values in the finite state space $\mathcal{I} = \{0, 1\}$ with the transition probability matrix

$$\Lambda = \begin{bmatrix} p_{00} & p_{01} \\ p_{10} & p_{11} \end{bmatrix} = \begin{bmatrix} 0.7 & 0.3 \\ 0.8 & 0.2 \end{bmatrix},$$

and other parameters are chosen as $d = 1$, $b = 2$, $\mu_{13} = 0.8$, $\mu_{21} = 0.8$, $\mu_{32} = 0.8$, $\Upsilon_1 = 1$, $\Upsilon_2 = 1$, $\Upsilon_3 = 1$, $\Upsilon_4 = 1$, and $\gamma = 0.5$. By using the Matlab software, the filter gains $H_{ij}(\theta_{ji}(t_k))$ can be obtained as

$$H_{ij}(\theta_{ji}(t_k)) = \begin{cases} H_{ij}(\theta_{ji}(t_k)), & (i,j) \in \mathscr{E} \\ 0, & \text{otherwise} \end{cases}$$

where

$$H_{12}(0) = \begin{bmatrix}0.0099 & 0.0297 & 0.0732 & 0.0542\end{bmatrix}^T,$$
$$H_{12}(1) = \begin{bmatrix}0.0168 & 0.0337 & 0.0979 & 0.0970\end{bmatrix}^T,$$
$$H_{23}(0) = \begin{bmatrix}0.0129 & 0.0330 & 0.0761 & 0.0580\end{bmatrix}^T,$$
$$H_{23}(1) = \begin{bmatrix}0.0190 & 0.0385 & 0.0995 & 0.0963\end{bmatrix}^T,$$
$$H_{31}(0) = \begin{bmatrix}0.0150 & 0.0322 & 0.0818 & 0.0661\end{bmatrix}^T,$$
$$H_{31}(1) = \begin{bmatrix}0.0200 & 0.0367 & 0.1013 & 0.1009\end{bmatrix}^T.$$

The filter gains $O_{i,r}$ are omitted for saving space.

In the simulation, the initial values of the state are set as $\phi(-1) = \phi(0) = \text{col}\{0, 0\}$, the process noise and the measurement noises are chosen as $w(T_k) = 0.1\sin(T_k)/(8T_k + 1)$ and $v_i(t_k) = \cos(t_k)/(10t_k + i)$ $(i = 1, 2, 3)$, respectively. Simulation results are shown in Figs. 2.1 and 2.2. The output $z(T_k)$ and its estimates are shown in Fig. 2.1, and the filtering errors of the filters are shown in Fig. 2.2 from which we can see that the filtering errors converge to zero at about $T_k = 56h$. It can be seen that the proposed filter design scheme is indeed effective.

Example 2.2

To show the superiority of the proposed filtering algorithm, in this example, a comparison of the filtering performance is conducted between the proposed filter and the filter developed without considering the packet dropouts and filter gain variations (denoted as FWPF) on the proposed system model.

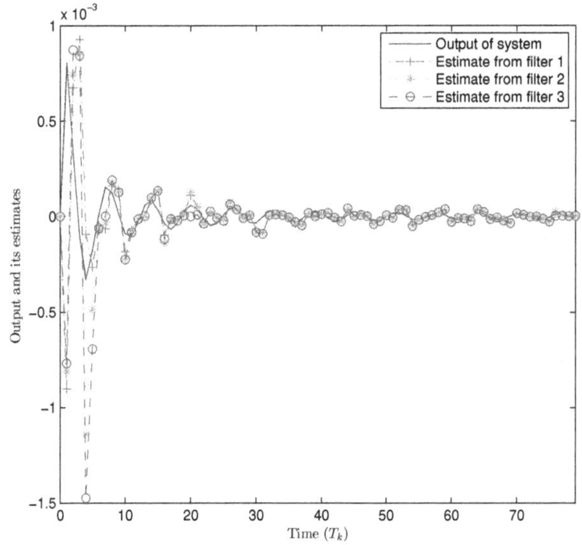

FIGURE 2.1: Output $z(T_k)$ and its estimates.

In this example, the probabilities of the packet dropouts are chosen as $\mu_{13} = 0.2$, $\mu_{21} = 0.2$, $\mu_{32} = 0.2$, the time-delay is chosen as $d = 1$, the gain variations matrices Υ_1, Υ_2, Υ_3, Υ_4 are set as $\Upsilon_1 = \Upsilon_2 = 3$ and $\Upsilon_3 = \Upsilon_4 = 5$, and the initial values of the state are set as $\phi(-1) = \phi(0) = \text{col}\{1, 1\}$. Other

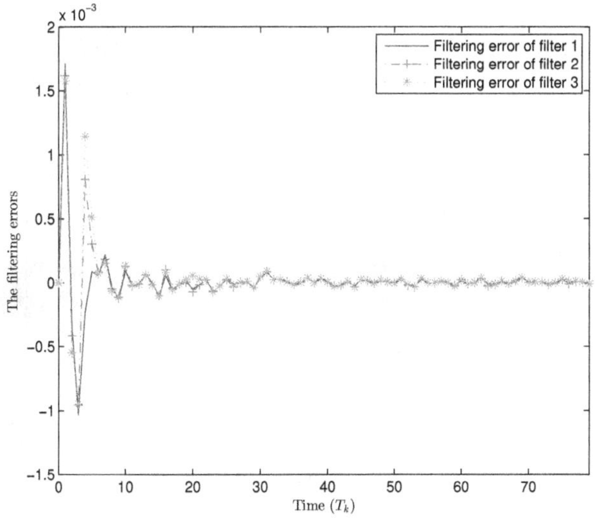

FIGURE 2.2: The filtering errors.

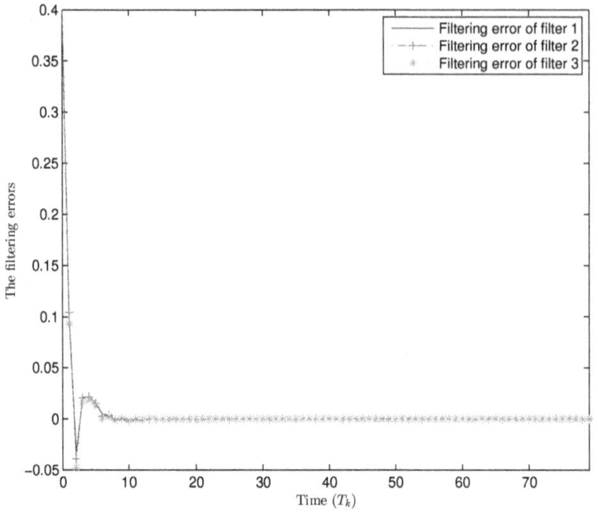

FIGURE 2.3: The filtering errors of the proposed filters.

parameters are chosen as those corresponding ones in Example 2.1. Simulation results are given in Figs. 2.3–2.4. It can be seen from Fig. 2.3 that the proposed filtering algorithm performs well when the system suffers packet dropouts and filter gain variations. Nevertheless, the FWPF fails to generate accurate

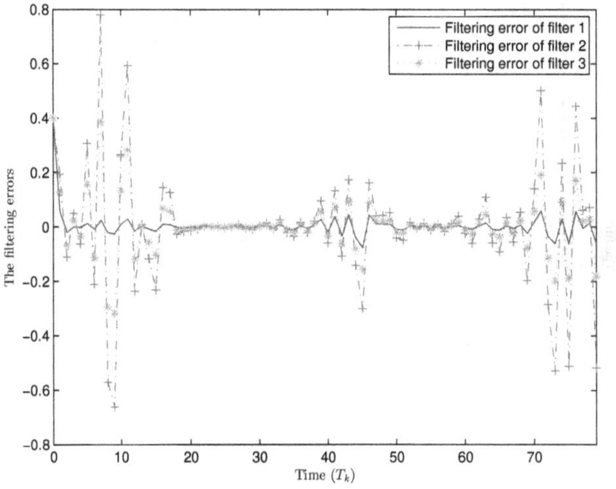

FIGURE 2.4: The filtering errors of the FWPF.

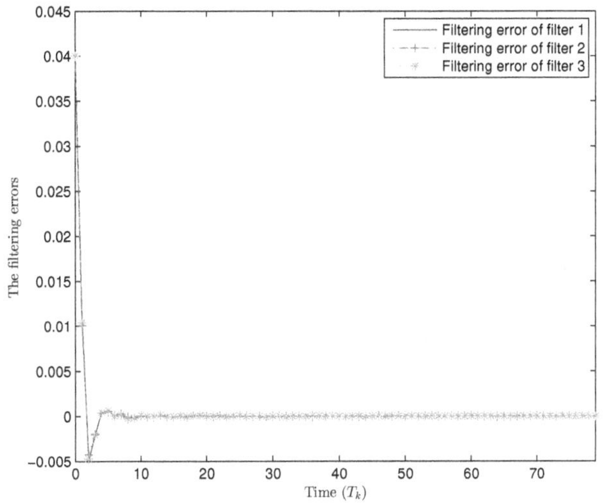

FIGURE 2.5: The filtering errors of the DFF without gain variations.

estimates to the system state. As such, the proposed filtering algorithm is indeed effective in dealing with the non-fragile filtering problem for multi-rate time-delayed systems with packet dropouts.

Example 2.3

In this example, we are going to show the influence of the filter gain variations on the filtering performance. First, an H_∞ filter (denoted as FWF) is designed for the proposed system (2.1) with measurement output (2.8) without considering the filter gain variations. Then, the developed FWF with filter gain variations and without filter gain variations are, respectively, applied to the proposed systems. The system parameters are chosen as the same with the corresponding ones in Example 2.2.

The simulation results are shown in Figs. 2.5–2.6. It can be seen from Fig. 2.5 that the FWF accurately estimate the system state when there does not exist filter gain variations. Nevertheless, when there exist filter gain variations in the FWF, as shown in Fig. 2.6, the filtering performance is seriously degraded. As such, the existence of the filter gain variations degrade the performance of the filtering algorithm.

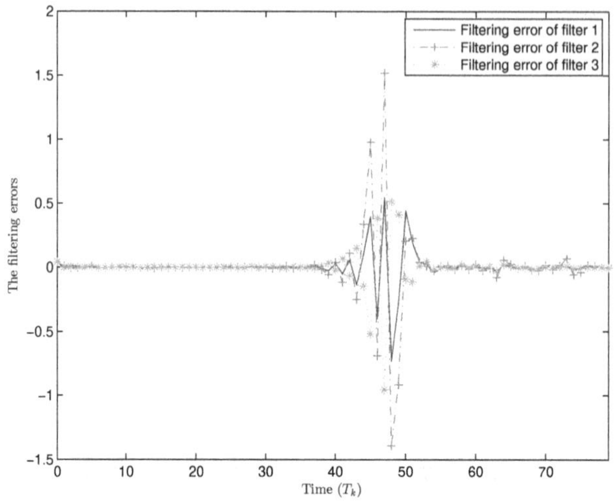

FIGURE 2.6: The filtering errors of the DFF with gain variations.

2.5 Conclusion

In this chapter, we have investigated the non-fragile H_∞ filtering problem for a class of discrete multi-rate time-delayed systems over sensor networks. The filter gains are assumed to subject to fluctuations in order to cater for real engineering applications. Both the exponential mean-square stability and the H_∞ performance of the filtering error system have been analyzed and the filter gains have been obtained by solving a set of matrix inequalities. Finally, a simulation example has been provided to illustrate the effectiveness of the proposed filtering scheme.

3

H_∞ Filtering for Multi-Rate Artificial Neural Networks with Integral Measurements

The filtering theory has been widely used in various practical applications such as target tracking, environmental monitoring, and power systems. Up to now, many filtering techniques have been developed that include H_∞ filtering, l_2-l_∞ filtering, Kalman-type filtering as well as set-membership filtering. Among others, the H_∞ filtering has gained particular research attention because its ability in attenuating the influence of the exogenous disturbances on the filtering error. The H_∞ filtering is effective in handling different kinds of systems such as fuzzy systems [147] and switched systems [172].

Recently, the ANNs have been found wide applications in different areas such as image processing and pattern recognition. In certain applications of the ANNs, the states of some primary neurons are needed to be known [194]. Unfortunately, the neuron states are difficult to be directly accessed and only partial information on the states can be obtained through sensor observation due to the large scale of the ANNs and the limited observation resources. In this case, an effective way to obtain the neuron state is using the filtering methods to estimate the neuron states based on the limited observations. Until now, plenty of research results have been obtained on the filtering problems for the ANNs.

In the filtering problems for MRSs, the lifting technique proposed in [113] is widely used to convert the MRSs into single-rate ones. Nevertheless, for multi-rate ANNs with inherent nonlinearity, the lifting technique is no longer applicable. Therefore, two interesting problems arise that: 1) how to convert the multi-rate ANNs into single-rate ones? and 2) how to design a filter with which the H_∞ requirement is ensured on the filtering error?

Due to the time spent on the data collection and processing, the measurement outputs of the sensor may not only rely on the current system state but also rely on the system states over a certain past period. Such a phenomenon, known as integral measurements, exists in practical engineering such as chemical processes [55] and nuclear reaction processes [13, 154]. Nevertheless, the integral measurements has been often overlooked in the area of signal processing and the filtering problem for systems with integral measurements has only obtained some initial attentions. As for the ANNs, the filtering problem with

DOI: 10.1201/9781032619507-3

integral measurements has not been studied yet, not to mention the case that the multi-rate sampling is also considered.

Motivated by the above discussions, in this chapter, we are going to consider the H_∞ filtering problem for multi-rate ANNs with integral measurements. The main contributions of this chapter are that: 1) the filtering problem is, for the first time, considered for multi-rate ANNs with integral measurements; 2) a novel method, which has better applicability than the lifting technique, is developed to convert the multi-rate ANNs into single-rate switched ones; and 3) sufficient conditions are obtained that guarantee the exponential stability and the H_∞ performance of the filtering error dynamics.

3.1 Problem Formulation

Consider the following ANN consists of n_x neurons:

$$\begin{cases} x_i(T_{k+1}) = a_i x_i(T_k) + \sum_{j=1}^{n_x} w_{ij} g_j(x_j(T_k)) \\ \qquad\qquad + b_i \omega_i(T_k), \quad i = 1, 2, \ldots, n_x \\ z_q(T_k) = \sum_{j=1}^{n_x} h_{qj} x_j(T_k), \quad q = 1, 2, \ldots, n_z \end{cases} \tag{3.1}$$

where $x_i(T_k) \in \mathbb{R}$ is the i-th neuron state, $z_q(T_k) \in \mathbb{R}$ is the signal to be estimated, $\omega_i(T_k) \in l_2([0,\infty);\mathbb{R})$ is the process noise, and $g_j(\cdot)$ is the non-linear activation function of the j-th neuron. a_i is the state feedback coefficient and w_{ij} is the connection weight. b_i and h_{qj} are known real scalars. The state update period of (3.1) is $h \triangleq T_{k+1} - T_k$.

Assumption 3.1 *The non-linear activation function $g_i(\cdot) : \mathbb{R} \to \mathbb{R}$ ($i = 1, 2, \ldots, n_x$) satisfies*

$$|g_i(a) - g_i(b)| \le \varepsilon_i |a - b|, \quad \forall\, a, b \in \mathbb{R} \tag{3.2}$$

where ε_i ($i = 1, 2, \ldots, n_x$) are known scalars.

The ANNs (3.1) is rewritten as

$$\begin{cases} x(T_{k+1}) = Ax(T_k) + Wg(x(T_k)) + B\omega(T_k), \\ z(T_k) = Hx(T_k) \end{cases} \tag{3.3}$$

where

$$\begin{aligned}
x(T_k) &\triangleq \text{col}\{x_1(T_k), x_2(T_k), \cdots, x_{n_x}(T_k)\},\\
z(T_k) &\triangleq \text{col}\{z_1(T_k), z_2(T_k), \cdots, z_{n_z}(T_k)\},\\
g(x(T_k)) &\triangleq \text{col}\{g_1(x_1(T_k)), g_2(x_2(T_k)), \cdots, g_{n_x}(x_{n_x}(T_k))\},\\
\omega(T_k) &\triangleq \text{col}\{\omega_1(T_k), \omega_2(T_k), \cdots, \omega_{n_x}(T_k)\},\\
A &\triangleq \text{diag}\{a_1, a_2, \cdots, a_{n_x}\},\\
B &\triangleq \text{diag}\{b_1, b_2, \cdots, b_{n_x}\},\\
W &\triangleq \left[w_{ij}\right]_{n_x \times n_x}, \quad H \triangleq \left[h_{qj}\right]_{n_z \times n_x}.
\end{aligned}$$

Considering the integral measurements, the measurement of the sensor with a sampling period $mh \triangleq t_{k+1} - t_k$ ($m \in \mathbb{N}^+$) is described as

$$y(t_k) = C\sum_{i=0}^{d} x(t_k - ih) + Ev(t_k) \tag{3.4}$$

where $y(t_k) \in \mathbb{R}^{n_y}$ is the measurement output and $v(t_k) \in l_2([0,\infty);\mathbb{R}^{n_v})$ is the measurement noise. C and E are known matrices with appropriate dimensions. dh ($d \in \mathbb{N}$) is the time interval to collect and process the data.

By denoting

$$\lambda(T_k) = \begin{cases} 1, & \text{if } \text{mod}(T_k, mh) = 0;\\ 0, & \text{otherwise} \end{cases} \tag{3.5}$$

and letting $y(T_k) = 0$ and $v(T_k) = 0$ for $\lambda(T_k) = 0$, (3.4) is rewritten as

$$y(T_k) = \tilde{C}_{\lambda(T_k)}\sum_{i=0}^{d} x(T_{k-i}) + \tilde{E}_{\lambda(T_k)}v(T_k) \tag{3.6}$$

where

$$\tilde{C}_{\lambda(T_k)} \triangleq \lambda(T_k)C, \quad \tilde{E}_{\lambda(T_k)} \triangleq \lambda(T_k)E.$$

As in [164], for convenience of presentation, all possible transitions from $\lambda(T_k)$ to $\lambda(T_{k+1})$ are represented by the set

$$\Omega \triangleq \big\{(s,t)|\lambda(T_k) = s, \lambda(T_{k+1}) = t, st \neq 1, s,t \in \{0,1\}\big\}.$$

The filter of the following form is used in this chapter:

$$\begin{cases} \hat{x}(T_{k+1}) = A\hat{x}(T_k) + Wg(\hat{x}(T_k)) \\ \qquad\qquad + K_{\lambda(T_k)}\Big(y(T_k) - \tilde{C}_{\lambda(T_k)}\sum_{i=0}^{d}\hat{x}(T_{k-i})\Big), \\ \hat{z}(T_k) = H\hat{x}(T_k) \end{cases} \tag{3.7}$$

where $\hat{x}(T_k)$ is the estimate of $x(T_k)$, $\hat{z}(T_k)$ is the estimate of $z(T_k)$, and $K_{\lambda(T_k)}$ is the filter gain to be determined.

In the following, for simplification, a matrix $X_{\lambda(T_k)}$ is simply denoted as X_s for each possible $\lambda(T_k) = s$. Denoting

$$\begin{aligned} e(T_k) &\triangleq x(T_k) - \hat{x}(T_k), \\ \tilde{z}(T_k) &\triangleq z(T_k) - \hat{z}(T_k), \end{aligned}$$

the filtering error dynamics is obtained as

$$\begin{cases} e(T_{k+1}) = \bar{A}_s e(T_k) + W\bar{g}(T_k) + B\omega(T_k) \\ \qquad\qquad - \sum_{i=1}^{d} \bar{C}_s e(T_{k-i}) - \bar{E}_s v(T_k), \\ \tilde{z}(T_k) = He(T_k) \end{cases} \tag{3.8}$$

where

$$\begin{aligned} \bar{A}_s &\triangleq A - K_s\tilde{C}_s, \quad \bar{C}_s \triangleq K_s\tilde{C}_s, \quad \bar{E}_s \triangleq K_s\tilde{E}_s, \\ \bar{g}(T_k) &\triangleq g(x(T_k)) - g(\hat{x}(T_k)). \end{aligned}$$

Definition 3.1 *The system (3.8) with $\omega(T_k) = 0$ and $v(T_k) = 0$ is said to be exponentially stable if there are scalars $\mu > 0$ and $0 < \nu < 1$ such that*

$$\|e(T_k)\|^2 \leq \mu\nu^k \|e(T_0)\|^2.$$

The objective of the addressed H_∞ filtering problem is to design a filter (3.7) such that:

1. the filtering error dynamic (3.8) with $\omega(T_k) = 0$ and $v(T_k) = 0$ is exponentially stable;
2. under the zero initial condition, for a given disturbance attenuation level $\gamma > 0$ and all non-zero $\omega(T_k)$ and $v(T_k)$, the filtering error $\tilde{z}(T_k)$ satisfies the following H_∞ performance requirement:

$$\sum_{k=0}^{\infty} \|\tilde{z}(T_k)\|^2 < \gamma^2 \sum_{k=0}^{\infty} \|\omega(T_k)\|^2 + \gamma^2 \sum_{k=0}^{\infty} \|v(T_k)\|^2. \tag{3.9}$$

3.2 H_∞ Filter Design

In this section, the H_∞ filtering problem is discussed for multi-rate ANNs with integral measurements. The exponential stability of the filtering error dynamic system (3.8) is firstly analyzed. Then, the H_∞ performance of the filtering error $\tilde{z}(T_k)$ is investigated. Finally, the H_∞ filtering scheme for multi-rate ANNs with integral measurements is developed.

In the following theorem, we are going to present sufficient conditions that guarantee the exponential stability of the filtering error dynamic system (3.8).

Theorem 3.1 *Let the filter gains K_s $(s \in \{0,1\})$ be given. The system (3.8) is exponentially stable with $\omega(T_k) = 0$ and $v(T_k) = 0$ if there exist positive-definite matrices P_s $(s \in \{0,1\})$ and Q such that*

$$\Xi_s \triangleq \begin{bmatrix} \Xi_s^{11} & -\bar{A}_s^T P_t \bar{C}_s L & \bar{A}_s^T P_t W \\ * & \Xi_s^{22} & -L^T \bar{C}_s^T P_t W \\ * & * & \Xi_s^{33} \end{bmatrix} < 0 \tag{3.10}$$

hold for all $(s,t) \in \Omega$, where

$$\begin{aligned} \Xi_s^{11} &\triangleq \bar{A}_s^T P_t \bar{A}_s - P_s + G + dQ, \\ \Xi_s^{22} &\triangleq L^T \bar{C}_s^T P_t \bar{C}_s L - \bar{Q}, \\ \Xi_s^{33} &\triangleq W^T P_t W - I, \ L \triangleq \begin{bmatrix} I & I & \cdots & I \end{bmatrix}, \\ \bar{Q} &\triangleq \mathrm{diag}\{Q, Q, \cdots, Q\}, \\ G &\triangleq \mathrm{diag}\{\varepsilon_1^2, \varepsilon_2^2, \cdots, \varepsilon_{n_x}^2\}. \end{aligned}$$

Proof *Choose the following Lyapunov-Krasovskii functional:*

$$V(T_k) = V_1(T_k) + V_2(T_k) \tag{3.11}$$

where

$$\begin{aligned} V_1(T_k) &\triangleq e^T(T_k) P_s e(T_k), \\ V_2(T_k) &\triangleq \sum_{i=1}^{d} \sum_{j=k-i}^{k-1} e^T(T_j) Q e(T_j). \end{aligned}$$

Note that the set Ω represents all possible transitions from $\lambda(T_k)$ to $\lambda(T_{k+1})$. Calculating the differences of $V_1(T_k)$ and $V_2(T_k)$ along the trajectory of (3.8) with $\omega(T_k) = 0$ and $v(T_k) = 0$, we have

$$\begin{aligned} \Delta V_1(T_k) =& e^T(T_{k+1}) P_t e(T_{k+1}) - e^T(T_k) P_s e(T_k) \\ =& \Big(\bar{A}_s e(T_k) + W\bar{g}(T_k) - \bar{C}_s L \bar{e}(T_k)\Big)^T P_t \\ & \times \Big(\bar{A}_s e(T_k) + W\bar{g}(T_k) - \bar{C}_s L \bar{e}(T_k)\Big) - e^T(T_k) P_s e(T_k) \\ =& e^T(T_k) \bar{A}_s^T P_t \bar{A}_s e(T_k) - e^T(T_k) P_s e(T_k) + 2e^T(T_k) \bar{A}_s^T P_t W \bar{g}(T_k) \\ & - 2e^T(T_k) \bar{A}_s^T P_t \bar{C}_s L \bar{e}(T_k) + \bar{g}^T(T_k) W^T P_t W \bar{g}(T_k) \\ & - 2\bar{g}^T(T_k) W^T P_t \bar{C}_s L \bar{e}(T_k) + \bar{e}^T(T_k) L^T \bar{C}_s^T P_t \bar{C}_s L \bar{e}(T_k) \\ =& \eta^T(T_k) \Gamma_s \eta(T_k) \end{aligned}$$

and

$$\begin{aligned}\Delta V_2(T_k) =& \sum_{i=1}^{d}\sum_{j=k+1-i}^{k} e^T(T_j)Qe(T_j) - \sum_{i=1}^{d}\sum_{j=k-i}^{k-1} e^T(T_j)Qe(T_j)\\ =& \sum_{i=1}^{d} e^T(T_k)Qe(T_k) - \sum_{i=1}^{d} e^T(T_{k-i})Qe(T_{k-i})\\ =& de^T(T_k)Qe(T_k) - \bar{e}^T(T_k)\bar{Q}\bar{e}(T_k)\end{aligned}$$

where

$$\begin{aligned}\bar{e}(T_k) \triangleq& \mathrm{col}\{e(T_{k-1}), e(T_{k-2}), \cdots, e(T_{k-d})\},\\ \eta(T_k) \triangleq& \mathrm{col}\{e(T_k), \bar{e}(T_k), \bar{g}(T_k)\},\\ \Gamma_s \triangleq& \begin{bmatrix} \bar{A}_s^T P_t \bar{A}_s - P_s & -\bar{A}_s^T P_t \bar{C}_s L & \bar{A}_s^T P_t W \\ * & L^T \bar{C}_s^T P_t \bar{C}_s L & -L^T \bar{C}_s^T P_t W \\ * & * & W^T P_t W \end{bmatrix}.\end{aligned}$$

Moreover, from Assumption 3.1, we know that

$$\bar{g}^T(T_k)\bar{g}(T_k) - e^T(T_k)Ge(T_k) \leq 0.$$

With the above discussions, we have

$$\begin{aligned}\Delta V(T_k) =& \Delta V_1(T_k) + \Delta V_2(T_k)\\ \leq& \eta^T(T_k)\Gamma_s\eta(T_k) + de^T(T_k)Qe(T_k) - \bar{e}^T(T_k)\bar{Q}\bar{e}(T_k)\\ &- \bar{g}^T(T_k)\bar{g}(T_k) + e^T(T_k)Ge(T_k)\\ =& \eta^T(T_k)\Xi_s\eta(T_k).\end{aligned}$$

It is known from (3.10) *that there exists a sufficiently small scalar* $\zeta > 0$ *such that*

$$\Delta V(T_k) + \zeta\|e(T_k)\|^2 < 0.$$

By following the similar analysis as in [165], it can be proved that the system (3.8) *is exponentially stable with* $\omega(T_k) = 0$ *and* $v(T_k) = 0$. *The proof is thus complete.*

Based on the result obtained in Theorem 3.1, we are now ready to present the following theorem.

Theorem 3.2 *Let the disturbance attenuation level* $\gamma > 0$ *and the filter gains* K_s $(s \in \{0, 1\})$ *be given. The system* (3.8) *is exponentially stable with* $\omega(T_k) = 0$ *and* $v(T_k) = 0$ *and satisfies the* H_∞ *performance requirement* (3.9) *for all non-zero* $\omega(T_k)$ *and* $v(T_k)$ *if there exist positive-definite matrices* P_s $(s \in$

$\{0,1\}$) *and* Q *such that*

$$\bar{\Xi}_s \triangleq \begin{bmatrix} \bar{\Xi}_s^{11} & \bar{\Xi}_s^{12} & \bar{A}_s^T P_t W & \bar{A}_s^T P_t B & \bar{\Xi}_s^{15} \\ * & \bar{\Xi}_s^{22} & \bar{\Xi}_s^{23} & \bar{\Xi}_s^{24} & \bar{\Xi}_s^{25} \\ * & * & \bar{\Xi}_s^{33} & W^T P_t B & \bar{\Xi}_s^{35} \\ * & * & * & \bar{\Xi}_s^{44} & \bar{\Xi}_s^{45} \\ * & * & * & * & \bar{\Xi}_s^{55} \end{bmatrix} < 0 \tag{3.12}$$

hold for all $(s,t) \in \Omega$, *where*

$$\begin{aligned}
\bar{\Xi}_s^{11} \triangleq & \bar{A}_s^T P_t \bar{A}_s - P_s + G + dQ + H^T H, \\
\bar{\Xi}_s^{12} \triangleq & - \bar{A}_s^T P_t \bar{C}_s L, \quad \bar{\Xi}_s^{15} \triangleq -\bar{A}_s^T P_t \bar{E}_s, \\
\bar{\Xi}_s^{23} \triangleq & - L^T \bar{C}_s^T P_t W, \quad \bar{\Xi}_s^{24} \triangleq -L^T \bar{C}_s^T P_t B, \\
\bar{\Xi}_s^{25} \triangleq & L^T \bar{C}_s^T P_t \bar{E}_s, \quad \bar{\Xi}_s^{35} \triangleq -W^T P_t \bar{E}_s, \\
\bar{\Xi}_s^{44} \triangleq & B^T P_t B - \gamma^2 I, \quad \bar{\Xi}_s^{45} \triangleq -B^T P_t \bar{E}_s, \\
\bar{\Xi}_s^{55} \triangleq & \bar{E}_s^T P_t \bar{E}_s - \gamma^2 I.
\end{aligned}$$

Proof *It is known from* (3.12) *that* (3.10) *is true, and therefore the system* (3.8) *is exponentially stable with* $\omega(T_k) = 0$ *and* $v(T_k) = 0$. *Next, we are going to prove that the system* (3.8) *satisfies the* H_∞ *performance requirement for all non-zero* $\omega(T_k)$ *and* $v(T_k)$.

Choose the Lyapunov-Krasovskii functional as (3.11). *Calculating the differences of* $V_1(T_k)$ *and* $V_2(T_k)$ *along the trajectory of* (3.8), *we have*

$$\begin{aligned}
\Delta V(T_k) =& \Delta V_1(T_k) + \Delta V_2(T_k) \\
=& \Big(\bar{A}_s e(T_k) + W\bar{g}(T_k) + B\omega(T_k) - \bar{C}_s L \bar{e}(T_k) - \bar{E}_s v(T_k)\Big)^T P_t \\
& \times \Big(\bar{A}_s e(T_k) + W\bar{g}(T_k) + B\omega(T_k) - \bar{C}_s L \bar{e}(T_k) - \bar{E}_s v(T_k)\Big) \\
& - e^T(T_k) P_s e(T_k) + d e^T(T_k) Q e(T_k) - \bar{e}^T(T_k) \bar{Q} \bar{e}(T_k) \\
\leq & \eta^T(T_k) \Xi_s \eta(T_k) + 2e^T(T_k) \bar{A}_s^T P_t B\omega(T_k) - 2e^T(T_k) \bar{A}_s^T P_t \bar{E}_s v(T_k) \\
& + 2\bar{g}^T(T_k) W^T P_t B\omega(T_k) - 2\bar{g}^T(T_k) W^T P_t \bar{E}_s v(T_k) \\
& + \omega^T(T_k) B^T P_t B\omega(T_k) - 2\omega^T(T_k) B^T P_t \bar{C}_s L \bar{e}(T_k) \\
& - 2\omega^T(T_k) B^T P_t \bar{E}_s v(T_k) + 2\bar{e}^T(T_k) L^T \bar{C}_s^T P_t \bar{E}_s v(T_k) \\
& + v^T(T_k) \bar{E}_s^T P_t \bar{E}_s v(T_k) \\
=& \bar{\eta}^T(T_k) \bar{\Gamma}_s \bar{\eta}(T_k)
\end{aligned}$$

where

$$\begin{aligned}
\bar{\eta}(T_k) \triangleq & \mathrm{col}\{\eta(T_k), \omega(T_k), v(T_k)\}, \\
\bar{\Gamma}_s \triangleq & \begin{bmatrix} \Xi_s & \Pi_1 & \Pi_2 \\ * & B^T P_t B & -B^T P_t \bar{E}_s \\ * & * & \bar{E}_s^T P_t \bar{E}_s \end{bmatrix},
\end{aligned}$$

$$\Pi_1 \triangleq \begin{bmatrix} \bar{A}_s^T P_t B \\ -L^T \bar{C}_s^T P_t B \\ W^T P_t B \end{bmatrix}, \quad \Pi_2 \triangleq \begin{bmatrix} -\bar{A}_s^T P_t \bar{E}_s \\ L^T \bar{C}_s^T P_t \bar{E}_s \\ -W^T P_t \bar{E}_s \end{bmatrix}.$$

Then, we have

$$\begin{aligned} &\Delta V(T_k) + \|\tilde{z}(T_k)\|^2 - \gamma^2 \|\omega(T_k)\|^2 - \gamma^2 \|v(T_k)\|^2 \\ &= \Delta V_1(T_k) + \Delta V_2(T_k) + e^T(T_k) H^T H e(T_k) \\ &\quad - \gamma^2 \omega^T(T_k)\omega(T_k) - \gamma^2 v^T(T_k) v(T_k) \\ &\le \bar{\eta}^T(T_k)\bar{\Gamma}_s \bar{\eta}(T_k) + e^T(T_k) H^T H e(T_k) \\ &\quad - \gamma^2 \omega^T(T_k)\omega(T_k) - \gamma^2 v^T(T_k) v(T_k) \\ &= \bar{\eta}^T(T_k)\bar{\Xi}_s \bar{\eta}(T_k) < 0. \end{aligned} \tag{3.13}$$

Noting $\tilde{z}(T_0) = 0$ and summing up (3.13) from 0 to ∞ with respect to k, one has

$$\sum_{k=0}^{\infty} \|\tilde{z}(T_k)\|^2 < \sum_{k=0}^{\infty} \gamma^2 \|\omega(T_k)\|^2 + \sum_{k=0}^{\infty} \gamma^2 \|v(T_k)\|^2.$$

Then, the H_∞ performance requirement is satisfied and the proof is complete.

Now, we are ready to solve the H_∞ filter design problem.

Theorem 3.3 *Let the disturbance attenuation level $\gamma > 0$ be given. The system (3.8) is exponentially stable with $\omega(T_k) = 0$ and $v(T_k) = 0$ and satisfies the H_∞ performance requirement (3.9) for all non-zero $\omega(T_k)$ and $v(T_k)$ if there exist positive-definite matrices P_s ($s \in \{0,1\}$) and Q, matrices Y_s ($s \in \{0,1\}$), and non-singular matrix S such that*

$$\begin{bmatrix} \Psi_s & \bar{\Phi}_s^T \\ * & P_t - S - S^T \end{bmatrix} < 0 \tag{3.14}$$

hold for all $(s,t) \in \Omega$, where

$$\begin{aligned} \Psi_s &\triangleq \mathrm{diag}\{\Psi_s^{11}, -\bar{Q}, -I, -\gamma^2 I, -\gamma^2 I\}, \\ \Psi_s^{11} &\triangleq -P_s + G + dQ + H^T H, \\ \bar{\Phi}_s &\triangleq \begin{bmatrix} SA - Y_s\tilde{C}_s & -Y_s\tilde{C}_s L & SW & SB & -Y_s\tilde{E}_s \end{bmatrix}. \end{aligned}$$

Moreover, the filter gains K_s ($s \in \{0,1\}$) are obtained by

$$K_s = S^{-1} Y_s. \tag{3.15}$$

Proof *It is obvious that $\bar{\Xi}_s$ can be rewritten as*

$$\bar{\Xi}_s = \Psi_s + \Phi_s^T P_t \Phi_s$$

where

$$\Phi_s \triangleq \begin{bmatrix} \bar{A}_s & -\bar{C}_s L & W & B & -\bar{E}_s \end{bmatrix}.$$

According to Lemma 2.2, $\bar{\Xi}_s < 0$ *is equivalent to*

$$\begin{bmatrix} \Psi_s & \Phi_s^T \\ * & -P_t^{-1} \end{bmatrix} < 0. \tag{3.16}$$

Pre- and post-multiplying (3.14) *by* $\mathrm{diag}\{I, I, I, I, I, S^{-1}\}$ *and noting* $Y_s = SK_s$*, we arrive at*

$$\begin{bmatrix} \Psi_s & \Phi_s^T \\ * & S^{-T}P_tS^{-1} - S^{-1} - S^{-T} \end{bmatrix} < 0.$$

From

$$(S^{-1} - P_t^{-1})^T P_t (S^{-1} - P_t^{-1}) \geq 0,$$

we have

$$-P_t^{-1} \leq S^{-T}P_tS^{-1} - S^{-1} - S^{-T}.$$

Then, it can be seen that

$$\begin{bmatrix} \Psi_s & \Phi_s^T \\ * & -P_t^{-1} \end{bmatrix} \leq \begin{bmatrix} \Psi_s & \Phi_s^T \\ * & S^{-T}P_tS^{-1} - S^{-1} - S^{-T} \end{bmatrix} < 0$$

which means $\bar{\Xi}_s < 0$*. Therefore, the system* (3.8) *is exponentially stable with* $\omega(T_k) = 0$ *and* $v(T_k) = 0$ *and satisfies the* H_∞ *performance constraint* (3.9) *for all non-zero* $\omega(T_k)$ *and* $v(T_k)$*. The filter gains* K_s *can be derived by solving* (3.15)*. The proof is complete.*

3.3 Simulation Examples

In this section, we give two examples to verify the proposed H_∞ filtering scheme. Moreover, the influence of the parameter m on the filtering performance is also discussed.

Consider an ANN consists of 3 neurons with $h = 1$ and

$$\begin{aligned} a_1 &= 0.85,\ a_2 = 0.7,\ a_3 = 0.3, \\ b_1 &= 0.14,\ b_2 = 0.23,\ b_3 = 0.1, \\ W &= \begin{bmatrix} 0.67 & -0.21 & 0.19 \\ 0.13 & 0.49 & -0.21 \\ 0.21 & -0.20 & 0.88 \end{bmatrix}, \\ H &= \begin{bmatrix} -0.61 & -0.44 & 0.29 \\ -0.19 & 0.2 & -0.23 \end{bmatrix}. \end{aligned}$$

The non-linear activation functions $g_i(\cdot)$ $(i = 1, 2, 3)$ are set as

$$\begin{aligned} g_1(x_1(T_k)) &= \tanh(0.39x_1(T_k)), \\ g_2(x_2(T_k)) &= \tanh(-0.26x_2(T_k)), \\ g_3(x_3(T_k)) &= \tanh(0.29x_3(T_k)). \end{aligned}$$

It is known that the conditions (3.2) are satisfied with $\varepsilon_1 = 0.39$, $\varepsilon_2 = 0.26$, and $\varepsilon_3 = 0.29$.

In the following, two examples with different m are provided to show the influence of the parameter m on the filtering performance.

Example 3.1

In this example, a sensor with a sampling period 3, i.e., $m = 3$, is employed to measure the ANN. The parameters of the sensor are

$$C = \begin{bmatrix} 0.26 & 0.24 & 0.18 \\ 0.18 & 0.34 & 0.2 \end{bmatrix}, \quad E = \begin{bmatrix} 0.2 \\ 0.18 \end{bmatrix}.$$

Moreover, we set $d = 1$.

From $m = 3$, we know that the set Ω representing all possible transitions from $\lambda(T_k)$ to $\lambda(T_{k+1})$ is

$$\Omega = \{(0, 1), (0, 0), (1, 0)\}.$$

With the given parameters, by using the Matlab LMI toolbox, the filter gains K_s $(s = 0, 1)$ are derived as:

$$K_0 = \begin{bmatrix} 0 & 0 \\ 0 & 0 \\ 0 & 0 \end{bmatrix}, \quad K_1 = \begin{bmatrix} 1.38 & -1.12 \\ -1.19 & 1.45 \\ 0.03 & 0.03 \end{bmatrix}. \tag{3.17}$$

In the simulation, the disturbance inputs $w_i(T_k)$ $(i = 1, 2, 3)$ and the measurement noise $v(t_k)$ are respectively chosen as

$$\begin{aligned} w_1(T_k) &= \frac{0.5\cos(T_k)}{0.4T_k + 1}, \quad w_2(T_k) = \frac{\cos(T_k)}{0.3T_k + 1}, \\ w_3(T_k) &= \frac{\cos(T_k)}{0.2T_k + 1}, \quad v(t_k) = \frac{2\cos(t_k)}{0.5t_k + 1}. \end{aligned}$$

The initial condition is set as $x(T_0) = \begin{bmatrix} 0.2 & 0.2 & -0.2 \end{bmatrix}^T$.

The simulation results of Example 3.1 are given in Figs. 3.1–3.4. The states of the neurons and the corresponding estimates are presented in Figs. 3.1–3.3. Figure 3.4 depicts the output filtering error $\tilde{z}(t_k)$. It can be seen from the simulation results that the proposed filtering scheme performs well.

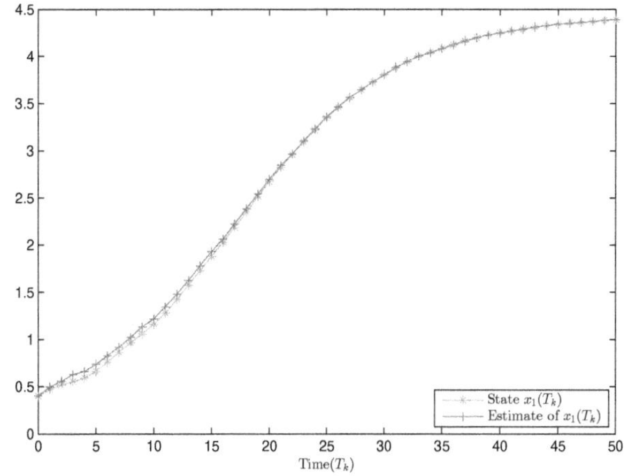

FIGURE 3.1: State $x_1(T_k)$ and its estimate with $m = 3$.

Example 3.2

In this example, the sampling period of the sensor is set as 10, that is, $m = 10$. Other parameters of the sensor are chosen as those in Example 3.1. From $m = 10$, we know that the set Ω is also

$$\Omega = \{(0,1),(0,0),(1,0)\}.$$

Note that the LMIs (3.14) do not include the parameter m, and other parameters and the set Ω are all the same with those in Example 3.1. The obtained

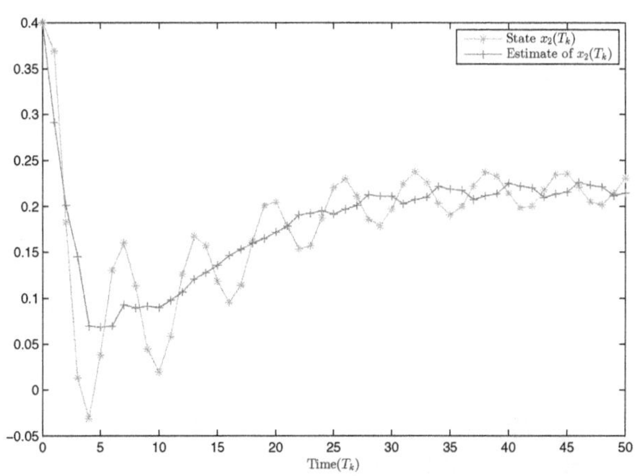

FIGURE 3.2: State $x_2(T_k)$ and its estimate with $m = 3$.

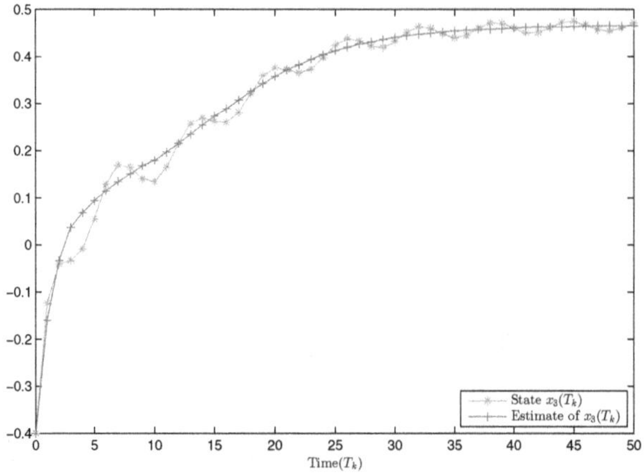

FIGURE 3.3: State $x_3(T_k)$ and its estimate with $m = 3$.

filter gains K_s ($s = 0, 1$) are also as (3.17). With the same conditions as in Example 3.1, simulation results are obtained and shown in Figs. 3.5–3.8.

It can be seen from the simulation results that the filtering performance with $m = 10$ is worse than the filtering performance with $m = 3$. Such a phenomenon is reasonable. With the increases of m, the number of $\lambda(T_k)$ that equals zero also increases, which inevitably leads to a degraded filtering performance.

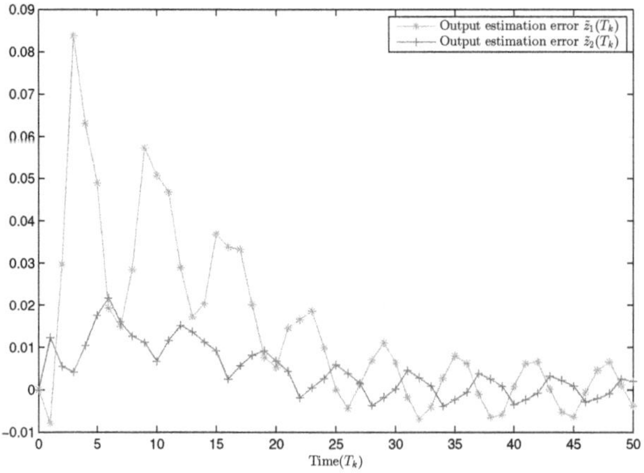

FIGURE 3.4: Output filtering error $\tilde{z}(T_k)$ with $m = 3$.

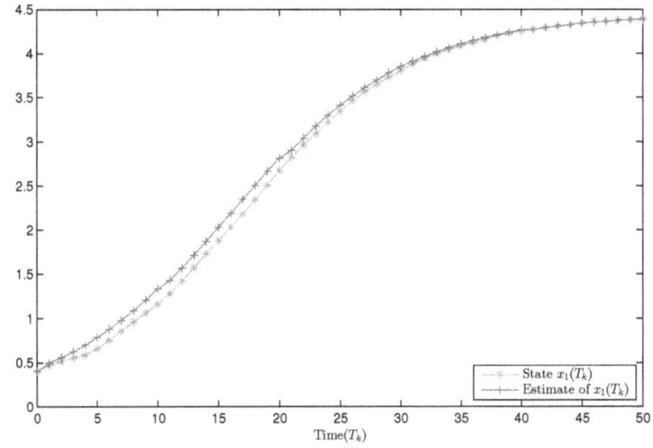

FIGURE 3.5: State $x_1(T_k)$ and its estimate with $m = 10$.

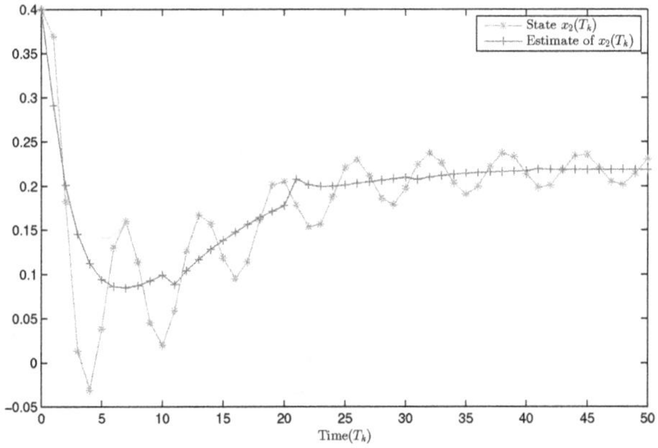

FIGURE 3.6: State $x_2(T_k)$ and its estimate with $m = 10$.

3.4 Conclusion

In this chapter, the H_∞ filtering problem has been studied for a class of multi-rate ANNs with integral measurements. A novel method has been put forward to convert the multi-rate ANNs to single-rate switched ones. By constructing a Lyapunov-Krasovskii functional and applying the switched system approach, sufficient conditions have been derived such that the filtering error dynamic system is exponentially stable and the H_∞ performance constraint is met.

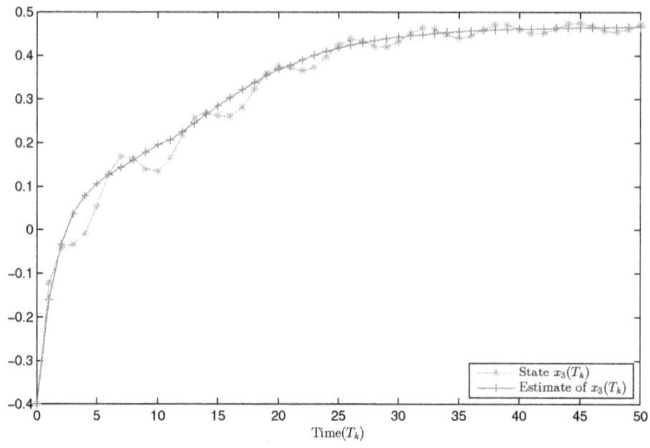

FIGURE 3.7: State $x_3(T_k)$ and its estimate with $m = 10$.

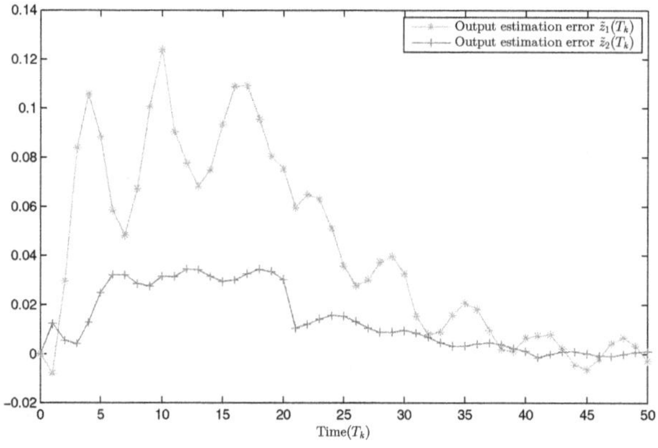

FIGURE 3.8: Output filtering error $\tilde{z}(T_k)$ with $m = 10$.

Then, the desired filter gains have been obtained by solving a set of LMIs. Finally, simulation examples have been conducted to verify the effectiveness of the developed H_∞ filtering algorithm and reveal the influence of the multi-rate sampling on the filtering performance.

4

Recursive State Estimation for Multi-Rate Systems with Sensor Resolutions

So far, many state estimation algorithms have been developed in the literature for a variety of systems. Among others, the recursive state estimation method, whose main aim is to minimize the estimation error covariance at each instant, has proven to be one of the most popular ones due primarily to its advantages in easy implementation and online computation. It is worth mentioning that, up to now, the focus of almost all recursive state estimation problems has been paid on the single-rate systems. Despite the significant engineering background of the multi-rate sampling mechanism, the recursive state estimation problems for MRSs still need extra research attentions.

The multi-rate sampling, though practically appealing, does bring multiple time sequences to the system and would invalidate the state estimation algorithms designed specifically for single-rate systems. Therefore, it is theoretically important to make dedicated efforts in dealing with the multi-rate sampling issue in the recursive state estimation problems, see [39,193] for some latest results. In particular, the filtering problem has been tackled in [193] for MRSs with asynchronous sensors and the lifting technique has been used to transform the MRS into a single-rate one.

Multiplicative noises, also known as state-dependent noises, have recently gained considerable research interest. Note that many practical plants can be modeled by systems with multiplicative noises, and the corresponding recursive state estimation problem for such kind of systems demands extra care, see e.g. [64,101] for some representative results on the single-rate systems with multiplicative noises. When it comes to the MRSs, the relevant recursive state estimation results have been really scattered because of the essential difficulties in handling the coupling between the multi-rate sampling and the multiplicative noises. For example, the traditional lifting technique is no longer directly applicable for MRSs because of the existence of the multiplicative noises. Therefore, there is a practical need to develop a novel method to solve the recursive state estimation problem for MRSs subject to multiplicative noises.

In engineering practice, no sensors could detect arbitrarily small changes of the measurement. The smallest change that a sensor can detect, known as the *sensor resolution*, is one of the important specifications of the sensor. Due to inherent limit of the sensor resolution, the actual measurement output

DOI: 10.1201/9781032619507-4

from the sensor is most likely to deviate from the true measurement output. Obviously, estimating the system state by using the deviated measurement would lead to a poor estimation performance. As such, in the state estimation problem, it is vitally important to take the sensor resolution into serious consideration.

The consideration of the phenomenon of sensor resolution is clearly a nontrivial task with some additional challenges outlined as follows: 1) how to establish a model to accurately characterize the sensor resolution? 2) how to mitigate the performance deterioration of the state estimation caused by the deviation of the actual measurement from the true measurement? and 3) how to design a state estimator with guaranteed estimation accuracy in spite of the sensor-resolution-induced measurement distortions? Note that preliminary result has been obtained in [187] on the state estimation problem for single-rate systems with non-logarithmic sensor resolutions. Unfortunately, for MRSs, the corresponding state estimation result has been really scattered.

To tackle the aforementioned challenges, in this chapter, we endeavor to develop an effective recursive state estimation scheme for MRSs with multiplicative noises and sensor-resolution-induced measurement distortions. The main novelties of this chapter are stressed as follows: 1) a novel method is put forward to handle the difficulties resulting from the coupling of the multi-rate sampling and the multiplicative noises; 2) a novel state estimation scheme, which is of acceptable computational complexity, is developed for systems undergoing sensor-resolution-induced measurement distortions; and 3) a locally minimized upper bound is guaranteed on the estimation error covariance.

4.1 Problem Formulation

Consider the following class of discrete-time systems with multiplicative noises:

$$x(k+1) = A(k)x(k) + \varepsilon(k)B(k)x(k) + E(k)w(k) \tag{4.1}$$

where $x(k) \in \mathbb{R}^{n_x}$ is the system state, $\varepsilon(k) \in \mathbb{R}$ is the zero-mean Gaussian multiplicative noise with unity covariance, and $w(k) \in \mathbb{R}^{n_w}$ is the zero-mean process noise with covariance $W(k) > 0$. $A(k)$, $B(k)$ and $E(k)$ are known time-varying matrices with compatible dimensions. The initial value $x(0)$ of the system state is a random variable with mean $\chi(0)$ and covariance $X(0)$.

In practical engineering, a sensor can only detect the change of the measurement that is larger than a certain value. Such a certain value is known as *sensor resolution* and defined as follows.

Definition 4.1 *[187] Let y_i $(i = 1, 2, \ldots, n_y)$ be the i-th element of the measurement output of a sensor. If y_i takes value in the set $\{jr_i | j = 0, \pm 1, \ldots, \pm z\}$*

where z is a given positive integer, then $R \triangleq \text{col}\{r_1, r_2, \cdots, r_{n_y}\}$ is the resolution of the sensor.

In this chapter, a sensor with the sampling period

$$b \triangleq s_{k+1} - s_k$$

and the sensor resolution

$$R \triangleq \text{col}\{r_1, r_2, \cdots, r_{n_y}\}$$

is deployed to measure the system. Without considering the sensor resolution, the *ideal* measurement output of the sensor is

$$y^{id}(s_k) = C(s_k)x(s_k) + D(s_k)v(s_k) \tag{4.2}$$

where $y^{id}(s_k) \in \mathbb{R}^{n_y}$ is the ideal measurement output of the sensor, $v(s_k) \in \mathbb{R}^{n_v}$ is the zero-mean measurement noise with covariance $V(s_k) > 0$, and $C(s_k)$ and $D(s_k)$ are known time-varying matrices with compatible dimensions.

Assumption 4.1 *The random variables $x(0)$, $\varepsilon(k)$, $w(k)$, and $v(s_k)$ are mutually uncorrelated.*

By taking the sensor resolution into consideration, the *actual* measurement from the sensor with sensor resolution R is

$$y_i^{ac}(s_k) = \begin{cases} \left\lfloor \dfrac{y_i^{id}(s_k)}{r_i} \right\rfloor r_i, & y_i^{id}(s_k) \geq r_i \\ 0, & y_i^{id}(s_k) \in (-r_i, r_i) \\ \left\lceil \dfrac{y_i^{id}(s_k)}{r_i} \right\rceil r_i, & y_i^{id}(s_k) \leq -r_i \end{cases} \tag{4.3}$$

where $y_i^{ac}(s_k)$ is the i-th element of the actual measurement $y^{ac}(s_k)$ from the sensor, $y_i^{id}(s_k)$ is the i-th element of $y^{id}(s_k)$, and r_i is the i-th element of the resolution R.

Noting that the system under consideration is a MRS, we are going to transform the MRS into a single-rate one. First, we rewrite the system (4.1) as

$$x(k+1) = F(k)x(k) + E(k)w(k)$$

where $F(k) \triangleq A(k) + \varepsilon(k)B(k)$. Then, setting

$$\bar{x}(s_k) \triangleq \text{col}\{x(s_{k-1}+1), \cdots, x(s_k - 1), x(s_k)\},$$

we have

$$\bar{x}(s_{k+1}) = \bar{A}(s_k)\bar{x}(s_k) + \bar{E}(s_k)\bar{w}(s_k) \tag{4.4}$$

where

$$
\begin{aligned}
\bar{w}(s_k) &\triangleq \mathrm{col}\{w(s_k), w(s_k+1), \cdots, w(s_{k+1}-1)\}, \\
\bar{A}(s_k) &\triangleq \Big[\underbrace{0_{bn_x \times n_x} \quad \cdots \quad 0_{bn_x \times n_x}}_{b-1} \quad \mathscr{A}(s_k)\Big], \\
\bar{E}(s_k) &\triangleq \mathscr{B}(s_k)\mathscr{E}(s_k), \ \mathscr{F}_m^n(s_k) \triangleq \prod_{i=m}^{n} F(s_{k+1}-i), \\
\mathscr{A}(s_k) &\triangleq \mathrm{col}\{\mathscr{F}_b^b(s_k), \mathscr{F}_{b-1}^b(s_k), \cdots, \mathscr{F}_1^b(s_k)\}, \\
\mathscr{B}(s_k) &\triangleq \begin{bmatrix} I & 0 & \cdots & 0 \\ \mathscr{F}_{b-1}^{b-1}(s_k) & I & \cdots & 0 \\ \mathscr{F}_{b-2}^{b-1}(s_k) & \mathscr{F}_{b-2}^{b-2}(s_k) & \cdots & 0 \\ \vdots & \vdots & \cdots & 0 \\ \mathscr{F}_1^{b-1}(s_k) & \mathscr{F}_1^{b-2}(s_k) & \cdots & I \end{bmatrix}, \\
\mathscr{E}(s_k) &\triangleq \mathrm{diag}\{E(s_k), E(s_k+1), \cdots, E(s_{k+1}-1)\}.
\end{aligned}
$$

Moreover, the ideal measurement model (4.2) is reformulated as

$$y^{id}(s_k) = \bar{C}(s_k)\bar{x}(s_k) + D(s_k)v(s_k)$$

where $\bar{C}(s_k) \triangleq \begin{bmatrix} 0 & \cdots & 0 & C(s_k) \end{bmatrix}$.

Due to the sensor resolution, the measurement received by the estimator is $y^{ac}(s_k)$ (instead of the ideal measurement $y^{id}(s_k)$). In this chapter, the state estimator is of the following form

$$\hat{x}(s_{k+1}) = \mathbb{E}\{\bar{A}(s_k)\}\hat{x}(s_k) + K(s_k)\big(y^{ac}(s_k) - \bar{C}(s_k)\hat{x}(s_k)\big) \tag{4.5}$$

where $\hat{x}(s_k)$ is the estimate of $\bar{x}(s_k)$ and $K(s_k)$ is the estimator gain matrix to be determined. The initial condition of the estimator is $\hat{x}(s_0) = \mathbb{E}\{\bar{x}(s_0)\}$.

Denoting the estimation error as

$$e(s_k) \triangleq \bar{x}(s_k) - \hat{x}(s_k)$$

and the difference between the actual measurement and the ideal measurement as

$$\Delta(s_k) \triangleq y^{ac}(s_k) - y^{id}(s_k),$$

we have the following estimation error dynamics:

$$
\begin{aligned}
e(s_{k+1}) = {} & \breve{A}(s_k)e(s_k) + \tilde{A}(s_k)\bar{x}(s_k) + \bar{E}(s_k)\bar{w}(s_k) \\
& - K(s_k)\Delta(s_k) - K(s_k)D(s_k)v(s_k)
\end{aligned} \tag{4.6}
$$

where

$$
\begin{aligned}
\tilde{A}(s_k) &\triangleq \bar{A}(s_k) - \mathbb{E}\{\bar{A}(s_k)\}, \\
\breve{A}(s_k) &\triangleq \mathbb{E}\{\bar{A}(s_k)\} - K(s_k)\bar{C}(s_k).
\end{aligned}
$$

Remark 4.1 *It can be seen from (4.4) that, due to the existence of the multiplicative noise $\varepsilon(k)$, the parameter matrices $\bar{A}(s_k)$ and $\bar{E}(s_k)$ of the augmented system (4.4) are essentially random matrices. Therefore, in the estimator (4.5), the expectation $\mathbb{E}\{\bar{A}(s_k)\}$ of the random matrix $\bar{A}(s_k)$ is used. It can be seen from the estimation error dynamic system (4.6) that the appearance of the random matrices $\bar{A}(s_k)$ and $\bar{E}(s_k)$ would bring additional difficulties in the derivation of the estimation error covariance (or its upper bound), and therefore a novel method is needed that can tackle the random matrices in an adequate way.*

The aim of the considered estimation problem is to develop a state estimator (4.5) such that the estimation error covariance

$$P(s_k) \triangleq \mathbb{E}\{e(s_k)e^T(s_k)\}$$

has a certain upper bound and, moreover, an appropriate gain matrix $K(s_k)$ is designed such that the derived upper bound is minimized.

4.2 Estimator Design

In this section, we will first derive an upper bound on the estimation error covariance, and then the estimator gain matrix will be characterized so as to minimize the obtained upper bound.

The following lemma will be useful in our later analysis.

Lemma 4.1 *Let a random matrix*

$$M(k) \triangleq \left[M_{ij}(k)\right]_{b\times b}, \quad M_{ij}(k) \in \mathbb{R}^{n_x\times n_y}$$

and a vector

$$x(k) \triangleq \mathrm{col}\{x_1(k), x_2(k), \cdots, x_b(k)\}, \quad x_i(k) \in \mathbb{R}^{n_y}$$

be given. The term $\mathbb{E}\{M(k)x(k)x^T(k)M^T(k)\}$ can be obtained by

$$\mathbb{E}\{M(k)x(k)x^T(k)M^T(k)\}_{mn} = \sum_{j=1}^{b}\sum_{i=1}^{b}\mathbb{E}\{M_{mi}(k)x_i(k)x_j^T(k)M_{nj}^T(k)\}$$

where $\mathbb{E}\{M(k)x(k)x^T(k)M^T(k)\}_{mn} \in \mathbb{R}^{n_x\times n_x}$ is the (m,n)-th submatrix of

$$\mathbb{E}\{M(k)x(k)x^T(k)M^T(k)\}.$$

Proof *The proof is easily accessible by matrix operations and is therefore omitted here.*

In the following, the recursion of the estimation error covariance is presented.

Lemma 4.2 *The estimation error covariance $P(s_{k+1})$ of the estimator* (4.5) *is recursively calculated by*

$$\begin{aligned}P(s_{k+1}) = & \breve{A}(s_k)P(s_k)\breve{A}^T(s_k) + \Gamma_1(s_k) + \Gamma_2(s_k) \\ & + K(s_k)\mathbb{E}\{\Delta(s_k)\Delta^T(s_k)\}K^T(s_k) \\ & + K(s_k)D(s_k)V(s_k)D^T(s_k)K^T(s_k) \\ & - \mathcal{A}(s_k) - \mathcal{A}^T(s_k) + \mathcal{B}(s_k) + \mathcal{B}^T(s_k)\end{aligned} \tag{4.7}$$

where

$$\begin{aligned}\Gamma_1(s_k) &\triangleq \mathbb{E}\{\tilde{A}(s_k)\bar{x}(s_k)\bar{x}^T(s_k)\tilde{A}^T(s_k)\}, \\ \Gamma_2(s_k) &\triangleq \mathbb{E}\{\bar{E}(s_k)\bar{w}(s_k)\bar{w}^T(s_k)\bar{E}^T(s_k)\}, \\ \mathcal{A}(s_k) &\triangleq \mathbb{E}\{\breve{A}(s_k)e(s_k)\Delta^T(s_k)K^T(s_k)\}, \\ \mathcal{B}(s_k) &\triangleq \mathbb{E}\{K(s_k)\Delta(s_k)v^T(s_k)D^T(s_k)K^T(s_k)\}.\end{aligned}$$

Proof *With the help of Assumption 4.1 and* (4.6), *it is obvious that* (4.7) *is true. The proof is complete.*

Note that the estimation error dynamics (4.6) contains the augmented state $\bar{x}(s_k)$. To facilitate the derivation of the $P(s_k)$, in the following, the covariance of the augmented state $\bar{x}(s_k)$ is given.

Lemma 4.3 *The state covariance $X(k) \triangleq \mathbb{E}\{x(k)x^T(k)\}$ is recursively calculated by*

$$\begin{aligned}X(k+1) = & A(k)X(k)A^T(k) + B(k)X(k)B^T(k) \\ & + E(k)W(k)E^T(k)\end{aligned} \tag{4.8}$$

with initial value $X(0)$. Moreover, the covariance of the augmented state $\bar{X}(s_k) \triangleq \mathbb{E}\{\bar{x}(s_k)\bar{x}^T(s_k)\}$ is derived by

$$\bar{X}(s_k) = \left[\bar{X}_{i,j}(s_k)\right]_{b\times b} \tag{4.9}$$

where

$$\bar{X}_{i,j}(s_k) \triangleq \mathbb{E}\{x(s_{k-1}+i)x^T(s_{k-1}+j)\} \in \mathbb{R}^{n_x\times n_x}$$

is obtained according to

$$\bar{X}_{i,j}(s_k) = \begin{cases}\prod_{l=b-i+1}^{b-j} A(s_k - l)X(s_{k-1}+j), & i > j \\ X(s_{k-1}+i), & i = j \\ X(s_{k-1}+i)\prod_{l=i}^{j-1} A^T(s_{k-1}+l), & i < j.\end{cases}$$

Proof *From* (4.1), *it is easily obtained that*

$$\begin{aligned}X(k+1) &= \mathbb{E}\{A(k)x(k)x^T(k)A^T(k) \\ &\quad + \varepsilon(k)B(k)x(k)x^T(k)B^T(k)\varepsilon^T(k) \\ &\quad + E(k)w(k)w^T(k)E^T(k)\} \\ &= A(k)X(k)A^T(k) + B(k)X(k)B^T(k) \\ &\quad + E(k)W(k)E^T(k).\end{aligned}$$

From the definition of $\bar{x}(s_k)$, *we know that* $\bar{X}_{i,j}(s_k)$ *is the* (i,j)*-th submatrix of* $\bar{X}(s_k)$. *For* $i = j$, *one has* $\bar{X}_{i,i}(s_k) = X(s_{k-1}+i)$. *For* $i > j$, *it is known from* (4.1) *that*

$$\begin{aligned}\bar{X}_{i,j}(s_k) &= A(s_k - b + i - 1)\bar{X}_{i-1,j}(s_k) \\ &= \prod_{l=b-i+1}^{b-j} A(s_k - l)\bar{X}_{j,j}(s_k) \\ &= \prod_{l=b-i+1}^{b-j} A(s_k - l)X(s_{k-1}+j).\end{aligned}$$

For $i < j$, *we have*

$$\begin{aligned}\bar{X}_{i,j}(s_k) &= \bar{X}_{i,j-1}(s_k)A^T(s_{k-1}+j-1) \\ &= \bar{X}_{i,i}(s_k)\prod_{l=i}^{j-1} A^T(s_{k-1}+l) \\ &= X(s_{k-1}+i)\prod_{l=i}^{j-1} A^T(s_{k-1}+l).\end{aligned}$$

Then, $\bar{X}(s_k)$ *is calculated by* (4.9). *The proof is complete.*

For simplification, the following notations are introduced:

$$\begin{aligned}\Theta_i(s_k) &\triangleq \mathbb{E}\left\{\mathscr{F}_i^b(s_k)x(s_k)x^T(s_k)(\mathscr{F}_i^b(s_k))^T\right\}, \\ \Omega_i^m(s_k) &\triangleq \mathbb{E}\{\bar{E}_{mi}(s_k)w(s_k+i-1)w^T(s_k+i-1)\bar{E}_{mi}^T(s_k)\}, \\ \bar{\mathscr{F}}_m^n(s_k) &\triangleq \prod_{i=m}^{n} A(s_{k+1}-i), \\ \bar{W}(s_k) &\triangleq E(s_k)W(s_k)E^T(s_k)\end{aligned}$$

where $\bar{E}_{mi}(s_k) \in \mathbb{R}^{n_x \times n_w}$ is the (m,i)-th submatrix of $\bar{E}(s_k)$.

In order to calculate the estimation error covariance, one also needs to calculate $\Gamma_1(s_k)$ and $\Gamma_2(s_k)$ with $\tilde{A}(s_k)$ and $\bar{E}(s_k)$ being random matrices. Obviously, the random matrices make the calculations non-trivial. In the following, a novel method is provided to handle such difficulties.

Lemma 4.4 *The term* $\Gamma_1(s_k)$ *is obtained by*

$$\Gamma_1(s_k) = \mathbb{E}\{\bar{A}(s_k)\bar{x}(s_k)\bar{x}^T(s_k)\bar{A}^T(s_k)\} - \mathbb{E}\{\bar{A}(s_k)\}\bar{X}(s_k)\mathbb{E}\{\bar{A}^T(s_k)\}$$

where

$$\mathbb{E}\{\bar{A}(s_k)\} = \Big[\underbrace{0_{bn_x\times n_x} \quad \cdots \quad 0_{bn_x\times n_x}}_{b-1} \quad \bar{\mathscr{A}}(s_k)\Big],$$

$$\bar{\mathscr{A}}(s_k) \triangleq \mathrm{col}\{\bar{\mathscr{F}}_b^b(s_k), \bar{\mathscr{F}}_{b-1}^b(s_k), \cdots, \bar{\mathscr{F}}_1^b(s_k)\}$$

and $\mathbb{E}\{\bar{A}(s_k)\bar{x}(s_k)\bar{x}^T(s_k)\bar{A}^T(s_k)\}$ *is derived according to*

$$\mathbb{E}\{\bar{A}(s_k)\bar{x}(s_k)\bar{x}^T(s_k)\bar{A}^T(s_k)\}_{mn} = \begin{cases} \bar{\mathscr{F}}_{b-m+1}^{b-n}(s_k)\Theta_{b-n+1}(s_k), & m > n \\ \Theta_{b-m+1}(s_k), & m = n \\ \Theta_{b-m+1}(s_k)\left(\bar{\mathscr{F}}_{b-n+1}^{b-m}(s_k)\right)^T, & m < n. \end{cases}$$

Here, $\mathbb{E}\{\bar{A}(s_k)\bar{x}(s_k)\bar{x}^T(s_k)\bar{A}^T(s_k)\}_{mn} \in \mathbb{R}^{n_x\times n_x}$ *is the* (m,n)*-th submatrix of*

$$\mathbb{E}\{\bar{A}(s_k)\bar{x}(s_k)\bar{x}^T(s_k)\bar{A}^T(s_k)\}$$

and $\Theta_i(s_k)$ *is calculated by repeating*

$$\Theta_{i-1}(s_k) = A(s_{k+1} - i + 1)\Theta_i(s_k)A^T(s_{k+1} - i + 1) + B(s_{k+1} - i + 1)\Theta_i(s_k)B^T(s_{k+1} - i + 1)$$

with

$$\Theta_b(s_k) = A(s_k)X(s_k)A^T(s_k) + B(s_k)X(s_k)B^T(s_k).$$

Proof *Noting the fact that* $\varepsilon(s_{k+1} - i)$ $(i = 1, 2, \ldots, b)$ *are uncorrelated with* $x(s_{k-1} + j)$ $(j = 1, 2, \ldots, b)$*, it is derived that*

$$\begin{aligned}
&\mathbb{E}\Big\{\tilde{A}(s_k)\bar{x}(s_k)\bar{x}^T(s_k)\tilde{A}^T(s_k)\Big\} \\
&= \mathbb{E}\{\bar{A}(s_k)\bar{x}(s_k)\bar{x}^T(s_k)\bar{A}^T(s_k)\} \\
&\quad + \mathbb{E}\Big\{\mathbb{E}\{\bar{A}(s_k)\}\bar{x}(s_k)\bar{x}^T(s_k)\mathbb{E}\{\bar{A}^T(s_k)\}\Big\} \\
&\quad - \mathbb{E}\Big\{\mathbb{E}\{\bar{A}(s_k)\}\bar{x}(s_k)\bar{x}^T(s_k)\bar{A}^T(s_k)\Big\} \\
&\quad - \mathbb{E}\Big\{\bar{A}(s_k)\bar{x}(s_k)\bar{x}^T(s_k)\mathbb{E}\{\bar{A}^T(s_k)\}\Big\} \\
&= \mathbb{E}\{\bar{A}(s_k)\bar{x}(s_k)\bar{x}^T(s_k)\bar{A}^T(s_k)\} \\
&\quad - \bar{\mathbb{E}}\{A(s_k)\}\mathbb{E}\{x(s_k)\bar{x}^T(s_k)\}\bar{\mathbb{E}}\{A^T(s_k)\}.
\end{aligned}$$

From the definition of $\bar{A}(s_k)$, by resorting to Lemma 4.1, the calculation of $\mathbb{E}\{\bar{A}(s_k)\bar{x}(s_k)\bar{x}^T(s_k)\bar{A}^T(s_k)\}$ reduces to the calculation of

$$\begin{aligned}&\mathbb{E}\{\bar{A}(s_k)\bar{x}(s_k)\bar{x}^T(s_k)\bar{A}^T(s_k)\}_{mn}\\&\quad=\mathbb{E}\{\mathscr{F}^b_{b-m+1}(s_k)x(s_k)x^T(s_k)(\mathscr{F}^b_{b-n+1}(s_k))^T\}.\end{aligned}$$

For $m=n$, it is obvious that

$$\mathbb{E}\{\bar{A}(s_k)\bar{x}(s_k)\bar{x}^T(s_k)\bar{A}^T(s_k)\}_{mm}=\Theta_{b-m+1}(s_k).$$

For $m>n$, we have

$$\begin{aligned}&\mathbb{E}\{\bar{A}(s_k)\bar{x}(s_k)\bar{x}^T(s_k)\bar{A}^T(s_k)\}_{mn}\\&\quad=\mathbb{E}\{\prod_{i=b-m+1}^{b-n}F(s_{k+1}-i)\\&\qquad\times\mathscr{F}^b_{b-n+1}(s_k)x(s_k)x^T(s_k)(\mathscr{F}^b_{b-n+1}(s_k))^T\}\\&\quad=\prod_{i=b-m+1}^{b-n}\mathbb{E}\{F(s_{k+1}-i)\}\Theta_{b-n+1}(s_k)\\&\quad=\bar{\mathscr{F}}^{b-n}_{b-m+1}(s_k)\Theta_{b-n+1}(s_k).\end{aligned}$$

Similarly, for $m<n$, we have

$$\begin{aligned}&\mathbb{E}\{\bar{A}(s_k)\bar{x}(s_k)\bar{x}^T(s_k)\bar{A}^T(s_k)\}_{mn}\\&\quad=\Theta_{b-m+1}(s_k)\left(\prod_{i=b-n+1}^{b-m}\mathbb{E}\{F(s_{k+1}-i)\}\right)^T\\&\quad=\Theta_{b-m+1}(s_k)\left(\bar{\mathscr{F}}^{b-m}_{b-n+1}(s_k)\right)^T.\end{aligned}$$

Since $\varepsilon(k)$ are mutually uncorrelated in time, we can derive the following relationship between $\Theta_{i-1}(s_k)$ and $\Theta_i(s_k)$:

$$\begin{aligned}\Theta_{i-1}(s_k)&=\mathbb{E}\{\mathscr{F}^b_{i-1}(s_k)x(s_k)x^T(s_k)(\mathscr{F}^b_{i-1}(s_k))^T\}\\&=\mathbb{E}\{F(s_{k+1}-i+1)\mathscr{F}^b_i(s_k)x(s_k)\\&\quad\times x^T(s_k)(\mathscr{F}^b_i(s_k))^TF^T(s_{k+1}-i+1)\}\\&=A(s_{k+1}-i+1)\mathbb{E}\{\mathscr{F}^b_i(s_k)x(s_k)\\&\quad\times x^T(s_k)(\mathscr{F}^b_i(s_k))^T\}A^T(s_{k+1}-i+1)\\&\quad+B(s_{k+1}-i+1)\mathbb{E}\{\mathscr{F}^b_i(s_k)x(s_k)\\&\quad\times x^T(s_k)(\mathscr{F}^b_i(s_k))^T\}B^T(s_{k+1}-i+1)\\&=A(s_{k+1}-i+1)\Theta_i(s_k)A^T(s_{k+1}-i+1)\\&\quad+B(s_{k+1}-i+1)\Theta_i(s_k)B^T(s_{k+1}-i+1).\end{aligned}$$

Moreover, it is known that

$$\begin{aligned}\Theta_b(s_k) &= \mathbb{E}\{F(s_k)x(s_k)x^T(s_k)F^T(s_k)\}\\ &= A(s_k)\mathbb{E}\{x(s_k)x^T(s_k)\}A^T(s_k)\\ &\quad + B(s_k)\mathbb{E}\{x(s_k)x^T(s_k)\}B^T(s_k)\\ &= A(s_k)X(s_k)A^T(s_k) + B(s_k)X(s_k)B^T(s_k).\end{aligned}$$

Then, $\Theta_{b-m+1}(s_k)$ and $\Theta_{b-n+1}(s_k)$ are calculated by repeating the above relationship. The proof is complete.

Lemma 4.5 *The term $\Gamma_2(s_k)$ is obtained according to*

$$\{\Gamma_2(s_k)\}_{mn} = \begin{cases} \bar{\mathscr{F}}^{b-n}_{b-m+1}(s_k)\sum_{i=1}^{n}\Omega_i^n(s_k), & m > n\\ \sum_{i=1}^{m}\Omega_i^m(s_k), & m = n\\ \sum_{i=1}^{m}\Omega_i^m(s_k)\left(\bar{\mathscr{F}}^{b-m}_{b-n+1}(s_k)\right)^T, & m < n.\end{cases}$$

Here, $\{\Gamma_2(s_k)\}_{mn} \in \mathbb{R}^{n_x \times n_x}$ is the (m,n)-th submatrix of $\Gamma_2(s_k)$. Moreover, $\Omega_i^m(s_k)$ is derived by repeating

$$\begin{aligned}\Omega_i^m(s_k) &= A(s_k+m-1)\Omega_i^{m-1}(s_k)A^T(s_k+m-1)\\ &\quad + B(s_k+m-1)\Omega_i^{m-1}(s_k)B^T(s_k+m-1)\end{aligned}$$

with

$$\Omega_i^i(s_k) = \bar{W}(s_k+i-1).$$

Proof *The proof of this lemma is similar to that of Lemma 4.4 and is therefore omitted here.*

Remark 4.2 *In Lemmas 4.4 and 4.5, a novel method is put forward to calculate $\Gamma_1(s_k) = \mathbb{E}\{\tilde{A}(s_k)\bar{x}(s_k)\bar{x}^T(s_k)\tilde{A}^T(s_k)\}$ and $\Gamma_2(s_k) = \mathbb{E}\{\bar{E}(s_k)\bar{w}(s_k)\bar{w}^T(s_k)\bar{E}^T(s_k)\}$. Note that the structures of the matrices $\tilde{A}(s_k)$ and $\bar{E}(s_k)$ complicate the calculations of $\Gamma_1(s_k)$ and $\Gamma_2(s_k)$, while the multiplicative noise $\varepsilon(k)$ makes the calculation even more complicated. Accordingly, it is quite difficult to directly calculate $\Gamma_1(s_k)$ and $\Gamma_2(s_k)$. With the help of Lemma 4.1, the calculations of $\Gamma_1(s_k)$ and $\Gamma_2(s_k)$ reduce to the calculation of the submatrices of $\Gamma_1(s_k)$ and $\Gamma_2(s_k)$, thereby reducing the computational complexity to a great extent.*

Due to the consideration of the sensor resolution, it is extremely hard (if not impossible) to obtain the exact estimation error covariance. As such, an alternative way is to obtain an upper bound on the estimation error covariance and then the gain matrix is characterized so as to minimize the obtained upper bound.

Lemma 4.6 *Denote*

$$\Delta_i(s_k) \triangleq y_i^{ac}(s_k) - y_i^{id}(s_k)$$

as the i-th element of $\Delta(s_k)$*. Then, one has*

$$|\Delta_i(s_k)| < r_i.$$

Proof *For a constant* $a \geq 1$*, from the definition of the floor function* $\lfloor \cdot \rfloor$*, we know that* $-1 < \lfloor a \rfloor - a \leq 0$ *and therefore*

$$-1 < \left\lfloor \frac{y_i^{id}(s_k)}{r_i} \right\rfloor - \frac{y_i^{id}(s_k)}{r_i} \leq 0, \quad y_i^{id}(s_k) \geq r_i$$

which is equivalent to $-r_i < y_i^{ac}(s_k) - y_i^{id}(s_k) \leq 0$ *for* $y_i^{id}(s_k) \geq r_i$*.*

Similarly, from the definition of the ceiling function $\lceil \cdot \rceil$*, we know that* $0 \leq \lceil b \rceil - b < 1$ *with* $b \leq -1$ *being a constant. Therefore, we have*

$$0 \leq y_i^{ac}(s_k) - y_i^{id}(s_k) < r_i, \quad y_i^{id}(s_k) \leq -r_i.$$

Moreover, it is easily known that $-r_i < y_i^{ac}(s_k) - y_i^{id}(s_k) < r_i$ *for* $-r_i < y_i^{id}(s_k) < r_i$*.*

Summarizing the above discussions, we have $|\Delta_i(s_k)| = \left|y_i^{ac}(s_k) - y_i^{id}(s_k)\right| < r_i$*, and the proof is complete.*

Theorem 4.1 *Let the positive scalars* $\gamma_1(s_k)$ *and* $\gamma_2(s_k)$ *be given. Denote*

$$\begin{aligned}
\bar{\gamma}_1(s_k) &\triangleq 1 + \gamma_1(s_k), \\
\bar{\gamma}_2(s_k) &\triangleq 1 + \gamma_1^{-1}(s_k) + \gamma_2(s_k), \\
\bar{\gamma}_3(s_k) &\triangleq 1 + \gamma_2^{-1}(s_k).
\end{aligned}$$

The solution $\bar{P}(s_{k+1})$ *for the following recursion*

$$\begin{aligned}
\bar{P}(s_{k+1}) = {} & \bar{\gamma}_1(s_k)\breve{A}(s_k)\bar{P}(s_k)\breve{A}^T(s_k) + \Gamma_1(s_k) + \Gamma_2(s_k) \\
& + \bar{\gamma}_2(s_k)\sum_{i=1}^{n_y} r_i^2 K(s_k)K^T(s_k) \\
& + \bar{\gamma}_3(s_k)K(s_k)D(s_k)V(s_k)D^T(s_k)K^T(s_k)
\end{aligned} \tag{4.10}$$

with initial value $\bar{P}(s_0) = P(s_0)$ *is an upper bound on the estimation error covariance* $P(s_{k+1})$*. Here,* $\Gamma_1(s_k)$ *and* $\Gamma_2(s_k)$ *can be obtained according to Lemmas 4.4 and 4.5, respectively. Moreover, the upper bound* $\bar{P}(s_{k+1})$ *is minimized with the estimator gain* $K(s_k)$ *chosen as*

$$K(s_k) = \Upsilon(s_k)\Sigma^{-1}(s_k) \tag{4.11}$$

and the minimized upper bound is

$$\begin{aligned}\bar{P}(s_{k+1}) =& -\Upsilon(s_k)\Sigma^{-1}(s_k)\Upsilon^T(s_k) + \Gamma_1(s_k) + \Gamma_2(s_k) \\ &+ \bar{\gamma}_1(s_k)\mathbb{E}\{\bar{A}(s_k)\}\bar{P}(s_k)\mathbb{E}^T\{\bar{A}(s_k)\}\end{aligned} \tag{4.12}$$

where

$$\begin{aligned}\Sigma(s_k) \triangleq& \bar{\gamma}_1(s_k)\bar{C}(s_k)\bar{P}(s_k)\bar{C}^T(s_k) \\ &+ \bar{\gamma}_2(s_k)\sum_{i=1}^{n_y} r_i^2 I + \bar{\gamma}_3(s_k)D(s_k)V(s_k)D^T(s_k), \\ \Upsilon(s_k) \triangleq& \bar{\gamma}_1(s_k)\mathbb{E}\{\bar{A}(s_k)\}\bar{P}(s_k)\bar{C}^T(s_k).\end{aligned}$$

Proof *The proof is completed by using the mathematical induction method. First, it is obvious that $P(s_0) \le \bar{P}(s_0)$ holds. Assuming that $P(s_k) \le \bar{P}(s_k)$ holds, we need to prove $P(s_{k+1}) \le \bar{P}(s_{k+1})$.*

With the help of the elementary inequality $ab^T + ba^T \le \delta aa^T + \delta^{-1}bb^T$ where a and b are vectors of appropriate dimensions and δ is a known scalar, it follows from Lemmas 4.2–4.4 that

$$\begin{aligned}P(s_{k+1}) \le& \bar{\gamma}_1(s_k)\breve{A}(s_k)P(s_k)\breve{A}^T(s_k) \\ &+ \Gamma_1(s_k) + \Gamma_2(s_k) \\ &+ \bar{\gamma}_2(s_k)K(s_k)\mathbb{E}\{\Delta(s_k)\Delta^T(s_k)\}K^T(s_k) \\ &+ \bar{\gamma}_3(s_k)K(s_k)D(s_k)V(s_k)D^T(s_k)K^T(s_k).\end{aligned}$$

Moreover, from Lemma 4.6, we know that

$$\begin{aligned}\mathbb{E}\{\Delta(s_k)\Delta^T(s_k)\} \le& \mathbb{E}\{\mathrm{tr}\{\Delta(s_k)\Delta^T(s_k)\}\}I \\ =& \sum_{i=1}^{n_y}\mathbb{E}\{\Delta_i^2(s_k)\}I \le \sum_{i=1}^{n_y} r_i^2 I.\end{aligned}$$

Accordingly, one has

$$\begin{aligned}P(s_{k+1}) \le& \bar{\gamma}_1(s_k)\breve{A}(s_k)P(s_k)\breve{A}^T(s_k) \\ &+ \Gamma_1(s_k) + \Gamma_2(s_k) \\ &+ \bar{\gamma}_2(s_k)\sum_{i=1}^{n_y} r_i^2 K(s_k)K^T(s_k) \\ &+ \bar{\gamma}_3(s_k)K(s_k)D(s_k)V(s_k)D^T(s_k)K^T(s_k).\end{aligned}$$

Therefore, $P(s_{k+1}) \le \bar{P}(s_{k+1})$ holds.

Now, we are going to characterize the estimator gain that minimizes $\bar{P}(s_{k+1})$. From the definition of $\breve{A}(s_k)$, (4.10) is rewritten as

$$\begin{aligned}\bar{P}(s_{k+1}) = & \bar{\gamma}_1(s_k)\mathbb{E}\{\bar{A}(s_k)\}\bar{P}(s_k)\mathbb{E}^T\{\bar{A}(s_k)\} \\ & + \bar{\gamma}_1(s_k)K(s_k)\bar{C}(s_k)\bar{P}(s_k)\bar{C}^T(s_k)K^T(s_k) \\ & - \bar{\gamma}_1(s_k)\mathbb{E}\{\bar{A}(s_k)\}\bar{P}(s_k)\bar{C}^T(s_k)K^T(s_k) \\ & - \bar{\gamma}_1(s_k)K(s_k)\bar{C}(s_k)\bar{P}(s_k)\mathbb{E}^T\{\bar{A}(s_k)\} \\ & + \Gamma_1(s_k) + \Gamma_2(s_k) \\ & + \bar{\gamma}_2(s_k)\sum_{i=1}^{n_y} r_i^2 K(s_k)K^T(s_k) \\ & + \bar{\gamma}_3(s_k)K(s_k)D(s_k)V(s_k)D^T(s_k)K^T(s_k).\end{aligned}$$

Noting that $\Gamma_1(s_k)$ and $\Gamma_2(s_k)$ do not contain the estimator gain $K(s_k)$, one has

$$\begin{aligned}\bar{P}(s_{k+1}) = & K(s_k)\Sigma(s_k)K^T(s_k) - \Upsilon(s_k)K^T(s_k) \\ & - K(s_k)\Upsilon^T(s_k) + \Gamma_1(s_k) + \Gamma_2(s_k) \\ & + \bar{\gamma}_1(s_k)\mathbb{E}\{\bar{A}(s_k)\}\bar{P}(s_k)\mathbb{E}^T\{\bar{A}(s_k)\} \\ = & (K(s_k) - \Upsilon(s_k)\Sigma^{-1}(s_k))\Sigma(s_k) \\ & \times (K(s_k) - \Upsilon(s_k)\Sigma^{-1}(s_k))^T \\ & - \Upsilon(s_k)\Sigma^{-1}(s_k)\Upsilon^T(s_k) \\ & + \bar{\gamma}_1(s_k)\mathbb{E}\{\bar{A}(s_k)\}\bar{P}(s_k)\mathbb{E}^T\{\bar{A}(s_k)\} \\ & + \Gamma_1(s_k) + \Gamma_2(s_k).\end{aligned}$$

Since $\Sigma(s_k) > 0$, the minimum of $\bar{P}(s_{k+1})$ is (4.12) with the estimator gain being (4.11). The proof is complete.

In the following, the proposed recursive estimation algorithm is summarized in Algorithm 1.

Remark 4.3 *In this chapter, the recursive state estimation problem is studied for MRSs with multiplicative noises and sensor resolution. In Lemmas 4.4–4.5, a novel method is used to handle the difficulties resulting from the coupling of the multi-rate sampling and the multiplicative noises. An upper bound is derived in Lemma 4.6 as a result of tackling the uncertainties caused by the sensor resolution. Based on the result given in Lemma 4.6, the estimator gain is characterized in Theorem 4.1 which minimizes the upper bound on the estimation error covariance and the corresponding minimized upper bound is presented. It is worth mentioning that the obtained minimal upper bound reflects all the system information including the multi-rate sampling, the multiplicative noises, and the sensor resolution.*

Algorithm 1 Recursive state estimation with the sensor resolution and multiplicative noises

Step 1. Give positive scalars $\gamma_1(s_k)$, $\gamma_2(s_k)$ and set the initial values $\hat{x}(s_0)$, $\bar{P}(s_0)$;

Step 2. At time instant s_k, calculate $\mathbb{E}\{\bar{A}(s_k)\}$, $\Gamma_1(s_k)$, and $\Gamma_2(s_k)$ according to Lemmas 4.4–4.5;

Step 3. Calculate $K(s_k)$ and $\bar{P}(s_{k+1})$ by using (4.11) and (4.12), respectively;

Step 4. Compute the estimate $\hat{x}(s_{k+1})$ with the estimator (4.5). Set $k = k + 1$;

Step 5. If $k < M$, then go to Step 2, else go to Step 6;

Step 6. Stop.

4.3 A Simulation Example

In this section, the usefulness of the proposed estimation algorithm is verified on the moving target tracking problem.

The dynamics of the moving target (modified from [145]) is formulated as

$$x(k+1) = A(k)x(k) + \varepsilon(k)B(k)x(k) + E(k)w(k)$$

where

$$x(k) \triangleq \begin{bmatrix} p_x^T(k) & \nu_x^T(k) & p_y^T(k) & \nu_y^T(k) \end{bmatrix}^T$$

with $(p_x(k), p_y(k))$ being the position of the target and $(\nu_x(k), \nu_y(k))$ being the velocity of the target. $\varepsilon(k)$ is the multiplicative noise with zero mean and unity covariance. $w(k)$ is the zero-mean process noise with covariance matrix

$$W(k) = \Lambda \begin{bmatrix} T^3/3 & T^2/2 & 0 & 0 \\ T^2/2 & T & 0 & 0 \\ 0 & 0 & T^3/3 & T^2/2 \\ 0 & 0 & T^2/2 & T \end{bmatrix}$$

where T is the sampling period and Λ is the acceleration variance. Moreover, the parameter matrices are

$$A(k) = \begin{bmatrix} 1 & T & 0 & 0 \\ 0 & 1 & 0 & 0 \\ 0 & 0 & 1 & T \\ 0 & 0 & 0 & 1 \end{bmatrix}, \quad E(k) = \begin{bmatrix} 1 & 0 & 0 & 0 \\ 0 & 1 & 0 & 0 \\ 0 & 0 & 1 & 0 \\ 0 & 0 & 0 & 1 \end{bmatrix},$$

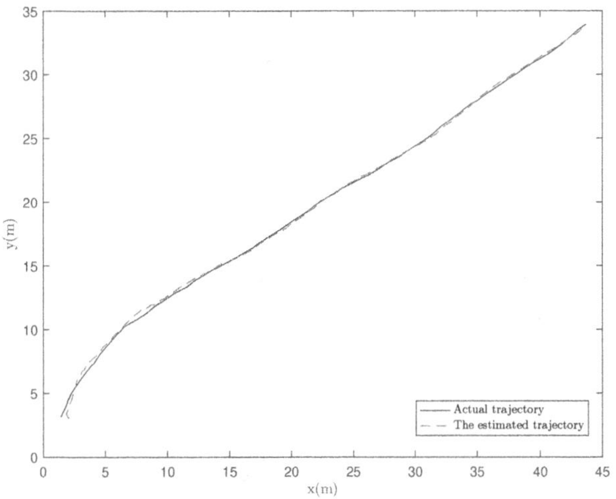

FIGURE 4.1: The trajectory of the moving target and its estimate.

$$B(k) = \begin{bmatrix} 0 & 0.01 & 0 & 0 \\ 0 & 0 & 0 & 0 \\ 0 & 0 & 0 & 0.01 \\ 0 & 0.01 & 0 & 0 \end{bmatrix}.$$

In this example, a sensor with the sampling period $b = 2T$ and the resolution $R = \text{col}\{0.1, 0.01\}$ is deployed to collect the information. The ideal measurement model is given as

$$y^{id}(s_k) = C(s_k)x(s_k) + D(s_k)v(s_k)$$

where $v(s_k)$ is the measurement noise with zero mean and covariance 0.05. The parameter matrices are

$$C(s_k) = \begin{bmatrix} 1 & 0 & 0 & 0 \\ 0 & 0 & 1 & 0 \end{bmatrix}, \; D(s_k) = \begin{bmatrix} 1 \\ 1 \end{bmatrix}.$$

In the simulation, we set $T = 1\text{s}$ and $\Lambda = 0.04$. The initial condition of the state is $x(0) = \begin{bmatrix} 2\text{m} & 0.1\text{m/s} & 3\text{m} & 0.2\text{m/s} \end{bmatrix}^T$.

The simulation results are given in Figs. 4.1–4.2. Figure 4.1 shows the actual trajectory of the moving target and the estimate of the trajectory. Figure 4.2 gives the trace of the minimal upper bound and the mean square error of the proposed estimation algorithm. The mean square error, denoted as MSE(k), is defined as

$$\text{MSE}(k) \triangleq \frac{1}{N}\left(x(k) - \vec{x}(k)\right)^T \left(x(k) - \vec{x}(k)\right)$$

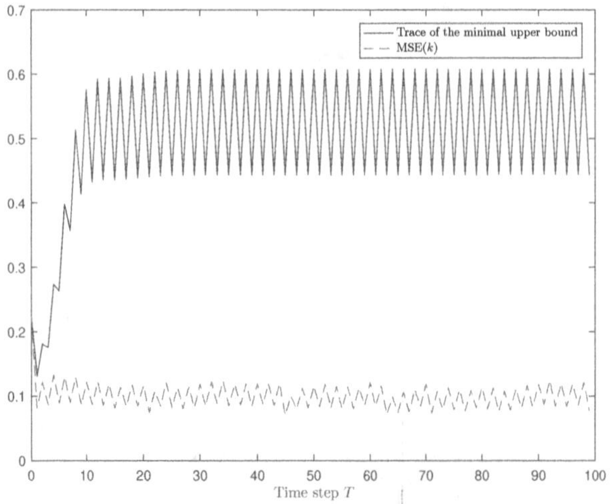

FIGURE 4.2: Trace of the minimal upper bound and the MSE.

with $N = 500$ and $\vec{x}(k)$ being the estimate of $x(k)$. The estimate $\vec{x}(k)$ is obtained by applying matrix operations to $\hat{x}(k)$. The simulation results verify that the developed estimation algorithm is effective in the moving target tracking problem and the derived minimal upper bound is indeed an upper bound of the mean square error.

To show the monotonicity of the minimal upper bound with respect to the resolution of the sensor, in the following, simulation results with different sensor resolutions are presented. Figures 4.3–4.4 give the minimal upper bounds and the MSEs with the sensor resolutions $R_1 = \text{col}\{0.5, 0.01\}$ and $R_2 = \text{col}\{0.1, 0.05\}$, respectively. From the simulation results, we know that the minimal upper bound and the MSE are increasing when the sensor resolution increases, which is in agreement with the engineering practice.

4.4 Conclusion

In this chapter, the recursive state estimation problem has been investigated for MRSs with multiplicative noises and sensor resolution. By applying the lifting technique, the MRS with multiplicative noises has been converted into a single-rate system with random parameter matrices. A novel method has been put forward to handle the difficulties from the random parameter matrices. A novel state estimation scheme with low mathematical complexity has also been developed which is robust to the uncertainty caused by the sensor

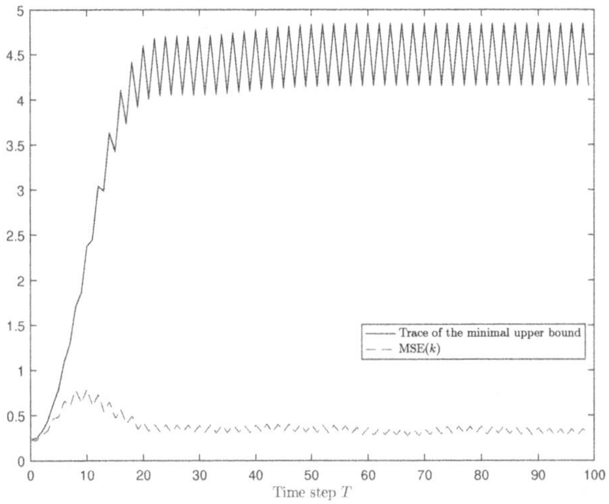

FIGURE 4.3: Estimation performance with resolution $R_1 = \text{col}\{0.5, 0.01\}$.

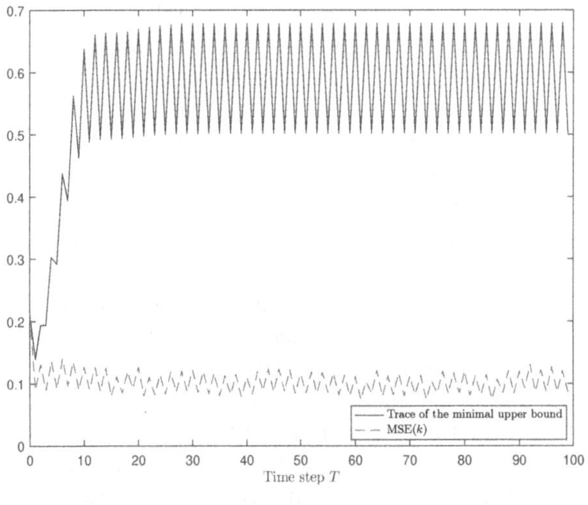

FIGURE 4.4: Estimation performance with resolution $R_2 = \text{col}\{0.1, 0.05\}$.

resolution. With the developed state estimation scheme, an upper bound has been obtained on the estimation error covariance and the estimator gain has been characterized with which the obtained upper bound is minimized. Finally, the effectiveness of the proposed estimation algorithm has been verified on the moving target tracking problem.

5

Minimum-Variance State and Fault Estimation for Multi-Rate Systems with Dynamical Bias

In engineering practice, the measurement output of the sensors are often subject to abrupt changes due to a variety of reasons such as sensor aging and random sensor failure. Such a phenomenon, customarily known as sensor fault, would largely degrade the estimation performance and it is therefore necessary to acquire the information of the sensor fault with hope to mitigate the impact from the sensor fault. Recently, the fault estimation problems for single-rate systems with sensor fault have been widely investigated. In [45], the joint estimation problem for the state and the sensor fault has been studied for discrete-time systems and, subsequently, a fault-tolerant controller design scheme has been proposed. Unfortunately, the joint state and fault estimation problem for MRSs has received little attention despite its practical significance, and this gives rise to the main motivation of this chapter.

In practical systems, it is quite common that the system noises consist of white noises and the strongly correlated noises, where the latter are called *random biases* that could be either constant or dynamic [100]. As early as in 1990, the random bias has been characterized by a dynamical equation in [71] where the joint state and random bias estimation problem has been considered. Thereafter, the state estimation problems for single-rate systems with random bias have received considerable research attention. For example, in [157], the state estimation problem has been studied for a class of two-dimensional systems with random bias and measurement quantization, and a recursive state estimation algorithm has been designed. Note that the corresponding state estimation problems for MRSs with random bias have not been considered yet, and this constitutes another motivation of this chapter.

Motivated by the above discussions, in this chapter, we aim to solve the joint state and fault estimation problem for MRSs subject to dynamical bias. The main contributions of this chapter are: 1) the joint state and fault estimation estimation problem is, for the first time, studied for MRSs with dynamical bias where the considered fault model is quite general; 2) different from the traditional lifting technique that often leads to high computational burden, a novel method is put forward to convert the MRSs into single-rate systems through the introduction of a time-varying delay into the measurement

DOI: 10.1201/9781032619507-5

equation; and 3) the proposed joint state and fault estimation algorithm is in the recursive form and therefore suitable for online application.

5.1 Problem Formulation

Consider the following class of discrete-time systems:

$$x(s+1) = A(s)x(s) + B(s)w(s) + E(s)h(s) \tag{5.1}$$

where $x(s) \in \mathbb{R}^{n_x}$ is the system state and $w(s) \in \mathbb{R}^{n_w}$ is the process noise. The initial value $x(0)$ is a random variable with the mean $\bar{x}(0)$ and the covariance $X(0)$. $h(s) \in \mathbb{R}^{n_h}$ stands for the random bias with the following dynamics:

$$h(s+1) = H(s)h(s) + \lambda(s) \tag{5.2}$$

where $\lambda(s) \in \mathbb{R}^{n_h}$ is a zero-mean Gaussian sequence with the covariance $\Lambda(s) > 0$. The initial value $h(0)$ of the random bias is a zero-mean Gaussian random variable with the covariance $\Pi(0)$. $H(s)$, $A(s)$, $B(s)$, and $E(s)$ are known time-varying matrices with compatible dimensions.

In this chapter, the sampling period of the sensor is $b \triangleq l_{m+1} - l_m$ where $b \geq 1$ is a positive integer. l_m is the sampling instant of the sensor with m being the order number of the sampling instant. The measurement model of the sensor is

$$y(l_m) = C(l_m)x(l_m) + D(l_m)v(l_m) + F(l_m)f(l_m) \tag{5.3}$$

where $y(l_m) \in \mathbb{R}^{n_y}$ is the measurement output and $v(l_m) \in \mathbb{R}^{n_v}$ is the measurement noise. $C(l_m)$, $D(l_m)$, and $F(l_m)$ are known time-varying matrices with compatible dimensions. $f(l_m)$ is the unknown sensor fault evolving according to [62]

$$f(l_{m+1}) = G(l_m)f(l_m) \tag{5.4}$$

where $G(l_m)$ is a known time-varying matrix with compatible dimensions.

The process noise $w(s)$ and the measurement noise $v(l_m)$ satisfy

$$\begin{aligned}
&\mathbb{E}\{w(s)\} = 0, \ \mathbb{E}\{w(s)w^T(t)\} = W(s)\delta(s,t), \\
&\mathbb{E}\{v(l_m)\} = 0, \ \mathbb{E}\{v(l_m)v^T(l_n)\} = V(l_m)\delta(l_m,l_n)
\end{aligned}$$

where $W(s)$ and $V(l_m)$ are known time-varying matrices with compatible dimensions.

Assumption 5.1 *The random variables $x(0)$, $h(0)$, $\lambda(s)$, $w(s)$, and $v(l_m)$ are mutually independent.*

In literature, a typical approach to dealing with MRSs is to use the lifting technique to obtain an augmented state equation with an increased state update period, which gives rise to heavy computational load. In this chapter, instead of utilizing the lifting technique, we aim to reconstruct a new measurement equation with a decreased sampling period. Here, the zero-order hold strategy is adopted to compensate the measurements at the non-sampling instants of the sensor.

With the zero-order hold strategy, the actual measurement used by the estimator is

$$\bar{y}(s) \triangleq y(\bar{l}_n), \quad \bar{l}_n \leq s < \bar{l}_{n+1}$$

with $\bar{l}_n$ being the largest measurement sampling instant that is not larger than s and $\bar{l}_{n+1} \triangleq \bar{l}_n + b$.

Define $\rho_s \triangleq s-\bar{l}_n$ ($\bar{l}_n \leq s < \bar{l}_{n+1}$). Then, the measurement $\bar{y}(s)$ is rewritten as

$$\begin{aligned}\bar{y}(s) = {} & C(s-\rho_s)x(s-\rho_s) + D(s-\rho_s)v(s-\rho_s) \\ & + F(s-\rho_s)f(s-\rho_s).\end{aligned} \tag{5.5}$$

In the following, we introduce a new notation

$$\bar{f}(s) \triangleq f(\bar{l}_n), \quad \bar{l}_n \leq s < \bar{l}_{n+1}. \tag{5.6}$$

Similarly, one has $\bar{f}(s) = f(s-\rho_s)$ for $\bar{l}_n \leq s < \bar{l}_{n+1}$. From the definition of $\bar{f}(s)$, it is obvious that

$$\bar{f}(s+1) = \begin{cases} \bar{f}(s), & \text{if } \bar{l}_n < s+1 < \bar{l}_{n+1} \\ G(s-b+1)\bar{f}(s), & \text{if } s+1 = \bar{l}_{n+1}. \end{cases}$$

Then, $\bar{f}(s+1)$ is further rewritten as

$$\bar{f}(s+1) = \bar{G}(s)\bar{f}(s)$$

where

$$\bar{G}(s) \triangleq (1-\delta(s+1,\bar{l}_{n+1}))I + \delta(s+1,\bar{l}_{n+1})G(s-b+1)$$

and $\delta(\cdot,\cdot)$ is the Kronecker delta function. Furthermore, with (5.6), $\bar{y}(s)$ is rewritten as

$$\begin{aligned}\bar{y}(s) = {} & C(s-\rho_s)x(s-\rho_s) + D(s-\rho_s)v(s-\rho_s) \\ & + F(s-\rho_s)\bar{f}(s).\end{aligned} \tag{5.7}$$

Remark 5.1 *In this chapter, to estimate the fault, the dynamics of the fault are required to be known [62]. Nevertheless, noting that the update period of the fault is b, it is impossible to obtain the relationship between $f(s+1)$ and $f(s)$. To solve such a problem, in this chapter, we introduce a new notation $\bar{f}(s)$*

that satisfies (5.6), *and the relationship between* $\bar{f}(s+1)$ *and* $\bar{f}(s)$ *is known. Then, the dynamics of* $\bar{f}(s)$ *is obtained and an estimator can be designed to estimate* $\bar{f}(s)$. *Note that, with the help of* (5.6), *the estimate of the fault* $f(l_m)$ *can be easily obtained from the estimate of* $\bar{f}(s)$.

Remark 5.2 *Motivated by [135], in this chapter, a novel method is put forward to transform the MRSs into single-rate systems. By introducing* ρ_s, *the measurement* (5.3) *with a sampling period* b *is transformed into* (5.5) *with a sampling period* 1. *Compared with the lifting technique, the proposed method avoids the augmentation of the system state, and therefore the computational complexity is reduced.*

From (5.1) and (5.7), it is obvious that the MRS is now transformed into a single-rate system with the time-varying delay ρ_s. To tackle the addressed joint state and fault estimation problem, we design the estimator of the following form:

$$\begin{cases} \hat{x}(s+1) = A(s)\hat{x}(s) + E(s)\hat{h}(s) + K_1(s)\Big(\bar{y}(s) - \\ \qquad\qquad C(s-\rho_s)\hat{x}(s-\rho_s) - F(s-\rho_s)\hat{f}(s)\Big) \\ \hat{h}(s+1) = H(s)\hat{h}(s) + K_2(s)\Big(\bar{y}(s) - \\ \qquad\qquad C(s-\rho_s)\hat{x}(s-\rho_s) - F(s-\rho_s)\hat{f}(s)\Big) \\ \hat{f}(s+1) = \bar{G}(s)\hat{f}(s) + K_3(s)\Big(\bar{y}(s) - \\ \qquad\qquad C(s-\rho_s)\hat{x}(s-\rho_s) - F(s-\rho_s)\hat{f}(s)\Big) \end{cases} \tag{5.8}$$

where $\hat{x}(s)$ is the estimate of the state $x(s)$, $\hat{h}(s)$ is the estimate of the bias $h(s)$, $\hat{f}(s)$ is the estimate of $\bar{f}(s)$, and $K_1(s)$, $K_2(s)$, and $K_3(s)$ are the estimator gain matrices to be designed. Moreover, we set $\hat{x}(0) = \bar{x}(0)$ and $\hat{h}(0) = \hat{f}(0) = 0$.

Denoting the state estimation error as $e_x(s) \triangleq x(s) - \hat{x}(s)$, the bias estimation error as $e_h(s) \triangleq h(s) - \hat{h}(s)$, and the fault estimation error as $e_f(s) \triangleq \bar{f}(s) - \hat{f}(s)$, we have

$$\begin{aligned} e_x(s+1) =& A(s)e_x(s) + B(s)w(s) + E(s)e_h(s) \\ & - K_1(s)\Big(C(s-\rho_s)e_x(s-\rho_s) \\ & + D(s-\rho_s)v(s-\rho_s) + F(s-\rho_s)e_f(s)\Big) \\ e_h(s+1) =& H(s)e_h(s) + \lambda(s) \\ & - K_2(s)\Big(C(s-\rho_s)e_x(s-\rho_s) \\ & + D(s-\rho_s)v(s-\rho_s) + F(s-\rho_s)e_f(s)\Big) \end{aligned}$$

$$e_f(s+1) = \bar{G}(s)e_f(s) - K_3(s)\Big(C(s-\rho_s)e_x(s-\rho_s) + D(s-\rho_s)v(s-\rho_s) + F(s-\rho_s)e_f(s)\Big).$$

Denoting $e(s) \triangleq \begin{bmatrix} e_x^T(s) & e_h^T(s) & e_f^T(s)\end{bmatrix}^T$, we have the following augmented system

$$\begin{aligned} e(s+1) = {} & \tilde{A}(s)e(s) - \bar{K}(s)\bar{C}(s-\rho_s)e(s-\rho_s) + \bar{I}\lambda(s) \\ & - \bar{K}(s)D(s-\rho_s)v(s-\rho_s) + \bar{B}(s)w(s) \end{aligned} \tag{5.9}$$

where

$$\begin{aligned} \tilde{A}(s) &\triangleq \bar{A}(s) - \bar{K}(s)\bar{F}(s-\rho_s), \\ \bar{A}(s) &\triangleq \begin{bmatrix} A(s) & E(s) & 0 \\ 0 & H(s) & 0 \\ 0 & 0 & \bar{G}(s) \end{bmatrix}, \\ \bar{K}(s) &\triangleq \begin{bmatrix} K_1^T(s) & K_2^T(s) & K_3^T(s) \end{bmatrix}^T, \\ \bar{F}(s-\rho_s) &\triangleq \begin{bmatrix} 0 & 0 & F(s-\rho_s) \end{bmatrix}, \ \bar{I} \triangleq \begin{bmatrix} 0 & I & 0 \end{bmatrix}^T \\ \bar{B}(s) &\triangleq \begin{bmatrix} B^T(s) & 0 & 0 \end{bmatrix}^T, \quad \bar{C}(s-\rho_s) \triangleq \begin{bmatrix} C(s-\rho_s) & 0 & 0 \end{bmatrix}. \end{aligned}$$

The aim of this chapter is to design the estimator (5.8) such that the estimation error covariance $P(s) \triangleq \mathbb{E}\{e(s)e^T(s)\}$ is minimized.

5.2 Joint State and Fault Estimator Design

In this section, we are going to solve the joint state and fault estimation problem for MRSs with dynamical bias. In the following lemma, the estimation error covariance is first derived.

Lemma 5.1 *The estimation error covariance $P(s+1)$ is calculated by the following recursion:*

$$\begin{aligned} P(s+1) = {} & \tilde{A}(s)P(s)\tilde{A}^T(s) + \bar{B}(s)W(s)\bar{B}^T(s) + \bar{I}\Lambda(s)\bar{I}^T \\ & + \bar{K}(s)\bar{C}(s-\rho_s)P(s-\rho_s)\bar{C}^T(s-\rho_s)\bar{K}^T(s) \\ & + \bar{K}(s)D(s-\rho_s)V(s-\rho_s)D^T(s-\rho_s)\bar{K}^T(s) \\ & - \mathscr{P}_1(s) - \mathscr{P}_1^T(s) - \mathscr{P}_2(s) - \mathscr{P}_2^T(s) \end{aligned} \tag{5.10}$$

where

$$\begin{aligned} \mathscr{P}_1(s) &\triangleq \tilde{A}(s)\mathbb{E}\{e(s)e^T(s-\rho_s)\}\bar{C}^T(s-\rho_s)\bar{K}^T(s), \\ \mathscr{P}_2(s) &\triangleq \tilde{A}(s)\mathbb{E}\{e(s)v^T(s-\rho_s)\}D^T(s-\rho_s)\bar{K}^T(s). \end{aligned}$$

Proof *It is easily known from Assumption 5.1 and* (5.9) *that* (5.10) *is true. Therefore, the proof is omitted here.*

From Lemma 5.1, we know that the calculation of the estimation error covariance needs the calculations of $\mathscr{P}_1(s)$ and $\mathscr{P}_2(s)$, for which some preliminary results are presented as follows.

Lemma 5.2 $\Omega_1(s) \triangleq \mathbb{E}\{e(s)v^T(s-\rho_s)\}$ *is calculated by*

$$\Omega_1(s) = \begin{cases} 0, & \text{for } \rho_s = 0 \\ -\Upsilon_1(s), & \text{for } \rho_s > 0 \end{cases} \tag{5.11}$$

where

$$\begin{aligned}\Upsilon_1(s) \triangleq (1-\delta(1,\rho_s)) &\sum_{i=2}^{\rho_s}\prod_{j=1}^{i-1}\tilde{A}(s-j) \\ &\times \bar{K}(s-i)D(s-\rho_s)V(s-\rho_s) \\ + \bar{K}(s-1)&D(s-\rho_s)V(s-\rho_s).\end{aligned}$$

Proof *From the definition of* ρ_s*, we know that* ρ_s *takes values in the set* $\{0,1,2,\ldots,b-1\}$ *and*

$$\rho_s = \begin{cases} 0, & \text{for } s = \bar{l}_n \\ \rho_{s-1}+1, & \text{for } \bar{l}_n < s < \bar{l}_{n+1}. \end{cases}$$

The proof of this lemma is divided in the following two cases.

Case 5.1 $\underline{\rho_s = 0}$. *For* $\rho_s = 0$*, it is obvious that*

$$\mathbb{E}\{e(s)v^T(s-\rho_s)\} = \mathbb{E}\{e(s)v^T(s)\} = 0.$$

Case 5.2 $\underline{\rho_s > 0}$. *By introducing*

$$\begin{aligned}\Delta(t) \triangleq \mathbb{E}\{&\bar{K}(t)\bar{C}(t-\rho_t)e(t-\rho_t)v^T(s-\rho_s) \\ &+ \bar{K}(t)D(t-\rho_t)v(t-\rho_t)v^T(s-\rho_s)\},\end{aligned}$$

we have

$$\begin{aligned}&\mathbb{E}\{e(s)v^T(s-\rho_s)\} \\ &= \tilde{A}(s-1)\mathbb{E}\{e(s-1)v^T(s-\rho_s)\} - \Delta(s-1) \\ &= \prod_{i=1}^{\rho_s}\tilde{A}(s-i)\mathbb{E}\{e(s-\rho_s)v^T(s-\rho_s)\} \\ &\quad - (1-\delta(1,\rho_s))\sum_{i=2}^{\rho_s}\prod_{j=1}^{i-1}\tilde{A}(s-j)\Delta(s-i) - \Delta(s-1).\end{aligned}$$

Accordingly, what we need to do is to calculate $\Delta(s-i)$ $(1 \leq i \leq \rho_s)$. *From the definition of* ρ_s, *we know that* $s - \rho_s = \bar{l}_n$. *For* $1 \leq i < \rho_s$, *one has*

$$\bar{l}_n = s - \rho_s < s - i \leq s - 1 = \bar{l}_n + \rho_s - 1 < \bar{l}_{n+1},$$

and therefore $s - i - \rho_{s-i} = s - \rho_s$. *For* $i = \rho_s$, *it is obvious that* $s - i - \rho_{s-i} = s - \rho_s$. *Accordingly,* $\Delta(s-i)$ $(1 \leq i \leq \rho_s)$ *is rewritten as*

$$\begin{aligned}\Delta(s-i) =& \mathbb{E}\{\bar{K}(s-i)\bar{C}(s-\rho_s)e(s-\rho_s)v^T(s-\rho_s) \\ & + \bar{K}(s-i)D(s-\rho_s)v(s-\rho_s)v^T(s-\rho_s)\} \\ =& \bar{K}(s-i)D(s-\rho_s)V(s-\rho_s).\end{aligned}$$

Therefore, one has

$$\begin{aligned}\mathbb{E}\{e(s)v^T(s-\rho_s)\} =& -(1-\delta(1,\rho_s))\sum_{i=2}^{\rho_s}\prod_{j=1}^{i-1}\tilde{A}(s-j) \\ & \times \bar{K}(s-i)D(s-\rho_s)V(s-\rho_s) \\ & - \bar{K}(s-1)D(s-\rho_s)V(s-\rho_s).\end{aligned}$$

The proof is complete.

Remark 5.3 *In Lemma 5.2, instead of simply applying the elementary equality to avoid the calculation of* $\mathbb{E}\{e(s)v^T(s-\rho_s)\}$, *we have derived the exact form of* $\mathbb{E}\{e(s)v^T(s-\rho_s)\}$. *It is worth mentioning that the calculation of* $\mathbb{E}\{e(s)v^T(s-\rho_s)\}$ *is non-trivial due to the existence of the time-varying delay* ρ_s.

Lemma 5.3 *The term* $\Omega_2(s) \triangleq \mathbb{E}\{e(s)e^T(s-\rho_s)\}$ *is recursively calculated by*

$$\Omega_2(s) = \begin{cases} P(s), & \text{for } \rho_s = 0 \\ \Upsilon_2(s), & \text{for } \rho_s > 0 \end{cases} \tag{5.12}$$

where

$$\begin{aligned}\Upsilon_2(s) \triangleq& \tilde{A}(s-1)\Omega_2(s-1) \\ & - \bar{K}(s-1)\bar{C}(s-\rho_{s-1}-1)P(s-\rho_{s-1}-1).\end{aligned}$$

Proof *The proof of this lemma is similar to that of Lemma 5.2 and is therefore omitted here.*

Theorem 5.1 *The estimator gains that minimize the estimation error covariance* $P(s)$ *are given as follows:*

$$K_1(s) = \begin{bmatrix} I & 0 & 0 \end{bmatrix} \bar{K}(s), \tag{5.13}$$
$$K_2(s) = \begin{bmatrix} 0 & I & 0 \end{bmatrix} \bar{K}(s), \tag{5.14}$$
$$K_3(s) = \begin{bmatrix} 0 & 0 & I \end{bmatrix} \bar{K}(s) \tag{5.15}$$

where

$$
\begin{aligned}
\bar{K}(s) &\triangleq \Psi(s)\Theta^{-1}(s), \\
\Theta(s) &\triangleq \bar{F}(s-\rho_s)P(s)\bar{F}^T(s-\rho_s) \\
&\quad + \bar{C}(s-\rho_s)P(s-\rho_s)\bar{C}^T(s-\rho_s) \\
&\quad + D(s-\rho_s)V(s-\rho_s)D^T(s-\rho_s) \\
&\quad + \bar{F}(s-\rho_s)\Omega_2(s)\bar{C}^T(s-\rho_s) \\
&\quad + \bar{C}(s-\rho_s)\Omega_2^T(s)\bar{F}^T(s-\rho_s) \\
&\quad + \bar{F}(s-\rho_s)\Omega_1(s)D^T(s-\rho_s) \\
&\quad + D(s-\rho_s)\Omega_1^T(s)\bar{F}^T(s-\rho_s), \\
\Psi(s) &\triangleq \bar{A}(s)P(s)\bar{F}^T(s-\rho_s) + \bar{A}(s)\Omega_2(s)\bar{C}^T(s-\rho_s) \\
&\quad + \bar{A}(s)\Omega_1(s)D^T(s-\rho_s).
\end{aligned}
$$

Moreover, the minimal estimation error covariance is given by

$$
\begin{aligned}
P(s+1) = &-\Psi(s)\Theta^{-1}(s)\Psi^T(s) + \bar{A}(s)P(s)\bar{A}^T(s) \\
&+ \bar{B}(s)W(s)\bar{B}^T(s) + \bar{I}\Lambda(s)\bar{I}^T.
\end{aligned}
$$

Proof *With the help of Lemmas 5.1–5.3, one has*

$$
\begin{aligned}
P(s+1) =&\bar{K}(s)\Theta(s)\bar{K}^T(s) + \bar{A}(s)P(s)\bar{A}^T(s) \\
&- \bar{K}(s)\bar{F}(s-\rho_s)P(s)\bar{A}^T(s) \\
&- \bar{A}(s)P(s)\bar{F}^T(s-\rho_s)\bar{K}^T(s) \\
&+ \bar{B}(s)W(s)\bar{B}^T(s) + \bar{I}\Lambda(s)\bar{I}^T \\
&- \bar{A}(s)\Omega_2(s)\bar{C}^T(s-\rho_s)\bar{K}^T(s) \\
&- \bar{K}(s)\bar{C}(s-\rho_s)\Omega_2^T(s)\bar{A}^T(s) \\
&- \bar{A}(s)\Omega_1(s)D^T(s-\rho_s)\bar{K}^T(s) \\
&- \bar{K}(s)D(s-\rho_s)\Omega_1^T(s)\bar{A}^T(s).
\end{aligned}
$$

We are now ready to derive the estimator gains that minimize the estimation error covariance. The estimation error covariance $P(s+1)$ is rewritten as

$$
\begin{aligned}
P(s+1) =&\bar{K}(s)\Theta(s)\bar{K}^T(s) - \Psi(s)\bar{K}^T(s) \\
&- \bar{K}(s)\Psi^T(s) + \tilde{A}(s)P(s)\tilde{A}^T(s) \\
&+ \bar{B}(s)W(s)\bar{B}^T(s) + \bar{I}\Lambda(s)\bar{I}^T \\
=&\big(\bar{K}(s) - \Psi(s)\Theta^{-1}(s)\big)\Theta(s) \\
&\quad \times \big(\bar{K}(s) - \Psi(s)\Theta^{-1}(s)\big)^T \\
&- \Psi(s)\Theta^{-1}(s)\Psi^T(s) + \bar{A}(s)P(s)\bar{A}^T(s) \\
&+ \bar{B}(s)W(s)\bar{B}^T(s) + \bar{I}\Lambda(s)\bar{I}^T.
\end{aligned}
$$

It is easily known that the estimation error covariance $P(s+1)$ is minimized by choosing $\bar{K}(s)$ as $\Psi(s)\Theta^{-1}(s)$. Noting the definition of $\bar{K}(s)$, the estimator gains that minimize $P(s+1)$ are derived by (5.13)–(5.15). The proof is complete.

Remark 5.4 *In this chapter, the joint state and fault estimation problem is concerned for a class of MRSs with dynamical bias. A novel method is put forward to convert the MRSs into single-rate systems and the proposed method has less computation complexity as compared to the lifting technique. First, in Lemma 5.1, the recursion of the estimation error covariance is derived. Then, with the help of Lemmas 5.2–5.3, the estimator gains that minimize the estimation error covariance as well as the minimal estimation error covariance are given in Theorem 5.1. It is worth noting that both the state and the fault are well estimated with the proposed estimation algorithm.*

5.3 An Illustrative Example

In the simulation example, we consider the joint state and fault estimation problem for a DC servo system [213] subject to random bias where the system parameters in (5.1)–(5.2) are given as follows:

$$A(s) = \begin{bmatrix} 1.12 + 0.3\sin(s) & 0.213 & -0.333 \\ 1 & 0 & 0 \\ 0 & 1 & 0 \end{bmatrix}, \quad b = 2,$$

$$E(s) = \begin{bmatrix} 0.45 & 0.26 & 0.12 + 0.2\sin(s) \\ 0.43 & 0.33 + 0.2\cos(s) & 0.28 \\ 0.33 & 0.34 & 0.25 \end{bmatrix},$$

$$H(s) = \begin{bmatrix} 0.31 & 0.12 & 0.26 \\ 0.37 & 0.21 & 0.34 \\ 0.52 & 0.15 & 0.25 \end{bmatrix}, \quad B(s) = \begin{bmatrix} 0.8 \\ 0 \\ 0 \end{bmatrix}, \quad F(l_m) = \begin{bmatrix} 1 \\ 1 \end{bmatrix},$$

$$C(l_m) = \begin{bmatrix} 1 & 2 + \sin(l_m) & 1 \\ 2 & 1 & 2 \end{bmatrix}, \quad D(l_m) = \begin{bmatrix} 0.43 \\ 0.51 \end{bmatrix}, \quad \Lambda(s) = 0.15I.$$

The covariances of the process noise $w(s)$ and the measurement noise $v(l_m)$ are 0.1 and 0.15, respectively. The initial conditions are given as $\hat{x}(0) = \begin{bmatrix} 0.52 & -0.56 & 0.55 \end{bmatrix}^T$.

In the simulation, the following sensor fault is considered:

$$f(l_{m+1}) = G(l_m)f(l_m)$$

with $G(l_m) = 1.8\sin(l_m)$. With the given parameters, the estimator gains $K_i(s)$ $(i = 1, 2, 3)$ and the estimation error covariance $P(s)$ are derived according to the proposed estimation algorithm. The simulation results are shown

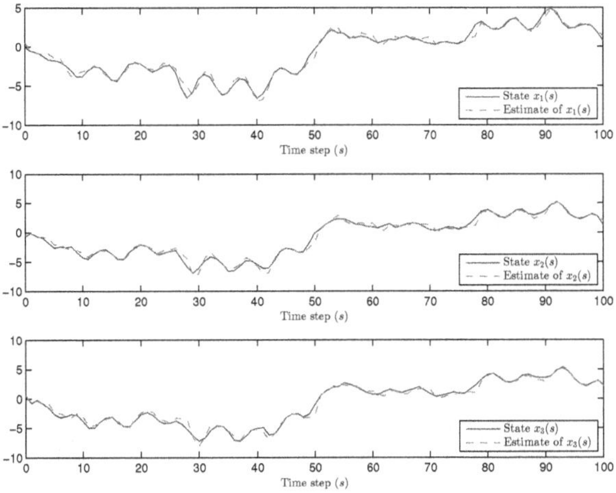

FIGURE 5.1: State $x(s)$ and the estimate.

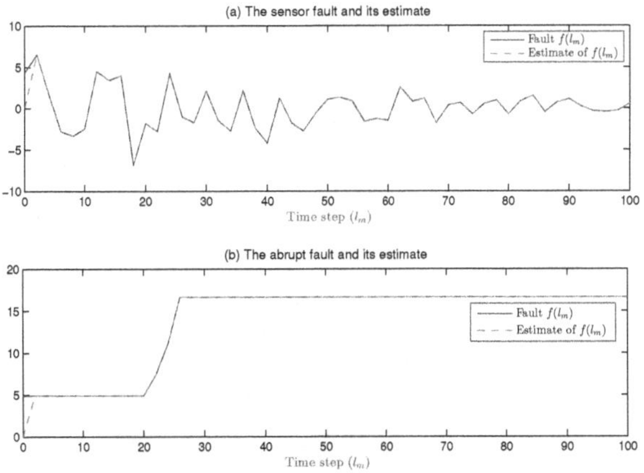

FIGURE 5.2: The fault estimation performance.

in Figs. 5.1–5.3. Figure 5.1 shows $x_i(s)$ $(i = 1, 2, 3)$ and the corresponding estimates where $x_i(s)$ denotes the ith element of the state $x(s)$. Figure 5.2(a) depicts the sensor fault $f(l_m)$ and its estimate. It is known from Figure 5.1 and Fig. 5.2(a) that the proposed estimation scheme can estimate the system state and the sensor fault with a satisfactory accuracy. Let $\text{MSE}_i(s)$ $(i = 1, 2, 3)$ denote the mean-square error of the estimation of $x_i(s)$, i.e.,

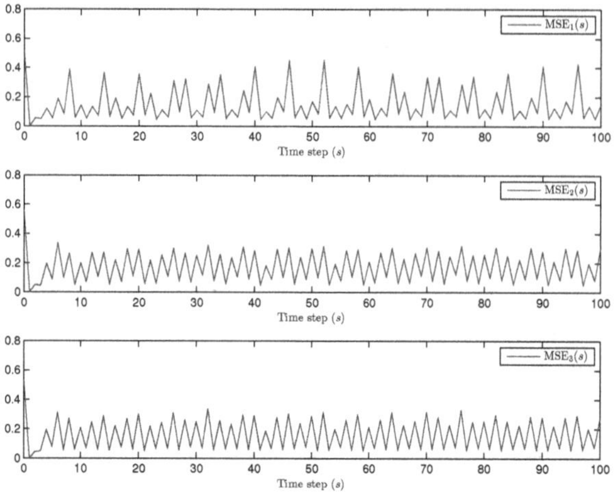

FIGURE 5.3: The mean-square error of the estimation.

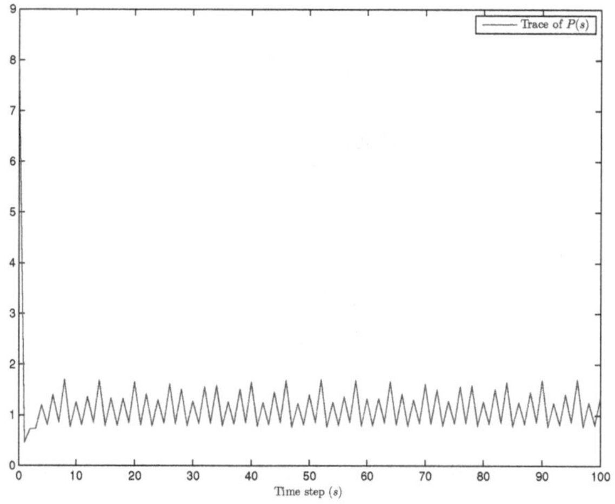

FIGURE 5.4: The trace of $P(s)$.

$\text{MSE}_i(s) = \frac{1}{N}\sum_{j=1}^{N}\left(x_i(s) - \hat{x}_i(s)\right)^2$. The $\text{MSE}_i(s)$ $(i = 1, 2, 3)$ are plotted in Fig. 5.3 which further verifies the estimation accuracy of the developed fault estimation algorithm. In Fig. 5.4, the trace of the estimation error covariance $P(s)$ is given. The simulation results verify that the proposed estimation scheme is indeed effective.

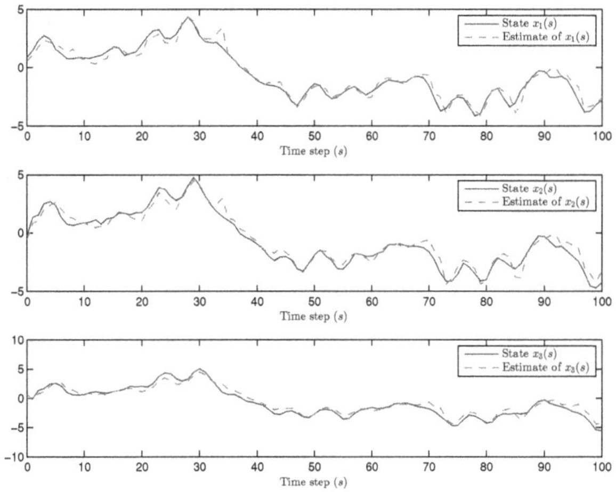

FIGURE 5.5: The estimate from the estimation scheme without considering the random bias.

To further verify the fault estimation performance, let us consider the abrupt fault described by (5.4) with

$$G(l_m) = \begin{cases} 1, & l_m \leq 20; \\ 1.5, & 20 < l_m \leq 26; \\ 1, & 27 < l_m. \end{cases}$$

The fault and its estimate are shown in Fig. 5.2(b), from which we can verify the effectiveness of the sensor fault estimation.

To show the superiority of the proposed estimation scheme, in the following, comparisons are conducted with the estimation scheme without considering the random bias. The simulation results are given in Figs. 5.5–5.6. The estimate from the estimation scheme without considering the random bias is shown in Fig. 5.5. The mean-square error of the estimation scheme without considering the random bias is plotted in Fig. 5.6. From the simulation results, the superiority of the developed estimation scheme is confirmed.

5.4 Conclusion

In this chapter, the joint state and fault estimation problem has been investigated for a class of MRSs with dynamical bias. To avoid the computational complexity from the lifting technique, a novel method has been developed

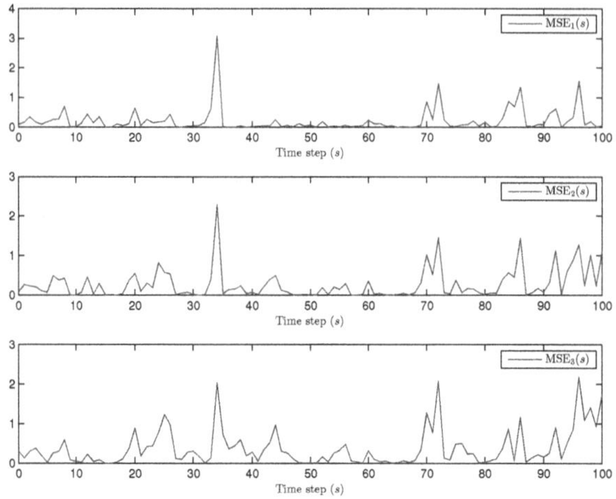

FIGURE 5.6: The mean-square error of the estimation scheme without considering the random bias.

to transform the MRSs into single-rate systems. Based on the transformed single-rate systems, the estimator has been designed that estimates the fault and the system state simultaneously. First, the recursion of the estimation error covariance has been derived. Then, the estimator gains have been characterized that minimize the estimation error covariance. Finally, a practical simulation has verified the usefulness of the proposed estimation scheme.

6

H_∞ Filtering for Multi-Rate Systems under p-Persistent CSMA Protocol

As one of the most frequently encountered network-induced phenomena, the sensor saturation is caused by many reasons such as physical limits of the sensors and imperfect communication environments. Sensor saturation is inherently a kind of non-linear phenomena which, if not adequately handled, would deteriorate the system performance or even result in the instability. In the available literature concerning sensor saturation, it has been often implicitly assumed that the occurrence of sensor saturations is deterministic, that is, the sensor always undergoes saturation. Such an assumption, however, does have its limitation when sensors are deployed in unattended environments such as power grids where sensors might frequently encounter some transient phenomena. Under this circumstance, the sensor saturations may occur in a probabilistic way and are randomly changeable in terms of their types and/or levels due to the random occurrence of network-induced phenomena such as random sensor failures, sensor aging, or sudden environment changes [163]. This gives rise to the so-called randomly occurring sensor saturation that has been largely overlooked in the literature in spite of its significant engineering background. Therefore, it is of great importance to examine the influence of randomly occurring sensor saturation on the corresponding filter performance, which constitutes partial motivation of this chapter.

Despite the fruitful results available for networked systems, the filter design problem for networked systems with communication protocols has received inadequate research attention. In real engineering practice, due to the limited capability of the shared communication network, data collisions occur frequently and not all the sensors are able to get the access to transmission simultaneously. As such, various communication protocols have been introduced to schedule the sensor accesses to certain shared networks according to some given priorities. Consequently, the control/filtering problems for networked systems with communication protocols have recently become quite hot research topics. For example, an observer-based controller has been designed in [212] for networked systems where the stochastic communication protocol is employed in both the sensor-to-controller network and the controller-to-actuator network. In [98], the exponential stability has been analyzed for discrete-time linear systems with the RR protocol.

DOI: 10.1201/9781032619507-6

The p-persistent CSMA protocol has been applied in communication networks such as Ethernet and 802.11a/b/g in many real-world applications. For example, the p-persistent CSMA protocol is used to regulate the communication between the train and the train control center in the communication-based train control systems [216]. The p-persistent CSMA protocol has a simple feature which is "listen before talk", that is, the sensors always check if the network is busy before transmitting measurement. p-persistent CSMA protocol is different from those protocols (e.g. the RR protocol) which schedule sensors with a global policy. Under the p-persistent CSMA protocol, all the sensors are assigned with identical priority and compete with each other for the access of transmission at each time instant. It should be pointed out that the sensors in multi-rate networked system have different sampling periods, and therefore the number of the sensors participating the competition for the transmission access at each time instant is *not* a constant. Consequently, the probability that a certain sensor gets the access of transmission is time-varying according to the p-persistent CSMA protocol. Such kind of dynamically changing probability brings substantial challenges to the filter design problem for multi-rate networked systems and it is the main motivation of this chapter to overcome these challenges by investigating the filtering problem for multi-rate networked systems under the p-persistent CSMA protocol.

In this chapter, the H_∞ filtering problem is studied for multi-rate networked systems with randomly occurring sensor saturations under the p-persistent CSMA protocol. Our purpose is to design an H_∞ filter such that the filtering error dynamics is exponentially mean-square stable and the H_∞ performance requirement is guaranteed. The main contributions of this chapter are highlighted as follows: 1) a combination of important factors contributing to the complexity of networked systems is investigated in a unified framework which comprises the multi-rate sampling, the randomly occurring sensor saturations and the p-persistent CSMA protocol; 2) a novel analytical method is developed to deal with the time-varying probability issue caused by the coupling of the multi-rate sampling and the p-persistent CSMA protocol; and 3) the stochastic analysis approach is conducted to derive sufficient condition that ensures the exponential mean-square stability as well as the H_∞ performance constraint.

6.1 Problem Formulation and Preliminaries

Consider the following discrete-time system

$$\begin{cases} x(T_{k+1}) = Ax(T_k) + B\omega(T_k) \\ z(T_k) = Mx(T_k) \end{cases} \tag{6.1}$$

with m-sensor measurements under randomly occurring saturations:

$$\begin{aligned} y_i(t_k^i) = {} & \beta_i(t_k^i)\sigma(C_i x(t_k^i)) + (1 - \beta_i(t_k^i))C_i x(t_k^i) \\ & + D_i v_i(t_k^i) \quad (i = 1, 2, \ldots, m) \end{aligned} \tag{6.2}$$

where $x(T_k) \in \mathbb{R}^{n_x}$ is the state vector, $y_i(t_k^i) \in \mathbb{R}^{n_y}$ is the measurement output from sensor i, $z(T_k) \in \mathbb{R}^{n_z}$ is the signal to be estimated, $\omega(T_k)$ is the process noise belonging to $l_2([0, \infty), \mathbb{R}^{n_\omega})$, $v_i(t_k^i)$ is the measurement noise for sensor i belonging to $l_2([0, \infty), \mathbb{R}^{n_v})$, and A, B, C_i, D_i, and M are known matrices with appropriate dimensions.

In this chapter, to cater for the engineering practice, it is also assumed that the state update period of the system and the sampling periods of the sensors are allowed to be different. The state update period for (6.1) is denoted by $h \triangleq T_{k+1} - T_k$, and the sampling periods for (6.2) are integer multiples of the state update period for (6.1), i.e. $t_{k+1}^i - t_k^i \triangleq b_i h$ where b_i $(i = 1, 2, \ldots, m)$ are positive integers. That is, the measurement from the ith sensor is generated at time instants $t_0^i, t_1^i, \cdots, t_k^i, \cdots$, while the state of the system updates at time instants $T_0, T_1, \cdots, T_k, \cdots$.

The saturation function $\sigma(\cdot) : \mathbb{R}^{n_y} \to \mathbb{R}^{n_y}$ is defined as

$$\sigma(s) \triangleq \begin{bmatrix} \sigma(s_1) & \sigma(s_2) & \cdots & \sigma(s_{n_y}) \end{bmatrix}^T \tag{6.3}$$

with

$$\sigma(s_l) \triangleq \mathrm{sign}(s_l) \min\{s_{l,\max}, |s_l|\}, \ l = 1, 2, \cdots, n_y$$

where $\mathrm{sign}(\cdot)$ is the signum function and $s_{l,\max}$ is the lth element of the saturation level.

The stochastic variables $\beta_i(t_k^i)$ $(i = 1, 2, \ldots, m)$ are mutually independent Bernoulli distributed white sequences taking values on 0 or 1 with

$$\begin{cases} \mathrm{Prob}\{\beta_i(t_k^i) = 1\} = \mu_i \\ \mathrm{Prob}\{\beta_i(t_k^i) = 0\} = 1 - \mu_i \end{cases}$$

where $\mu_i \in [0, 1]$ $(i = 1, 2, \ldots, m)$ are known constants.

In model (6.2), the stochastic variable $\beta_i(t_k^i)$ is used to describe the random occurrence of the sensor saturation on the ith sensor. $\beta_i(t_k^i) = 1$ represents that the sensor saturation occurs on the ith sensor and $\beta_i(t_k^i) = 0$ otherwise. Generally, the probability of the occurrence of the sensor saturation for a fixed sensor can be obtained via statistical experiments during a period of time.

Remark 6.1 *In reality, owing to the fast-growing and the wide application of the digital technologies, there are more and more systems that can be described as discrete-time system. As such, the discrete-time system can closely reflect the engineering practice. On the other hand, almost all practical systems are time-varying and contain nonlinearity which should be described as non-linear* *time-varying systems. Nevertheless, most of the non-linear time-varying systems can be approximately represented by the linear time-invariant systems with sufficient accuracy. Therefore, many practical systems can be described as the linear time-invariant systems. For example, the state-space model of the continuous stirred tank reactor, the most generally employed bioreactor for biohydrogen production, can be described as a discrete linear time-invariant system [192]. The phenomenon of randomly occurring sensor saturation exists extensively in the practical engineering and the measurement model* (6.2) *is capable of closely reflecting such a phenomenon. Moreover, the measurement model* (6.2) *is transformed to the commonly used measurement model in [9] when* $s_{l,\max}$ *approaches infinity. As such, the proposed measurement model in this chapter is more general.*

Remark 6.2 *In the real world, the multi-rate sampling strategy (sampling different signals with different rates) owns certain advantages from the engineering perspective and has been widely used in various practical applications. To cater for the engineering practice, in this chapter, it is assumed that the sampling rates of the sensors and the state update rate of the system are different. Note that, the developed filtering algorithm for MRSs can be easily extended to traditional single-rate systems by simply letting* $b_i = 1$.

In this chapter, the data transmissions from the sensors to the filter are realized by a communication network which, due to limited bandwidth, only allows one sensor to transmit its measurement at each time instant. The p-persistent CSMA protocol is used to schedule the m sensors with different measurement periods. Under the p-persistent CSMA protocol, when a sensor has measurement to transmit (i.e. at its sampling instant), it senses the communication network for idle or busy first. If the network is idle, the sensor begins to transmit the measurement with a probability p. If the network is busy, the sensor does not transmit the measurement. When more than one sensor senses that the network is idle and begins to transmit simultaneously, a collision occurs and, in this situation, all the sensors abort their transmissions immediately, which means that no sensor gets the access of transmission.

Let $\bar{y}_i(t_k^i)$ represent the received measurement of the filter from sensor i after transmitted through the communication network, then we have

$$\begin{aligned}\bar{y}_i(t_k^i) = {} & \alpha_i(t_k^i)\beta_i(t_k^i)\sigma(C_i x(t_k^i)) + \alpha_i(t_k^i)D_i v_i(t_k^i) \\ & + \alpha_i(t_k^i)(1-\beta_i(t_k^i))C_i x(t_k^i) \end{aligned} \tag{6.4}$$

where the stochastic variable $\alpha_i(t_k^i)$ is a Bernoulli distributed white sequence taking values on 0 or 1. $\alpha_i(t_k^i) = 1$ means that the transmission of measurement from sensor i at time instant t_k^i is successful, and $\alpha_i(t_k^i) = 0$ means that the transmission at time instant t_k^i fails.

According to the above description of the p-persistent CSMA protocol and noting that the communication channel is idle at the beginning of each sampling instant, it is easily known that the probability of a sensor successfully transmitting its measurement at a given time instant is $p(1-p)^{N-1}$ where N is the number of the sensors wanting to transmit data at this time instant [149]. Then, we know

$$\begin{cases} \text{Prob}\{\alpha_i(t_k^i) = 1\} = p(1-p)^{N(t_k^i)-1} \\ \text{Prob}\{\alpha_i(t_k^i) = 0\} = 1 - p(1-p)^{N(t_k^i)-1} \end{cases} \tag{6.5}$$

where $N(t_k^i) \in \{1, 2, \cdots, m\}$ represents the number of the sensors which try to transmit their measurements at time instant t_k^i. It can be derived from (6.5) that

$$\nu_i(t_k^i) \triangleq \mathbb{E}\{\alpha_i(t_k^i)\} = \bar{p} + \Delta_i(t_k^i)$$

where $\bar{p} \triangleq \frac{1}{2}(p + p(1-p)^{m-1})$ and $\Delta_i(t_k^i) \triangleq p(1-p)^{N(t_k^i)-1} - \bar{p}$. Then, we have

$$|\Delta_i(t_k^i)| \leq \bar{\Delta}, \quad |\nu_i(t_k^i)| \leq \bar{\nu} \tag{6.6}$$

where $\bar{\Delta} \triangleq \frac{1}{2}(p - p(1-p)^{m-1})$ and $\bar{\nu} \triangleq \bar{p} + \bar{\Delta}$.

According to the p-persistent CSMA protocol, we know that $\mathbb{E}\{\alpha_i(t_k^i)\alpha_j(t_k^j)\}$ $(i, j \in \{1, 2, \cdots, m\})$ satisfies

$$\mathbb{E}\{\alpha_i(t_k^i)\alpha_j(t_k^j)\} = \begin{cases} \nu_i(t_k^i)\nu_j(t_k^j), & \text{if } t_k^i \neq t_k^j, \\ 0, & \text{if } t_k^i = t_k^j,\ i \neq j, \\ \nu_i(t_k^i), & \text{if } t_k^i = t_k^j,\ i = j. \end{cases}$$

Noting (6.6), we have

$$\mathbb{E}\{\alpha_i(t_k^i)\alpha_j(t_k^j)\} - \nu_i(t_k^i)\nu_j(t_k^j) \leq \bar{\nu}. \tag{6.7}$$

Remark 6.3 *In networked systems, system components are connected through communication networks of limited communication capability. As such, data collisions may occur when more than one component transmits information through communication network simultaneously. Nevertheless, data collisions can be prevented through the utilization of communication protocols that decide which component can transmit its data at a certain time instant. The p-persistent CSMA protocol is often used in Ethernet, which is a widely employed communication network in industry, to prevent the data from collisions [149]. Therefore, the networked system with p-persistent CSMA protocol studied in this chapter caters for many industrial systems in a networked environment.*

Remark 6.4 *In this chapter, the sensor itself in the multi-rate networked system* (6.1)-(6.2) *has individual sampling period and, subsequently, the number of the sensors competing for the access of measurement transmission is time-varying (at different time instant) which, in turn, renders the time-varying* *probability that a certain sensor gets the access of transmission under the p-persistent CSMA protocol. This kind of time-varying probability (i.e. the mathematical expectation due to the fact that the stochastic variable is Bernoulli distributed) creates a great deal of difficulty in the subsequent filter design. In* (6.5)–(6.7)*, we have analytically given certain bounds on the time-varying probabilities, thereby facilitating the filter design in sequel.*

Recall that system (6.1) evolves with a period h, while the measurements (6.4) are sampled with periods $b_i h$. Accordingly, the system under consideration is essentially a MRS that largely complicates the system analysis/synthesis issues. An effective way to deal with the MRS is to convert it to a certain single-rate system by using the well-known lifting technique.

By defining the least common multiple of b_i $(i = 1, 2, \ldots, m)$ as L, we set $q_i \triangleq L/b_i$ and denote the sampling instant of the derived single-rate system as $\mathcal{T}_k$, i.e. $\mathcal{T}_{k+1} - \mathcal{T}_k = Lh$. Then, it is obtained from (6.1) and (6.4) that

$$\begin{cases} \bar{x}(\mathcal{T}_{k+1}) = \bar{A}\bar{x}(\mathcal{T}_k) + \bar{B}\bar{\omega}(\mathcal{T}_k) \\ \tilde{y}_i(\mathcal{T}_k) = \bar{\alpha}_i(\mathcal{T}_k)\bar{\beta}_i(\mathcal{T}_k)\bar{\sigma}(\bar{C}_i\bar{x}(\mathcal{T}_k)) \\ \qquad + \bar{\alpha}_i(\mathcal{T}_k)(I - \bar{\beta}_i(\mathcal{T}_k))\bar{C}_i\bar{x}(\mathcal{T}_k) \\ \qquad + \bar{\alpha}_i(\mathcal{T}_k)\bar{D}_i\bar{v}_i(\mathcal{T}_k) \\ \bar{z}(\mathcal{T}_k) = \bar{M}\bar{x}(\mathcal{T}_k) \end{cases} \tag{6.8}$$

where

$$\begin{aligned}
\bar{x}(\mathcal{T}_k) &\triangleq \mathrm{col}_{j=0}^{L-1}\{x(\mathcal{T}_k + jh)\}, \quad \bar{z}(\mathcal{T}_k) \triangleq \mathrm{col}_{j=0}^{L-1}\{z(\mathcal{T}_k + jh)\}, \\
\bar{v}_i(\mathcal{T}_k) &\triangleq \mathrm{col}_{j=0}^{q_i-1}\{v_i(\mathcal{T}_k + jb_ih)\}, \quad \bar{C}_i \triangleq \mathrm{col}_{j=0}^{q_i-1}\{C_{i,j}\}, \\
\tilde{y}_i(\mathcal{T}_k) &\triangleq \mathrm{col}_{j=0}^{q_i-1}\{\tilde{y}_i(\mathcal{T}_k + jb_ih)\}, \quad \bar{D}_i \triangleq \mathrm{diag}_{q_i}\{D_i\}, \\
\bar{\omega}(\mathcal{T}_k) &\triangleq \mathrm{col}_{j=L-1}^{2L-2}\{\omega(\mathcal{T}_k + jh)\}, \quad \bar{M} \triangleq \mathrm{diag}_L\{M\}, \\
\bar{\alpha}_i(\mathcal{T}_k) &\triangleq \mathrm{diag}_{j=0}^{q_i-1}\{\alpha_i(\mathcal{T}_k + jb_ih)\}, \\
\bar{\beta}_i(\mathcal{T}_k) &\triangleq \mathrm{diag}_{j=0}^{q_i-1}\{\beta_i(\mathcal{T}_k + jb_ih)\}, \\
\bar{\sigma}(\bar{C}_i\bar{x}(\mathcal{T}_k)) &\triangleq \mathrm{col}_{j=0}^{q_i-1}\{\sigma(C_ix(\mathcal{T}_k + jb_ih))\}, \\
\hat{A} &\triangleq \mathrm{col}\{A, A^2, \cdots, A^L\}, \quad \bar{A} \triangleq \begin{bmatrix} 0 & 0 & \cdots & \hat{A} \end{bmatrix}, \\
\bar{B} &\triangleq \mathrm{col}\{B_L, B_{L-1}, \cdots, B_1\}, \\
C_{i,j} &\triangleq \begin{bmatrix} \underbrace{0 \ \cdots \ 0}_{j \times b_i} & C_i & 0 & \cdots & 0 \end{bmatrix}, \\
B_j &\triangleq \begin{bmatrix} A^{L-j}B & A^{L-j-1}B & \cdots & B & 0 & \cdots & 0 \end{bmatrix}.
\end{aligned}$$

Next, by denoting

$$\begin{aligned}
\tilde{y}(\mathcal{T}_k) &\triangleq \mathrm{col}_{j=1}^m\{\tilde{y}_j(\mathcal{T}_k)\}, \quad \bar{v}(\mathcal{T}_k) \triangleq \mathrm{col}_{j=1}^m\{\bar{v}_j^T(\mathcal{T}_k)\}, \\
\bar{\alpha}(\mathcal{T}_k) &\triangleq \mathrm{diag}_{j=1}^m\{\bar{\alpha}_j(\mathcal{T}_k)\}, \quad \bar{\beta}(\mathcal{T}_k) \triangleq \mathrm{diag}_{j=1}^m\{\bar{\beta}_j(\mathcal{T}_k)\}, \\
\bar{C} &\triangleq \mathrm{col}_{j=1}^m\{\bar{C}_j\}, \quad \bar{D} \triangleq \mathrm{diag}_{j=1}^m\{\bar{D}_j\}, \\
\tilde{\sigma}(\bar{C}\bar{x}(\mathcal{T}_k)) &\triangleq \mathrm{col}_{j=1}^m\{\bar{\sigma}(\bar{C}_j\bar{x}(\mathcal{T}_k))\},
\end{aligned}$$

the system (6.8) can be rewritten in the following compact form

$$\begin{cases}
\bar{x}(\mathcal{T}_{k+1}) = \bar{A}\bar{x}(\mathcal{T}_k) + \bar{B}\bar{\omega}(\mathcal{T}_k) \\
\quad \tilde{y}(\mathcal{T}_k) = \bar{\alpha}(\mathcal{T}_k)(I - \bar{\beta}(\mathcal{T}_k))\bar{C}\bar{x}(\mathcal{T}_k) \\
\qquad\qquad + \bar{\alpha}(\mathcal{T}_k)\bar{\beta}(\mathcal{T}_k)\tilde{\sigma}(\bar{C}\bar{x}(\mathcal{T}_k)) \\
\qquad\qquad + \bar{\alpha}(\mathcal{T}_k)\bar{D}\bar{v}(\mathcal{T}_k) \\
\quad \bar{z}(\mathcal{T}_k) = \bar{M}\bar{x}(\mathcal{T}_k).
\end{cases} \tag{6.9}$$

In this chapter, the following filter is adopted

$$\begin{cases}
\hat{x}(\mathcal{T}_{k+1}) = K\hat{x}(\mathcal{T}_k) + H\tilde{y}(\mathcal{T}_k) \\
\quad \hat{z}(\mathcal{T}_k) = \bar{M}\hat{x}(\mathcal{T}_k)
\end{cases} \tag{6.10}$$

where H and K are the filter gains to be designed, $\hat{x}(\mathcal{T}_k)$ is the estimate of $\bar{x}(\mathcal{T}_k)$, and $\hat{z}(\mathcal{T}_k)$ is the estimate of $\bar{z}(\mathcal{T}_k)$.

Now, letting $\tilde{z}(\mathcal{T}_k) = \bar{z}(\mathcal{T}_k) - \hat{z}(\mathcal{T}_k)$ and denoting

$$\begin{aligned}
&\eta(\mathcal{T}_k) \triangleq \begin{bmatrix} \bar{x}(\mathcal{T}_k) \\ \hat{x}(\mathcal{T}_k) \end{bmatrix}, \quad \tilde{B} \triangleq \begin{bmatrix} \bar{B} & 0 \\ 0 & H\tilde{p}\bar{D} \end{bmatrix}, \quad \varpi(\mathcal{T}_k) \triangleq \begin{bmatrix} \bar{\omega}(\mathcal{T}_k) \\ \bar{v}(\mathcal{T}_k) \end{bmatrix}, \\
&\tilde{A} \triangleq \begin{bmatrix} \bar{A} & 0 \\ H\tilde{p}(I - \tilde{\beta})\bar{C} & K \end{bmatrix}, \quad \tilde{E} \triangleq \begin{bmatrix} 0 \\ H\tilde{p}\tilde{\beta} \end{bmatrix}, \quad \bar{F}_{i,j} \triangleq \begin{bmatrix} 0 \\ H\vec{F}_{i,j} \end{bmatrix}, \\
&\tilde{M} \triangleq \begin{bmatrix} \bar{M} & -\bar{M} \end{bmatrix}, \quad \Lambda \triangleq \begin{bmatrix} I & 0 \end{bmatrix}, \quad \bar{\Lambda} \triangleq \begin{bmatrix} 0 & I \end{bmatrix}, \\
&\tilde{p} \triangleq \mathrm{diag}_{\varrho}\{\bar{p}\}, \quad \bar{\mu}_i \triangleq \mathrm{diag}_{q_i}\{\mu_i\}, \quad \tilde{\beta} \triangleq \mathrm{diag}_{j=1}^m\{\bar{\mu}_j\}, \\
&s_{i,j} \triangleq \sum_{a=1}^{i} q_a - (q_i - j), \quad \varrho \triangleq \sum_{i=1}^{m} q_i, \\
&\vec{F}_{i,j} \triangleq \mathrm{diag}\{\underbrace{0, \cdots, 0}_{s_{i,j}}, 1, 0, \cdots, 0\},
\end{aligned}$$

an augmented system that governs the filtering error dynamics of the filter is obtained as

$$
\begin{cases}
\eta(\mathcal{T}_{k+1}) = \tilde{A}\eta(\mathcal{T}_k) + \tilde{B}\varpi(\mathcal{T}_k) + \tilde{E}\tilde{\sigma}(\bar{C}\Lambda\eta(\mathcal{T}_k)) \\
\qquad + \sum_{i,j} \Delta_i(\mathcal{T}_k + jb_ih)(1-\mu_i)\bar{F}_{i,j}\bar{C}\Lambda\eta(\mathcal{T}_k) \\
\qquad + \sum_{i,j} \Delta_i(\mathcal{T}_k + jb_ih)\mu_i\bar{F}_{i,j}\tilde{\sigma}(\bar{C}\Lambda\eta(\mathcal{T}_k)) \\
\qquad + \sum_{i,j} \Delta_i(\mathcal{T}_k + jb_ih)\bar{F}_{i,j}\bar{D}\bar{\Lambda}\varpi(\mathcal{T}_k) \\
\qquad + \sum_{i,j} (\alpha_i(\mathcal{T}_k + jb_ih)(1-\beta_i(\mathcal{T}_k + jb_ih)) \\
\qquad\quad -\nu_i(\mathcal{T}_k + jb_ih)(1-\mu_i))\bar{F}_{i,j}\bar{C}\Lambda\eta(\mathcal{T}_k) \\
\qquad + \sum_{i,j} (\alpha_i(\mathcal{T}_k + jb_ih)\beta_i(\mathcal{T}_k + jb_ih) \\
\qquad\quad -\nu_i(\mathcal{T}_k + jb_ih)\mu_i)\bar{F}_{i,j}\tilde{\sigma}(\bar{C}\Lambda\eta(\mathcal{T}_k)) \\
\qquad + \sum_{i,j} (\alpha_i(\mathcal{T}_k + jb_ih) \\
\qquad\quad -\nu_i(\mathcal{T}_k + jb_ih))\bar{F}_{i,j}\bar{D}\bar{\Lambda}\varpi(\mathcal{T}_k) \\
\tilde{z}(\mathcal{T}_k) = \tilde{M}\eta(\mathcal{T}_k).
\end{cases}
\tag{6.11}
$$

Definition 6.1 *The augmented system* (6.11) *with* $\varpi(\mathcal{T}_k) = 0$ *is said to be exponentially mean-square stable if there exist constants* $\lambda > 0$ *and* $0 < \hbar < 1$ *such that*

$$\mathbb{E}\{\|\eta(\mathcal{T}_k)\|^2\} \leq \lambda\hbar^{\mathcal{T}_k}\mathbb{E}\{\|\eta(\mathcal{T}_0)\|^2\}.$$

The main purpose of this chapter is to design the filter in the form of (6.10) such that the following requirements are satisfied simultaneously:

1. the zero-solution of the augmented system (6.11) with $\varpi(\mathcal{T}_k) = 0$ is exponentially mean-square stable; and
2. under zero initial condition, for a given disturbance attenuation level $\gamma > 0$ and all non-zero $\varpi(\mathcal{T}_k)$, the filtering error $\tilde{z}(\mathcal{T}_k)$ satisfies the following condition:

$$\sum_{k=0}^{\infty} \mathbb{E}\{\|\tilde{z}(\mathcal{T}_k)\|^2\} < \gamma^2 \sum_{k=0}^{\infty} \|\varpi(\mathcal{T}_k)\|^2. \tag{6.12}$$

6.2 H_∞ Filter Design

In this section, the exponentially mean-square stability and the H_∞ performance are first analyzed for system (6.11) and the filter design problem is then studied.

Along the similar line in [155], we know from (6.3) that there exists a diagonal matrix Ω satisfying $0 < \Omega < I$ and

$$\begin{aligned}&\left[\tilde{\sigma}(\bar{C}\Lambda\eta(\mathcal{T}_k)) - \Omega\bar{C}\Lambda\eta(\mathcal{T}_k)\right]^T\\ &\quad\times\left[\tilde{\sigma}(\bar{C}\Lambda\eta(\mathcal{T}_k)) - \bar{C}\Lambda\eta(\mathcal{T}_k)\right] \leq 0.\end{aligned} \tag{6.13}$$

The following lemma will be used in obtaining our main results.

Lemma 6.1 *[53] For any real matrices $X_{i,j}$ and $Y_{l,n}$, constants $g_{i,j}$ and $h_{l,n}$ $(1 \leq i,l \leq m,\ 0 \leq j \leq q_i - 1,\ 0 \leq n \leq q_l - 1)$, and a symmetric positive definite matrix S, we have*

$$\begin{aligned}&2\sum_{i=1}^{m}\sum_{l=1}^{m}\sum_{j=0}^{q_i-1}\sum_{n=0}^{q_l-1} g_{i,j}h_{l,n}X_{i,j}^T S Y_{l,n}\\ &\leq \varrho\sum_{i=1}^{m}\sum_{j=0}^{q_i-1} g_{i,j}^2 X_{i,j}^T S X_{i,j} + \varrho\sum_{i=1}^{m}\sum_{j=0}^{q_i-1} h_{i,j}^2 Y_{i,j}^T S Y_{i,j}.\end{aligned}$$

Theorem 6.1 *Let the filter gains H, K and the matrix Ω be given. The zero-solution of the augmented system (6.11) with $\varpi(\mathcal{T}_k) = 0$ is exponentially mean-square stable if there exist a positive definite matrix P and a positive scalar ε satisfying*

$$\bar{\Gamma} \triangleq \begin{bmatrix}\bar{\Gamma}_{11} & \tilde{A}^T P\tilde{E} + \varepsilon\Lambda^T\bar{C}^T(I+\Omega)/2\\ * & \bar{\Gamma}_{22}\end{bmatrix} < 0 \tag{6.14}$$

where

$$\begin{aligned}\bar{\Gamma}_{11} \triangleq& \sum_{i,j}\left(2\bar{\varsigma}_{i,j} + 2\varrho\bar{\varsigma}_{i,j}^2 + 2\bar{\varrho}\bar{\nu}(1-\mu_i)^2 + \bar{\vartheta}_{i,j}^1 + \bar{\vartheta}_{i,j}^2\right)\\ &\times\Lambda^T\bar{C}^T\bar{F}_{i,j}^T P\bar{F}_{i,j}\bar{C}\Lambda + \sum_{i,j}\left(\bar{\tau}_{i,j} + \bar{\varsigma}_{i,j}\right)\tilde{A}^T P\tilde{A}\\ &-\varepsilon\Lambda^T\bar{C}^T\Omega\bar{C}\Lambda + \tilde{A}^T P\tilde{A} - P,\\ \bar{\Gamma}_{22} \triangleq& \sum_{i,j}\left(2\bar{\tau}_{i,j} + 2\varrho\bar{\tau}_{i,j}^2 + 2\bar{\varrho}\bar{\nu}\mu_i^2 + \bar{\vartheta}_{i,j}^2 + \bar{\vartheta}_{i,j}^3\right)\bar{F}_{i,j}^T P\bar{F}_{i,j}\\ &+\sum_{i,j}\left(\bar{\tau}_{i,j} + \bar{\varsigma}_{i,j}\right)\tilde{E}^T P\tilde{E} + \tilde{E}^T P\tilde{E} - \varepsilon I,\\ \varsigma_{i,j} \triangleq& \Delta_i(\mathcal{T}_k + jb_ih)(1-\mu_i),\quad \bar{\varsigma}_{i,j} \triangleq \bar{\Delta}(1-\mu_i),\end{aligned}$$

$$\begin{aligned}
\tau_{i,j} &\triangleq \Delta_i(\mathcal{T}_k + jb_ih)\mu_i, \quad \bar{\tau}_{i,j} \triangleq \bar{\Delta}\mu_i, \quad \bar{\varrho} \triangleq \varrho - 1,\\
\vartheta^1_{i,j} &\triangleq \nu_i(\mathcal{T}_k + jb_ih)(1-\mu_i) - \nu_i^2(\mathcal{T}_k + jb_ih)(1-\mu_i)^2,\\
\bar{\vartheta}^1_{i,j} &\triangleq \bar{\nu}(1-\mu_i), \quad \vartheta^2_{i,j} \triangleq \nu_i^2(\mathcal{T}_k + jb_ih)\mu_i(\mu_i - 1),\\
\bar{\vartheta}^2_{i,j} &\triangleq \bar{\nu}^2\mu_i(\mu_i - 1), \quad \bar{\vartheta}^3_{i,j} \triangleq \bar{\nu}\mu_i,\\
\vartheta^3_{i,j} &\triangleq \nu_i(\mathcal{T}_k + jb_ih)\mu_i - \nu_i^2(\mathcal{T}_k + jb_ih)\mu_i^2.
\end{aligned}$$

Proof *Choosing the following Lyapunov function for system* (6.11)*:*

$$V(\eta(\mathcal{T}_k)) \triangleq \eta^T(\mathcal{T}_k)P\eta(\mathcal{T}_k), \tag{6.15}$$

the difference of the Lyapunov function can be written as

$$\Delta V(\eta(\mathcal{T}_k)) = \mathbb{E}\{V(\eta(\mathcal{T}_{k+1}))|\eta(\mathcal{T}_k)\} - V(\eta(\mathcal{T}_k)).$$

Calculating $\Delta V(\eta(\mathcal{T}_k))$ *along the trajectory of system* (6.11) *with* $\varpi(\mathcal{T}_k) = 0$*, we have*

$$\begin{aligned}
\mathbb{E}\{\Delta V(\eta(\mathcal{T}_k))\} =& \mathbb{E}\{\eta^T(\mathcal{T}_{k+1})P\eta(\mathcal{T}_{k+1}) - \eta^T(\mathcal{T}_k)P\eta(\mathcal{T}_k)\}\\
=& \mathbb{E}\{\eta^T(\mathcal{T}_k)(\tilde{A}^TP\tilde{A} - P)\eta(\mathcal{T}_k)\\
&+ 2\eta^T(\mathcal{T}_k)\tilde{A}^TP\tilde{E}\tilde{\sigma}(\bar{C}\Lambda\eta(\mathcal{T}_k))\\
&+ \tilde{\sigma}^T(\bar{C}\Lambda\eta(\mathcal{T}_k))\tilde{E}^TP\tilde{E}\tilde{\sigma}(\bar{C}\Lambda\eta(\mathcal{T}_k))\\
&+ 2\sum_{i,j}\varsigma_{i,j}\eta^T(\mathcal{T}_k)\tilde{A}^TP\bar{F}_{i,j}\bar{C}\Lambda\eta(\mathcal{T}_k)\\
&+ 2\sum_{i,j}\tau_{i,j}\eta^T(\mathcal{T}_k)\tilde{A}^TP\bar{F}_{i,j}\tilde{\sigma}(\bar{C}\Lambda\eta(\mathcal{T}_k))\\
&+ 2\sum_{i,j}\varsigma_{i,j}\tilde{\sigma}^T(\bar{C}\Lambda\eta(\mathcal{T}_k))\tilde{E}^TP\bar{F}_{i,j}\bar{C}\Lambda\eta(\mathcal{T}_k)\\
&+ 2\sum_{i,j}\tau_{i,j}\tilde{\sigma}^T(\bar{C}\Lambda\eta(\mathcal{T}_k))\tilde{E}^TP\bar{F}_{i,j}\tilde{\sigma}(\bar{C}\Lambda\eta(\mathcal{T}_k))\\
&+ \sum_{ij,ln}\varsigma_{i,j}\varsigma_{l,n}\eta^T(\mathcal{T}_k)\Lambda^T\bar{C}^T\bar{F}^T_{i,j}P\bar{F}_{l,n}\bar{C}\Lambda\eta(\mathcal{T}_k)\\
&+ 2\sum_{ij,ln}\varsigma_{i,j}\tau_{l,n}\eta^T(\mathcal{T}_k)\Lambda^T\bar{C}^T\bar{F}^T_{i,j}P\bar{F}_{l,n}\tilde{\sigma}(\bar{C}\Lambda\eta(\mathcal{T}_k))\\
&+ \sum_{ij,ln}\tau_{i,j}\tau_{l,n}\tilde{\sigma}^T(\bar{C}\Lambda\eta(\mathcal{T}_k))\bar{F}^T_{i,j}P\bar{F}_{l,n}\tilde{\sigma}(\bar{C}\Lambda\eta(\mathcal{T}_k))\\
&+ \sum_{ij\neq ln}\delta_{ij,ln}(1-\mu_i)(1-\mu_l)\\
&\qquad \times \eta^T(\mathcal{T}_k)\Lambda^T\bar{C}^T\bar{F}^T_{i,j}P\bar{F}_{l,n}\bar{C}\Lambda\eta(\mathcal{T}_k)\\
&+ 2\sum_{ij\neq ln}\delta_{ij,ln}(1-\mu_i)\mu_l
\end{aligned}$$

$$
\begin{aligned}
&\times \eta^T(\mathcal{T}_k)\Lambda^T\bar{C}^T\bar{F}_{i,j}^T P\bar{F}_{l,n}\tilde{\sigma}(\bar{C}\Lambda\eta(\mathcal{T}_k))\\
&+\sum_{ij\neq ln}\delta_{ij,ln}\mu_i\mu_l\tilde{\sigma}^T(\bar{C}\Lambda\eta(\mathcal{T}_k))\bar{F}_{i,j}^T P\bar{F}_{l,n}\tilde{\sigma}(\bar{C}\Lambda\eta(\mathcal{T}_k))\\
&+\sum_{i,j}\vartheta_{i,j}^1\eta^T(\mathcal{T}_k)\Lambda^T\bar{C}^T\bar{F}_{i,j}^T P\bar{F}_{i,j}\bar{C}\Lambda\eta(\mathcal{T}_k)\\
&+2\sum_{i,j}\vartheta_{i,j}^2\eta^T(\mathcal{T}_k)\Lambda^T\bar{C}^T\bar{F}_{i,j}^T P\bar{F}_{i,j}\tilde{\sigma}(\bar{C}\Lambda\eta(\mathcal{T}_k))\\
&+\sum_{i,j}\vartheta_{i,j}^3\tilde{\sigma}^T(\bar{C}\Lambda\eta(\mathcal{T}_k))\bar{F}_{i,j}^T P\bar{F}_{i,j}\tilde{\sigma}(\bar{C}\Lambda\eta(\mathcal{T}_k))
\end{aligned}
$$

where

$$
\begin{aligned}
\delta_{ij,ln} \triangleq &\mathbb{E}\{\alpha_i(\mathcal{T}_k+jb_ih)\alpha_l(\mathcal{T}_k+nb_lh)\}\\
&-\nu_i(\mathcal{T}_k+jb_ih)\nu_l(\mathcal{T}_k+nb_lh).
\end{aligned}
$$

By using the elementary inequality $2a^Tb\leq a^Ta+b^Tb$ *and Lemma 6.1, the following inequalities can be obtained:*

$$
\begin{aligned}
&2\sum_{i,j}\varsigma_{i,j}\eta^T(\mathcal{T}_k)\tilde{A}^T P\bar{F}_{i,j}\bar{C}\Lambda\eta(\mathcal{T}_k)\\
&\quad\leq\sum_{i,j}\varsigma_{i,j}\eta^T(\mathcal{T}_k)\tilde{A}^T P\tilde{A}\eta(\mathcal{T}_k)\\
&\quad\quad+\sum_{i,j}\varsigma_{i,j}\eta^T(\mathcal{T}_k)\Lambda^T\bar{C}^T\bar{F}_{i,j}^T P\bar{F}_{i,j}\bar{C}\Lambda\eta(\mathcal{T}_k),\\
&2\sum_{i,j}\tau_{i,j}\eta^T(\mathcal{T}_k)\tilde{A}^T P\bar{F}_{i,j}\tilde{\sigma}(\bar{C}\Lambda\eta(\mathcal{T}_k))\\
&\quad\leq\sum_{i,j}\tau_{i,j}\eta^T(\mathcal{T}_k)\tilde{A}^T P\tilde{A}\eta(\mathcal{T}_k)\\
&\quad\quad+\sum_{i,j}\tau_{i,j}\tilde{\sigma}^T(\bar{C}\Lambda\eta(\mathcal{T}_k))\bar{F}_{i,j}^T P\bar{F}_{i,j}\tilde{\sigma}(\bar{C}\Lambda\eta(\mathcal{T}_k)),\\
&2\sum_{i,j}\varsigma_{i,j}\tilde{\sigma}^T(\bar{C}\Lambda\eta(\mathcal{T}_k))\tilde{E}^T P\bar{F}_{i,j}\bar{C}\Lambda\eta(\mathcal{T}_k)\\
&\quad\leq\sum_{i,j}\varsigma_{i,j}\tilde{\sigma}^T(\bar{C}\Lambda\eta(\mathcal{T}_k))\tilde{E}^T P\tilde{E}\tilde{\sigma}(\bar{C}\Lambda\eta(\mathcal{T}_k))\\
&\quad\quad+\sum_{i,j}\varsigma_{i,j}\eta^T(\mathcal{T}_k)\Lambda^T\bar{C}^T\bar{F}_{i,j}^T P\bar{F}_{i,j}\bar{C}\Lambda\eta(\mathcal{T}_k),\\
&2\sum_{i,j}\tau_{i,j}\tilde{\sigma}^T(\bar{C}\Lambda\eta(\mathcal{T}_k))\tilde{E}^T P\bar{F}_{i,j}\tilde{\sigma}(\bar{C}\Lambda\eta(\mathcal{T}_k))\\
&\quad\leq\sum_{i,j}\tau_{i,j}\tilde{\sigma}^T(\bar{C}\Lambda\eta(\mathcal{T}_k))\tilde{E}^T P\tilde{E}\tilde{\sigma}(\bar{C}\Lambda\eta(\mathcal{T}_k))
\end{aligned}
$$

$$+\sum_{i,j}\tau_{i,j}\tilde{\sigma}^T(\bar{C}\Lambda\eta(\mathcal{T}_k))\bar{F}_{i,j}^TP\bar{F}_{i,j}\tilde{\sigma}(\bar{C}\Lambda\eta(\mathcal{T}_k)),$$

$$\sum_{ij\neq ln}\delta_{ij,ln}(1-\mu_i)(1-\mu_l)\eta^T(\mathcal{T}_k)\Lambda^T\bar{C}^T\bar{F}_{i,j}^TP\bar{F}_{l,n}\bar{C}\Lambda\eta(\mathcal{T}_k)$$
$$\leq\sum_{ij\neq ln}\delta_{ij,ln}(1-\mu_i)^2\eta^T(\mathcal{T}_k)\Lambda^T\bar{C}^T\bar{F}_{i,j}^TP\bar{F}_{i,j}\bar{C}\Lambda\eta(\mathcal{T}_k),$$

$$2\sum_{ij,ln}\varsigma_{i,j}\tau_{l,n}\eta^T(\mathcal{T}_k)\Lambda^T\bar{C}^T\bar{F}_{i,j}^TP\bar{F}_{l,n}\tilde{\sigma}(\bar{C}\Lambda\eta(\mathcal{T}_k))$$
$$\leq\varrho\sum_{i,j}\varsigma_{i,j}^2\eta^T(\mathcal{T}_k)\Lambda^T\bar{C}^T\bar{F}_{i,j}^TP\bar{F}_{i,j}\bar{C}\Lambda\eta(\mathcal{T}_k)$$
$$+\varrho\sum_{i,j}\tau_{i,j}^2\tilde{\sigma}^T(\bar{C}\Lambda\eta(\mathcal{T}_k))\bar{F}_{i,j}^TP\bar{F}_{i,j}\tilde{\sigma}(\bar{C}\Lambda\eta(\mathcal{T}_k)),$$

$$\sum_{ij,ln}\varsigma_{i,j}\varsigma_{l,n}\eta^T(\mathcal{T}_k)\Lambda^T\bar{C}^T\bar{F}_{i,j}^TP\bar{F}_{l,n}\bar{C}\Lambda\eta(\mathcal{T}_k)$$
$$\leq\varrho\sum_{i,j}\varsigma_{i,j}^2\eta^T(\mathcal{T}_k)\Lambda^T\bar{C}^T\bar{F}_{i,j}^TP\bar{F}_{i,j}\bar{C}\Lambda\eta(\mathcal{T}_k),$$

$$\sum_{ij\neq ln}\delta_{ij,ln}\mu_i\mu_l\tilde{\sigma}^T(\bar{C}\Lambda\eta(\mathcal{T}_k))\bar{F}_{i,j}^TP\bar{F}_{l,n}\tilde{\sigma}(\bar{C}\Lambda\eta(\mathcal{T}_k))$$
$$\leq\sum_{ij\neq ln}\delta_{ij,ln}\mu_i^2\tilde{\sigma}^T(\bar{C}\Lambda\eta(\mathcal{T}_k))\bar{F}_{i,j}^TP\bar{F}_{i,j}\tilde{\sigma}(\bar{C}\Lambda\eta(\mathcal{T}_k)),$$

$$\sum_{ij,ln}\tau_{i,j}\tau_{l,n}\tilde{\sigma}^T(\bar{C}\Lambda\eta(\mathcal{T}_k))\bar{F}_{i,j}^TP\bar{F}_{l,n}\tilde{\sigma}(\bar{C}\Lambda\eta(\mathcal{T}_k))$$
$$\leq\varrho\sum_{i,j}\tau_{i,j}^2\tilde{\sigma}^T(\bar{C}\Lambda\eta(\mathcal{T}_k))\bar{F}_{i,j}^TP\bar{F}_{i,j}\tilde{\sigma}(\bar{C}\Lambda\eta(\mathcal{T}_k)),$$

$$2\sum_{ij\neq ln}\delta_{ij,ln}(1-\mu_i)\mu_l\eta^T(\mathcal{T}_k)\Lambda^T\bar{C}^T\bar{F}_{i,j}^TP\bar{F}_{l,n}\tilde{\sigma}(\bar{C}\Lambda\eta(\mathcal{T}_k))$$
$$\leq\sum_{ij\neq ln}\delta_{ij,ln}(1-\mu_i)^2\eta^T(\mathcal{T}_k)\Lambda^T\bar{C}^T\bar{F}_{i,j}^TP\bar{F}_{i,j}\bar{C}\Lambda\eta(\mathcal{T}_k)$$
$$+\sum_{ij\neq ln}\delta_{ij,ln}\mu_i^2\tilde{\sigma}^T(\bar{C}\Lambda\eta(\mathcal{T}_k))\bar{F}_{i,j}^TP\bar{F}_{i,j}\tilde{\sigma}(\bar{C}\Lambda\eta(\mathcal{T}_k)),$$

$$2\sum_{i,j}\vartheta_{i,j}^2\eta^T(\mathcal{T}_k)\Lambda^T\bar{C}^T\bar{F}_{i,j}^TP\bar{F}_{i,j}\tilde{\sigma}(\bar{C}\Lambda\eta(\mathcal{T}_k))$$
$$\leq\sum_{i,j}\vartheta_{i,j}^2\eta^T(\mathcal{T}_k)\Lambda^T\bar{C}^T\bar{F}_{i,j}^TP\bar{F}_{i,j}\bar{C}\Lambda\eta(\mathcal{T}_k)$$
$$+\sum_{i,j}\vartheta_{i,j}^2\tilde{\sigma}^T(\bar{C}\Lambda\eta(\mathcal{T}_k))\bar{F}_{i,j}^TP\bar{F}_{i,j}\tilde{\sigma}(\bar{C}\Lambda\eta(\mathcal{T}_k)).$$

Then, it is obvious that

$$\mathbb{E}\{\Delta V(\eta(\mathcal{T}_k))\} \leq \mathbb{E}\{\zeta^T(\mathcal{T}_k)\Gamma\zeta(\mathcal{T}_k)\}$$

where

$$\begin{aligned}
\zeta(\mathcal{T}_k) &\triangleq \left[\eta^T(\mathcal{T}_k) \quad \tilde{\sigma}^T(\bar{C}\Lambda\eta(\mathcal{T}_k))\right]^T, \quad \Gamma \triangleq \begin{bmatrix} \Gamma_{11} & \tilde{A}^T P\tilde{E} \\ * & \Gamma_{22} \end{bmatrix}, \\
\Gamma_{11} &\triangleq \Big(\sum_{i,j}(2\varsigma_{i,j} + 2\varrho\varsigma_{i,j}^2 + \vartheta_{i,j}^1 + \vartheta_{i,j}^2) \\
&\quad + 2\sum_{ij\neq ln}\delta_{ij,ln}(1-\mu_i)^2\Big)\Lambda^T\bar{C}^T\bar{F}_{i,j}^T P\bar{F}_{i,j}\bar{C}\Lambda \\
&\quad + \sum_{i,j}(\tau_{i,j}+\varsigma_{i,j})\,\tilde{A}^T P\tilde{A} + \tilde{A}^T P\tilde{A} - P, \\
\Gamma_{22} &\triangleq \tilde{E}^T P\tilde{E} + \Big(\sum_{i,j}(2\tau_{i,j} + 2\varrho\tau_{i,j}^2 + \vartheta_{i,j}^2 + \vartheta_{i,j}^3) \\
&\quad + 2\sum_{ij\neq ln}\delta_{ij,ln}\mu_i^2\Big)\bar{F}_{i,j}^T P\bar{F}_{i,j} + \sum_{i,j}(\tau_{i,j}+\varsigma_{i,j})\,\tilde{E}^T P\tilde{E}.
\end{aligned}$$

Furthermore, to deal with the uncertainties $\Delta_i(\mathcal{T}_k+jb_ih)$*, it follows from* (6.6) *and* (6.7) *that*

$$\begin{aligned}
&\sum_{ij\neq ln}\delta_{ij,ln}(1-\mu_i)^2 \leq \bar{\varrho}\sum_{i,j}\bar{\nu}(1-\mu_i)^2, \\
&\sum_{ij\neq ln}\delta_{ij,ln}\mu_i^2 \leq \bar{\varrho}\sum_{i,j}\bar{\nu}\mu_i^2, \quad \varsigma_{i,j} \leq \bar{\varsigma}_{i,j}, \quad \tau_{i,j} \leq \bar{\tau}_{i,j}, \\
&\vartheta_{i,j}^1 \leq \bar{\vartheta}_{i,j}^1, \quad \vartheta_{i,j}^2 \leq \bar{\vartheta}_{i,j}^2, \quad \vartheta_{i,j}^3 \leq \bar{\vartheta}_{i,j}^3.
\end{aligned}$$

Moreover, it is found from (6.13) *that*

$$\begin{aligned}
\mathbb{E}\{\Delta V(\eta(\mathcal{T}_k))\} \leq&\, \mathbb{E}\{\zeta^T(\mathcal{T}_k)\Gamma\zeta(\mathcal{T}_k) - \varepsilon[\tilde{\sigma}(\bar{C}\Lambda\eta(\mathcal{T}_k)) - \Omega\bar{C}\Lambda\eta(\mathcal{T}_k)]^T \\
&\times[\tilde{\sigma}(\bar{C}\Lambda\eta(\mathcal{T}_k)) - \bar{C}\Lambda\eta(\mathcal{T}_k)]\} \\
\leq&\, \mathbb{E}\{\zeta^T(\mathcal{T}_k)\bar{\Gamma}\zeta(\mathcal{T}_k)\}.
\end{aligned}$$

From (6.14)*, we know that there exists a positive scalar* $\kappa > 0$ *such that* $\bar{\Gamma} < -\kappa\mathrm{diag}\{I,\ 0\}$ *and subsequently* $\mathbb{E}\{\Delta V(\eta(\mathcal{T}_k))\} < -\kappa\left\|\eta(\mathcal{T}_k)\right\|^2$*. Then, from Lemma 1 in [165], the augmented system* (6.11) *is exponentially mean-square stable. The proof is complete.*

Now, we are ready to discuss the H_∞ performance requirement. In the following theorem, a sufficient condition is given that guarantees the exponentially mean-square stability as well as the H_∞ performance requirement for the filtering error dynamics.

Theorem 6.2 *Let the disturbance attenuation level $\gamma > 0$, the filter gains H, K and the matrix Ω be given. The augmented system (6.11) is exponentially mean-square stable with $\varpi(\mathcal{T}_k) = 0$ and satisfies the H_∞ performance requirement (6.12) if there exist a positive definite matrix P and a positive scalar ε satisfying*

$$\hat{\Gamma} \triangleq \begin{bmatrix} \hat{\Gamma}_{11} & \tilde{A}^T P \tilde{E} + \varepsilon \Lambda^T \bar{C}^T (I + \Omega)/2 & \tilde{A}^T P \tilde{B} \\ * & \hat{\Gamma}_{22} & \tilde{E}^T P \tilde{B} \\ * & * & \hat{\Gamma}_{33} \end{bmatrix} < 0 \tag{6.16}$$

where

$$\begin{aligned}
\hat{\Gamma}_{11} \triangleq & \tilde{A}^T P \tilde{A} - P + \sum_{i,j} \mathcal{X}_{i,j}^{(1)} \Lambda^T \bar{C}^T \bar{F}_{i,j}^T P \bar{F}_{i,j} \bar{C} \Lambda \\
& + \sum_{i,j} \mathcal{Y}_{i,j} \tilde{A}^T P \tilde{A} + \tilde{M}^T \tilde{M} - \varepsilon \Lambda^T \bar{C}^T \Omega \bar{C} \Lambda, \\
\hat{\Gamma}_{22} \triangleq & \tilde{E}^T P \tilde{E} + \sum_{i,j} \mathcal{X}_{i,j}^{(2)} \bar{F}_{i,j}^T P \bar{F}_{i,j} + \sum_{i,j} \mathcal{Y}_{i,j} \tilde{E}^T P \tilde{E} - \varepsilon I, \\
\hat{\Gamma}_{33} \triangleq & \tilde{B}^T P \tilde{B} + \sum_{i,j} \mathcal{X}_{i,j}^{(3)} \bar{\Lambda}^T \bar{D}^T \bar{F}_{i,j}^T P \bar{F}_{i,j} \bar{D} \bar{\Lambda} - \gamma^2 I \\
& + \sum_{i,j} \mathcal{Y}_{i,j} \tilde{B}^T P \tilde{B}, \\
\mathcal{X}_{i,j}^{(1)} \triangleq & 3\bar{\varsigma}_{i,j} + 3\varrho \bar{\varsigma}_{i,j}^2 + (2\bar{\varrho} + \varrho) \bar{\nu} (1 - \mu_i)^2 + \bar{\vartheta}_{i,j}^1 + \bar{\vartheta}_{i,j}^2, \\
\mathcal{X}_{i,j}^{(2)} \triangleq & 3\bar{\tau}_{i,j} + 3\varrho \bar{\tau}_{i,j}^2 + (2\bar{\varrho} + \varrho) \bar{\nu} \mu_i^2 + \bar{\vartheta}_{i,j}^2 + \bar{\vartheta}_{i,j}^3, \\
\mathcal{X}_{i,j}^{(3)} \triangleq & 3\bar{\upsilon}_{i,j} + 3\varrho \bar{\upsilon}_{i,j}^2 + 3\varrho \bar{\nu}, \quad \mathcal{Y}_{i,j} \triangleq \bar{\tau}_{i,j} + \bar{\varsigma}_{i,j} + \bar{\upsilon}_{i,j}, \\
\upsilon_{i,j} \triangleq & \Delta_i (\mathcal{T}_k + j b_i h), \quad \bar{\upsilon}_{i,j} \triangleq \bar{\Delta}.
\end{aligned}$$

Proof *First, it is easy to see that the inequality (6.14) in the Theorem 6.1 can be ensured by the inequality (6.16), and then the augmented system (6.11) is exponentially mean-square stable with $\varpi(\mathcal{T}_k) = 0$. It now remains to show that the filtering error $\tilde{z}(\mathcal{T}_k)$ satisfies the H_∞ performance requirement (6.12) under zero initial condition.*

By choosing the Lyapunov function as (6.15), we obtain

$$\begin{aligned}
& \mathbb{E}\{\Delta V(\eta(\mathcal{T}_k))\} + \mathbb{E}\{\|\tilde{z}(\mathcal{T}_k)\|^2\} - \gamma^2 \|\varpi(\mathcal{T}_k)\|^2 \\
\leq & \mathbb{E}\{\zeta^T(\mathcal{T}_k) \Gamma \zeta(\mathcal{T}_k) + \eta^T(\mathcal{T}_k) \tilde{M}^T \tilde{M} \eta(\mathcal{T}_k) \\
& + 2\eta^T(\mathcal{T}_k) \tilde{A}^T P \tilde{B} \varpi(\mathcal{T}_k) + \varpi^T(\mathcal{T}_k) \tilde{B}^T P \tilde{B} \varpi(\mathcal{T}_k) \\
& + 2\varpi^T(\mathcal{T}_k) \tilde{B}^T P \tilde{E} \tilde{\sigma}(\bar{C} \Lambda \eta(\mathcal{T}_k)) - \gamma^2 \varpi^T(\mathcal{T}_k) \varpi(\mathcal{T}_k) \\
& + 2 \sum_{i,j} \upsilon_{i,j} \eta^T(\mathcal{T}_k) \tilde{A}^T P \bar{F}_{i,j} \bar{D} \bar{\Lambda} \varpi(\mathcal{T}_k) \\
& + 2 \sum_{i,j} \varsigma_{i,j} \varpi^T(\mathcal{T}_k) \tilde{B}^T P \bar{F}_{i,j} \bar{C} \Lambda \eta(\mathcal{T}_k)
\end{aligned}$$

$$
\begin{aligned}
&+2\sum_{i,j}\tau_{i,j}\varpi^T(\mathcal{T}_k)\tilde{B}^TP\bar{F}_{i,j}\tilde{\sigma}(\bar{C}\Lambda\eta(\mathcal{T}_k))\\
&+2\sum_{i,j}\upsilon_{i,j}\varpi^T(\mathcal{T}_k)\tilde{B}^TP\bar{F}_{i,j}\bar{D}\bar{\Lambda}\varpi(\mathcal{T}_k)\\
&+2\sum_{i,j}\upsilon_{i,j}\tilde{\sigma}^T(\bar{C}\Lambda\eta(\mathcal{T}_k))\tilde{E}^TP\bar{F}_{i,j}\bar{D}\bar{\Lambda}\varpi(\mathcal{T}_k)\\
&+2\sum_{ij,ln}\varsigma_{i,j}\upsilon_{l,n}\eta^T(\mathcal{T}_k)\Lambda^T\bar{C}^T\bar{F}_{i,j}^TP\bar{F}_{l,n}\bar{D}\bar{\Lambda}\varpi(\mathcal{T}_k)\\
&+2\sum_{ij,ln}\tau_{i,j}\upsilon_{l,n}\tilde{\sigma}^T(\bar{C}\Lambda\eta(\mathcal{T}_k))\bar{F}_{i,j}^TP\bar{F}_{l,n}\bar{D}\bar{\Lambda}\varpi(\mathcal{T}_k)\\
&+\sum_{ij,ln}\upsilon_{i,j}\upsilon_{l,n}\varpi^T(\mathcal{T}_k)\bar{\Lambda}^T\bar{D}^T\bar{F}_{i,j}^TP\bar{F}_{l,n}\bar{D}\bar{\Lambda}\varpi(\mathcal{T}_k)\\
&+2\sum_{ij,ln}\delta_{ij,ln}(1-\mu_i)\eta^T(\mathcal{T}_k)\Lambda^T\bar{C}^T\bar{F}_{i,j}^TP\bar{F}_{l,n}\bar{D}\bar{\Lambda}\varpi(\mathcal{T}_k)\\
&+2\sum_{ij,ln}\delta_{ij,ln}\mu_i\tilde{\sigma}^T(\bar{C}\Lambda\eta(\mathcal{T}_k))\bar{F}_{i,j}^TP\bar{F}_{l,n}\bar{D}\bar{\Lambda}\varpi(\mathcal{T}_k)\\
&+\sum_{ij,ln}\delta_{ij,ln}\varpi^T(\mathcal{T}_k)\bar{\Lambda}^T\bar{D}^T\bar{F}_{i,j}^TP\bar{F}_{l,n}\bar{D}\bar{\Lambda}\varpi(\mathcal{T}_k)\\
\leq&\mathbb{E}\{\bar{\zeta}^T(\mathcal{T}_k)\tilde{\Gamma}\bar{\zeta}(\mathcal{T}_k)\}
\end{aligned}
$$

where

$$
\begin{aligned}
\bar{\zeta}(\mathcal{T}_k)&\triangleq\begin{bmatrix}\eta^T(\mathcal{T}_k) & \tilde{\sigma}^T(\bar{C}\Lambda\eta(\mathcal{T}_k)) & \varpi^T(\mathcal{T}_k)\end{bmatrix}^T,\\
\tilde{\Gamma}&\triangleq\begin{bmatrix}\tilde{\Gamma}_{11} & \tilde{A}^TP\tilde{E} & \tilde{A}^TP\tilde{B}\\ * & \tilde{\Gamma}_{22} & \tilde{E}^TP\tilde{B}\\ * & * & \tilde{\Gamma}_{33}\end{bmatrix},\\
\tilde{\Gamma}_{11}&\triangleq\Gamma_{11}+\Big(\sum_{i,j}(\varsigma_{i,j}+\varrho\varsigma_{i,j}^2)+\sum_{ij,ln}\delta_{ij,ln}(1-\mu_i)^2\Big)\\
&\qquad\times\Lambda^T\bar{C}^T\bar{F}_{i,j}^TP\bar{F}_{i,j}\bar{C}\Lambda+\sum_{i,j}\upsilon_{i,j}\tilde{A}^TP\tilde{A}\\
&\quad+\tilde{M}^T\tilde{M},\\
\tilde{\Gamma}_{22}&\triangleq\Gamma_{22}+\Big(\sum_{i,j}(\tau_{i,j}+\varrho\tau_{i,j}^2)+\sum_{ij,ln}\delta_{ij,ln}\mu_i^2\Big)\bar{F}_{i,j}^TP\bar{F}_{i,j}\\
&\quad+\sum_{i,j}\upsilon_{i,j}\tilde{E}^TP\tilde{E},\\
\tilde{\Gamma}_{33}&\triangleq\Big(\sum_{i,j}(3\upsilon_{i,j}+3\varrho\upsilon_{i,j}^2)+3\sum_{ij,ln}\delta_{ij,ln}\Big)\\
&\qquad\times\bar{\Lambda}^T\bar{D}^T\bar{F}_{i,j}^TP\bar{F}_{i,j}\bar{D}\bar{\Lambda}+\tilde{B}^TP\tilde{B}-\gamma^2I
\end{aligned}
$$

$$+\sum_{i,j}(\upsilon_{i,j}+\tau_{i,j}+\varsigma_{i,j})\tilde{B}^T P\tilde{B}.$$

Using the similar treatment in the proof of Theorem 6.1, it follows that

$$\begin{aligned}&\mathbb{E}\{\Delta V(\eta(\mathcal{T}_k))\}+\mathbb{E}\{\|\tilde{z}(\mathcal{T}_k)\|^2\}-\gamma^2\|\varpi(\mathcal{T}_k)\|^2\\ \le&\mathbb{E}\{\bar{\zeta}^T(\mathcal{T}_k)\tilde{\Gamma}\bar{\zeta}(\mathcal{T}_k)-\varepsilon[\tilde{\sigma}(\bar{C}\Lambda\eta(\mathcal{T}_k))-\Omega\bar{C}\Lambda\eta(\mathcal{T}_k)]^T\\ &\times[\tilde{\sigma}(\bar{C}\Lambda\eta(\mathcal{T}_k))-\bar{C}\Lambda\eta(\mathcal{T}_k)]\}\\ \le&\mathbb{E}\{\bar{\zeta}^T(\mathcal{T}_k)\hat{\Gamma}\bar{\zeta}(\mathcal{T}_k)\}\end{aligned}$$

which, from (6.16), *implies that*

$$\mathbb{E}\{\Delta V(\eta(\mathcal{T}_k))\}+\mathbb{E}\{\|\tilde{z}(\mathcal{T}_k)\|^2\}-\gamma^2\|\varpi(\mathcal{T}_k)\|^2<0 \tag{6.17}$$

for all non-zero $\varpi(\mathcal{T}_k)$. By summing up both sides of (6.17) *from* 0 *to* ∞ *with respect to k, we derive that*

$$\sum_{k=0}^{\infty}\mathbb{E}\{\|\tilde{z}(\mathcal{T}_k)\|^2\}<\gamma^2\sum_{k=0}^{\infty}\|\varpi(\mathcal{T}_k)\|^2$$

under the zero initial condition. The proof is complete.

In the following theorem, the solution to the H_∞ filtering problem for MRSs with randomly occurring sensor saturations under the p-persistent CSMA protocol is obtained.

Theorem 6.3 *Let the disturbance attenuation level $\gamma>0$ and the matrix Ω be given. For the discrete-time system* (6.1) *with measurement model* (6.2) *undergoing randomly occurring sensor saturations, the augmented system* (6.11) *is exponentially mean-square stable with $\varpi(\mathcal{T}_k)=0$ and satisfies the H_∞ performance constraint* (6.12) *if there exist a positive definite matrix* $P=\text{diag}\{Q_1,\ Q_2\}$, *matrices X, Y and a positive scalar ε such that*

$$\begin{bmatrix}\bar{\Pi}_1 & \Pi_2 & \tilde{\Pi}\\ * & -Q_2 & 0\\ * & * & -\tilde{Q}_2\end{bmatrix}<0 \tag{6.18}$$

where

$$\bar{\Pi}_1\triangleq\begin{bmatrix}\bar{\Pi}_{111} & -\bar{M}^T\bar{M} & \bar{\Pi}_{112} & \bar{A}^TQ_1\bar{B} & 0\\ * & \bar{M}^T\bar{M}-Q_2 & 0 & 0 & 0\\ * & * & -\varepsilon I & 0 & 0\\ * & * & * & \bar{\Pi}_{144} & 0\\ * & * & * & * & -\gamma^2 I\end{bmatrix},$$

$$\begin{aligned}\tilde{Q}_2\triangleq&\text{diag}\{\bar{Q}_2,\ \bar{Q}_2,\ \bar{Q}_2,\ \bar{Q}_2,\ \bar{Q}_2,\ \bar{Q}_2\},\\ \tilde{\Pi}\triangleq&\begin{bmatrix}\Pi_3 & \Pi_4 & \Pi_5 & \Pi_6 & \Pi_7 & \Pi_8\end{bmatrix},\quad \bar{Q}_2\triangleq\text{diag}_\varrho\{Q_2\},\end{aligned}$$

$$
\begin{aligned}
\bar{\Pi}_{111} \triangleq & \bar{M}^T\bar{M} - Q_1 - \varepsilon\bar{C}^T\Omega\bar{C} + \bar{A}^TQ_1\bar{A} + \sum_{i,j}\mathcal{Y}_{i,j}\bar{A}^TQ_1\bar{A}, \\
\bar{\Pi}_{144} \triangleq & -\gamma^2 I + \bar{B}^TQ_1\bar{B} + \sum_{i,j}\mathcal{Y}_{i,j}\bar{B}^TQ_1\bar{B}, \\
\bar{\Pi}_{112} \triangleq & \varepsilon\bar{C}^T(I+\Omega)/2, \quad \Pi_4 = \begin{bmatrix} 0 & 0 & \bar{S}_1\tilde{p}\tilde{\beta} & 0 & 0 \end{bmatrix}^T, \\
\Pi_2 \triangleq & \begin{bmatrix} X\tilde{p}(I-\tilde{\beta})\bar{C} & Y & X\tilde{p}\tilde{\beta} & 0 & X\tilde{p}\bar{D} \end{bmatrix}^T, \\
\Pi_3 \triangleq & \begin{bmatrix} \bar{S}_1\tilde{p}(I-\tilde{\beta})\bar{C} & \bar{S}_2 & 0 & 0 & 0 \end{bmatrix}^T, \\
\Pi_5 \triangleq & \begin{bmatrix} 0 & 0 & 0 & 0 & \bar{S}_1\tilde{p}\bar{D} \end{bmatrix}^T, \quad \Pi_6 \triangleq \begin{bmatrix} \bar{S}_3\bar{C} & 0 & 0 & 0 & 0 \end{bmatrix}^T, \\
\Pi_7 \triangleq & \begin{bmatrix} 0 & 0 & \bar{S}_4 & 0 & 0 \end{bmatrix}^T, \quad \Pi_8 \triangleq \begin{bmatrix} 0 & 0 & 0 & 0 & \bar{S}_5\bar{D} \end{bmatrix}^T, \\
\bar{S}_j \triangleq & \begin{bmatrix} \bar{S}_{j,1} & \bar{S}_{j,2} & \cdots & \bar{S}_{j,m} \end{bmatrix}^T, \ j \in (1,2,3,4,5), \\
\bar{S}_{1,i} \triangleq & \begin{bmatrix} \sqrt{\mathcal{Y}_{i,0}}X^T & \cdots & \sqrt{\mathcal{Y}_{i,q_i-1}}X^T \end{bmatrix}, \\
\bar{S}_{2,i} \triangleq & \begin{bmatrix} \sqrt{\mathcal{Y}_{i,0}}Y^T & \cdots & \sqrt{\mathcal{Y}_{i,q_i-1}}Y^T \end{bmatrix}, \\
\bar{S}_{3,i} \triangleq & \begin{bmatrix} \sqrt{\mathcal{X}^{(1)}_{i,0}}\vec{F}_{i,0}X^T & \cdots & \sqrt{\mathcal{X}^{(1)}_{i,q_i-1}}\vec{F}_{i,q_i-1}X^T \end{bmatrix}, \\
\bar{S}_{4,i} \triangleq & \begin{bmatrix} \sqrt{\mathcal{X}^{(2)}_{i,0}}\vec{F}_{i,0}X^T & \cdots & \sqrt{\mathcal{X}^{(2)}_{i,q_i-1}}\vec{F}_{i,q_i-1}X^T \end{bmatrix}, \\
\bar{S}_{5,i} \triangleq & \begin{bmatrix} \sqrt{\mathcal{X}^{(3)}_{i,0}}\vec{F}_{i,0}X^T & \cdots & \sqrt{\mathcal{X}^{(3)}_{i,q_i-1}}\vec{F}_{i,q_i-1}X^T \end{bmatrix}.
\end{aligned}
$$

Moreover, if the above inequality is feasible, the desired filter gains can be determined by

$$
H = Q_2^{-1}X, \quad K = Q_2^{-1}Y. \tag{6.19}
$$

Proof $\hat{\Gamma}$ *can be rewritten as follow:*

$$
\begin{aligned}
\hat{\Gamma} = & \bar{\Pi}_1 + \bar{\Pi}_2Q_2\bar{\Pi}_2^T + \bar{\Pi}_3\bar{Q}_2\bar{\Pi}_3^T + \bar{\Pi}_4\bar{Q}_2\bar{\Pi}_4^T + \bar{\Pi}_5\bar{Q}_2\bar{\Pi}_5^T \\
& + \bar{\Pi}_6\bar{Q}_2\bar{\Pi}_6^T + \bar{\Pi}_7\bar{Q}_2\bar{\Pi}_7^T + \bar{\Pi}_8\bar{Q}_2\bar{\Pi}_8^T
\end{aligned}
$$

where

$$
\begin{aligned}
\bar{\Pi}_2 \triangleq & \begin{bmatrix} H\tilde{p}(I-\tilde{\beta})\bar{C} & K & H\tilde{p}\tilde{\beta} & 0 & H\tilde{p}\bar{D} \end{bmatrix}^T, \\
\bar{\Pi}_3 \triangleq & \begin{bmatrix} S_1\tilde{p}(I-\tilde{\beta})\bar{C} & S_2 & 0 & 0 & 0 \end{bmatrix}^T, \\
\bar{\Pi}_4 \triangleq & \begin{bmatrix} 0 & 0 & S_1\tilde{p}\tilde{\beta} & 0 & 0 \end{bmatrix}^T, \\
\bar{\Pi}_5 \triangleq & \begin{bmatrix} 0 & 0 & 0 & 0 & S_1\tilde{p}\bar{D} \end{bmatrix}^T, \\
\bar{\Pi}_6 \triangleq & \begin{bmatrix} S_3\bar{C} & 0 & 0 & 0 & 0 \end{bmatrix}^T, \\
\bar{\Pi}_7 \triangleq & \begin{bmatrix} 0 & 0 & S_4 & 0 & 0 \end{bmatrix}^T, \\
\bar{\Pi}_8 \triangleq & \begin{bmatrix} 0 & 0 & 0 & 0 & S_5\bar{D} \end{bmatrix}^T,
\end{aligned}
$$

$$
\begin{aligned}
S_j &\triangleq \begin{bmatrix} S_{j,1} & S_{j,2} & \cdots & S_{j,m} \end{bmatrix}^T, \ j \in (1,2,3,4,5), \\
S_{1,i} &\triangleq \begin{bmatrix} \sqrt{\mathcal{Y}_{i,0}}H^T & \cdots & \sqrt{\mathcal{Y}_{i,q_i-1}}H^T \end{bmatrix}, \\
S_{2,i} &\triangleq \begin{bmatrix} \sqrt{\mathcal{Y}_{i,0}}K^T & \cdots & \sqrt{\mathcal{Y}_{i,q_i-1}}K^T \end{bmatrix}, \\
S_{3,i} &\triangleq \begin{bmatrix} \sqrt{\mathcal{X}_{i,0}^{(1)}}\vec{F}_{i,0}H^T & \cdots & \sqrt{\mathcal{X}_{i,q_i-1}^{(1)}}\vec{F}_{i,q_i-1}H^T \end{bmatrix}, \\
S_{4,i} &\triangleq \begin{bmatrix} \sqrt{\mathcal{X}_{i,0}^{(2)}}\vec{F}_{i,0}H^T & \cdots & \sqrt{\mathcal{X}_{i,q_i-1}^{(2)}}\vec{F}_{i,q_i-1}H^T \end{bmatrix}, \\
S_{5,i} &\triangleq \begin{bmatrix} \sqrt{\mathcal{X}_{i,0}^{(3)}}\vec{F}_{i,0}H^T & \cdots & \sqrt{\mathcal{X}_{i,q_i-1}^{(3)}}\vec{F}_{i,q_i-1}H^T \end{bmatrix}.
\end{aligned}
$$

Then, by using the Schur Complement Lemma, $\hat{\Gamma} < 0$ holds if and only if

$$
\begin{bmatrix}
\bar{\Pi}_1 & \bar{\Pi}_2 Q_2 & \bar{\Pi}_3 \bar{Q}_2 & \bar{\Pi}_4 \bar{Q}_2 & \bar{\Pi}_5 \bar{Q}_2 & \bar{\Pi}_6 \bar{Q}_2 & \bar{\Pi}_7 \bar{Q}_2 & \bar{\Pi}_8 \bar{Q}_2 \\
* & -Q_2 & 0 & 0 & 0 & 0 & 0 & 0 \\
* & * & -\bar{Q}_2 & 0 & 0 & 0 & 0 & 0 \\
* & * & * & -\bar{Q}_2 & 0 & 0 & 0 & 0 \\
* & * & * & * & -\bar{Q}_2 & 0 & 0 & 0 \\
* & * & * & * & * & -\bar{Q}_2 & 0 & 0 \\
* & * & * & * & * & * & -\bar{Q}_2 & 0 \\
* & * & * & * & * & * & * & -\bar{Q}_2
\end{bmatrix} < 0.
$$

Noting (6.19), *we can obtain* (6.18) *readily. The proof is complete.*

Remark 6.5 *In this chapter, we aim to design an H_∞ filter for a class of discrete networked MRSs with randomly occurring sensor saturations under the p-persistent CSMA protocol. In Theorems 6.1–6.2, sufficient conditions are established to ensure the existence of the desired filter with which the filtering error dynamics is exponentially mean-square stable and the H_∞ performance requirement is guaranteed. Furthermore, in Theorem 6.3, the desired filter gains are characterized by means of solving some LMIs. Note that all the information about the system parameters, the multi-rates, the randomly occurring sensor saturations and the p-persistent CSMA protocol are reflected in our main results.*

Remark 6.6 *Our main results are based on the LMI condition and the explicit expression of the desired filter gains can be characterized in terms of the solutions to LMI* (6.18). *The LMI toolbox of Matlab provides the state-of-art interior-point LMI solvers which are efficient in solving LMIs. On the other hand, the procedure to obtain the desired filter is given as follows:*

- *give the system parameters, the H_∞ performance index γ, and the matrix Ω;*
- *obtain the positive definite matrix P, matrices X and Y, and positive scalar ε by solving the LMI* (6.18) *using the Matlab LMI toolbox;*
- *derive the filter parameter matrices H and K by solving* (6.19).

Furthermore, to show the feasibility of the proposed H_∞ filtering algorithm, a practical example is given in Section 6.3 concerning the estimate of the concentration of the product in a stirred tank reactor system.

6.3 Illustrative Examples

In this section, two illustrative examples are given to show the effectiveness of the proposed filtering scheme.

Example 6.1

Consider a system described by (6.1) with the following matrix parameters:

$$A = \begin{bmatrix} -0.89 & -0.28 \\ 0.6 & 0.7 \end{bmatrix}, \; B = \begin{bmatrix} 0.02 \\ 0.01 \end{bmatrix}, \; M = \begin{bmatrix} 1 & 1 \\ 1 & 1 \end{bmatrix}$$

and two sensors under randomly occurring sensor saturations with

$$C_1 = \begin{bmatrix} 0.2 & 0.3 \end{bmatrix}, \; C_2 = \begin{bmatrix} 0.3 & -0.2 \end{bmatrix},$$
$$D_1 = 0.1, \; D_2 = 0.1, \; b_1 = 1, \; b_2 = 2.$$

Other parameters are taken as $\mu_1 = 0.5, \; \mu_2 = 0.6, \; p = 0.5$, and the disturbance attenuation level is given as $\gamma = 1$.

With the above parameters, by using the Matlab LMI toolbox, the desired filter gains can be obtained as follows:

$$H = \begin{bmatrix} 0.3621 & -0.1314 & 0.5063 \\ 0.3621 & -0.1314 & 0.5063 \\ -0.1131 & 0.3797 & -0.5392 \\ -0.1131 & 0.3797 & -0.5392 \end{bmatrix},$$

$$K = \begin{bmatrix} 0.5451 & -0.1986 & -0.000 & -0.000 \\ -0.1986 & 0.5451 & -0.000 & -0.000 \\ -0.000 & -0.000 & 0.5452 & -0.1985 \\ -0.000 & -0.000 & -0.1985 & 0.5452 \end{bmatrix}.$$

In the simulation, the process noise $\omega(T_k)$ and the measurement noises $v_i(t_k^i)$ $(i = 1, 2)$ are selected as $\omega(T_k) = \frac{\sin(T_k)}{0.5k+1}$, $v_1(t_k^1) = \frac{\cos(t_k^1)}{k+1}$, and $v_2(t_k^2) = \frac{\sin(t_k^2)}{k+1}$, respectively. The initial value of the state is chosen as $x(0) = \begin{bmatrix} 0.62 & -0.51 \end{bmatrix}^T$.

The simulation results are shown in Figs. 6.1 and 6.2. The output $z(T_k)$ and its estimate are presented in Fig. 6.1, and Fig. 6.2 depicts the filtering error dynamics. The simulation results show the effectiveness of the designed filtering scheme.

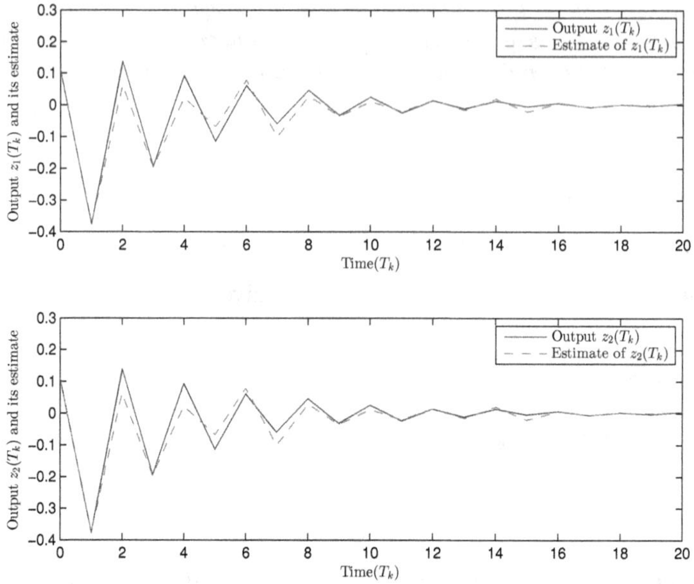

FIGURE 6.1: Output $z(T_k)$ and its estimate.

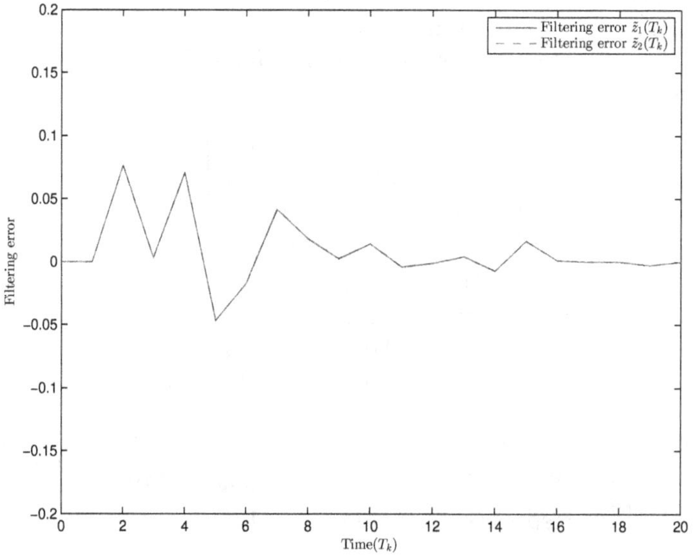

FIGURE 6.2: Filtering error dynamics.

TABLE 6.1: The model parameters and main operating point of the continuous stirred tank reactor.

Model Parameters	
$k_{0_{1,2}} = 1.287 * 10^{12}$ h^{-1}	$k_{0_3} = 9.043 * 10^9$ l/mol
$E_{A_{1,2}}/R = 9758.3$ K	$E_{A_3}/R = 8560.0$ K
$H_{AB} = 4.2$ kJ/mol	$H_{AD} = -41.85$ kJ/mol
$H_{BC} = -11.0$ kJ/mol	$\frac{F}{V} = 18.83$ h^{-1}
$\sigma = 0.9342 * 10^{-4}$ kg/l	$C_P = 3.01$ kJ/kg · K
$m_K = 0.9342 * 10^{-4}$ kg/l	$C_{PK} = 2.0$ kJ/kg · K
$A_R = 0.215$ m^2	$k_W = 4032$ kJ/kg · K
Main Operating Point	
$c_{As} = 1.235$ mol/l	$c_{Bs} = 0.9$ mol/l
$\varrho_s = 134.14$ °C	$\varpi_s = 128.95$ °C

Example 6.2

Consider a continuous stirred tank reactor with the following balance equations [192]:

$$
\begin{aligned}
\frac{dc_A}{dt} &= \frac{F}{V}(c_{A0} - c_A) - k_1 c_A - k_3 c_A^2, \\
\frac{dc_B}{dt} &= -\frac{F}{V}c_B + k_1 c_A - k_2 c_B, \\
\frac{d\varrho}{dt} &= \frac{F}{V}(\varrho_0 - \varrho) + \frac{k_W A_R}{\sigma C_P V}(\varpi - \varrho) \\
&\quad - \frac{k_1 c_A H_{AB} + k_2 c_B H_{BC} + k_3 c_A^2 H_{AD}}{\sigma C_P}, \\
\frac{d\varpi}{dt} &= \frac{1}{m_K C_{PK}}[Q_K + k_W A_R(\varrho - \varpi)]
\end{aligned}
$$

where c_A and c_B denote the concentration of the educt A and the product B, respectively. ϱ is the reactor temperature and ϖ is the coolant temperature. F, V, σ, and C_P are the normalized process stream inflow, the volume flow, the density and the heat capacity, respectively. H_{AB}, H_{BC}, and H_{AD} are the reaction enthalpy and k_1, k_2, and k_3 are rate coefficients that can be calculated by

$$
k_i = k_{0_i} e^{\frac{-E_{Ai}}{\varrho + 273.15}}, \; i = 1, 2, 3.
$$

Moreover, the model parameters and main operating point of the continuous stirred tank reactor are shown in Table 6.1.

By setting $x(t) \triangleq \begin{bmatrix} c_A(t) & c_B(t) & \varrho(t) & \varpi(t) \end{bmatrix}^T$, the linearized state-space model of the continuous stirred tank reactor near the operating point is obtained as

$$
\dot{x}(t) = \mathcal{A}x(t) + \mathcal{B}w(t)
$$

where $w(t)$ is the noise belonging to $L_2[0,\infty)$ and

$$\mathcal{A} \triangleq \begin{bmatrix} -86.0948 & 0 & -4.2075 & 0 \\ 50.6138 & -69.4438 & 0.9974 & 0 \\ 172.2205 & 197.9984 & -36.7414 & 30.7978 \\ 0 & 0 & 86.6880 & -86.6880 \end{bmatrix},$$

$$\mathcal{B} \triangleq \begin{bmatrix} 0 & 0 & 1 & 0 \end{bmatrix}^T.$$

In the continuous stirred tank reactor system, the concentration of the product B is an important value, but utilizing chemical methods to measure such a value is expensive. Consequently, the signal processing methods are used to estimate the concentration. In this simulation, the concentration is estimated only based on the measurements of the reactor temperature ϱ. Here, two sensors with the following parameters and randomly occurring sensor saturations are employed to measure the continuous stirred tank reactor system:

$$C_1 = \begin{bmatrix} 0 & 0 & 1 & 0 \end{bmatrix}, \; C_2 = \begin{bmatrix} 0 & 0 & 1 & 0 \end{bmatrix},$$
$$\mu_1 = 0.5, \;\; \mu_2 = 0.6, \;\; D_1 = 0.51, \;\; D_2 = 0.51,$$
$$M = \begin{bmatrix} 1 & 1 & 1 & 1 \end{bmatrix}.$$

The sensor 1 and sensor 2 have different sampling period h and $2h$ where $h = 3s$, respectively.

Discretizing the linearized state-space model of the continuous stirred tank reactor with period h, we have

$$A = \begin{bmatrix} -0.0752 & -0.0700 & -0.0226 & -0.0088 \\ 0.0228 & 0.0050 & -0.0111 & -0.0035 \\ 1.7660 & 1.3396 & 0.3882 & 0.1640 \\ 1.8159 & 1.4322 & 0.4618 & 0.1938 \end{bmatrix},$$
$$B = \begin{bmatrix} -0.0025 & 0.0135 & 0.0671 & 0.0506 \end{bmatrix}.$$

By choosing other parameters same as those in Example 6.1 and using the Matlab LMI toolbox, we obtain the desired filter gain matrices. In the simulation, the noises $\omega(T_k)$, $v_1(t_k^1)$, and $v_2(t_k^2)$ are chosen as $\omega(T_k) = \frac{10\sin(T_k)}{k+1}$, $v_1(t_k^1) = \frac{10\cos(t_k^1)}{k+1}$, and $v_2(t_k^2) = \frac{8\sin(t_k^2)}{k+1}$, respectively. The simulation results are shown in Figs. 6.3–6.4.

The output $z(T_k)$ of the continuous stirred tank reactor system and the estimate of the output are shown in Fig. 6.3. In Fig. 6.4, the filtering error $\tilde{z}(T_k)$ is depicted. It can be seen from the Fig. 6.3 that the estimate follows the real output well and from the Fig. 6.4 that the filtering error approaches zero as time increasing. From the simulation results, we know that the proposed filtering scheme is indeed effective in estimating the concentration of the product B in the continuous stirred tank reactor system.

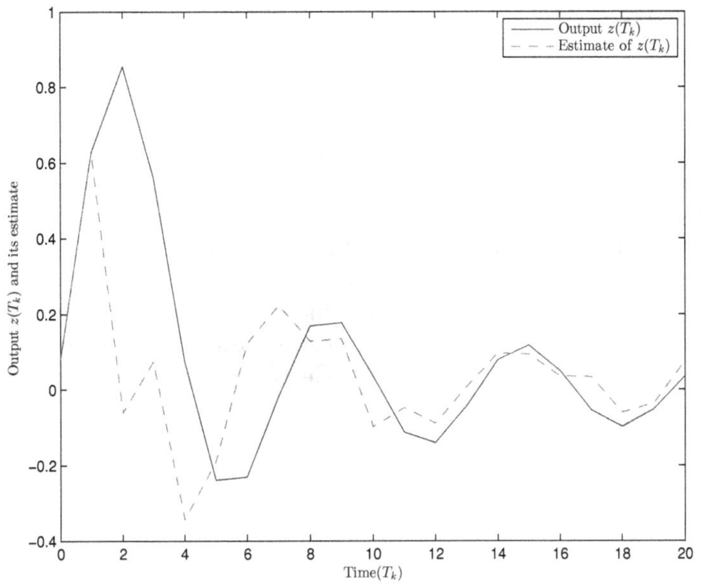

FIGURE 6.3: Output $z(T_k)$ and its estimate.

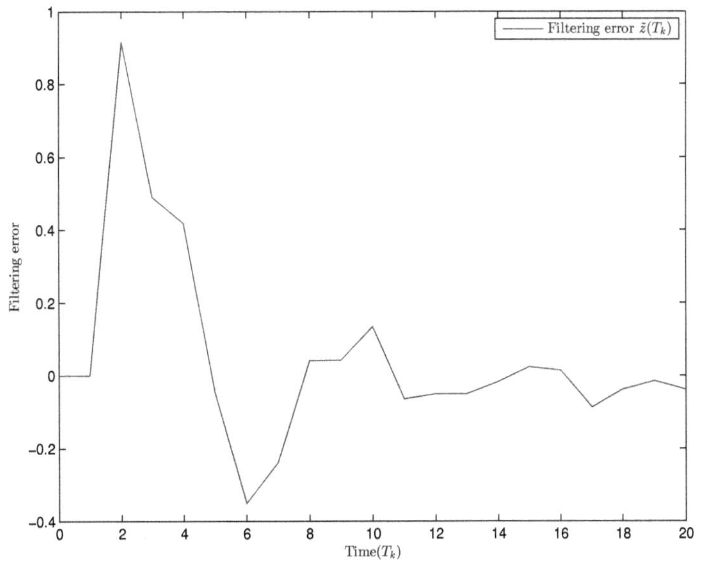

FIGURE 6.4: Filtering error dynamics.

6.4 Conclusion

This chapter has concerned the H_∞ filtering problem for a class of discrete-time networked MRSs with randomly occurring sensor saturations. The p-persistent CSMA protocol has been used to decide which sensor could get the access to transmission. The MRS has been converted into a single-rate system to facilitate the filter design. An H_∞ filter has been proposed such that the augmented system is exponentially mean-square stable and the H_∞ performance requirement is guaranteed simultaneously. By solving a LMI, the desired filter gain matrices have been derived. Finally, a numerical example has been provided to demonstrate the effectiveness of the proposed filtering approach.

7

l_2-l_∞ State Estimation for Artificial Neural Networks under High-Rate Channels with Round-Robin Protocol

In the late 1940s, inspired by the biological neural networks, the concept of ANNs has been put forward [44]. In the ANNs, the concept of neuron is extended to the node that owns adaptive weight and is capable of approximating non-linear function. Nowadays, due to the capability of self-learning and non-linear function approximation, the ANNs have been widely applied in many areas such as image processing, pattern recognition, and crack detection. As a foundation of certain applications of the ANNs, the actual states of some important neurons are required to be known. Unfortunately, the actual states of the neurons are often difficult to be directly acquired and only the measurement outputs are available through observation. Consequently, it is of vital importance to estimate the states of the neurons in the ANNs by using the available measurement outputs [21, 156]. On the other hand, time-delays are ubiquitous in the information transmissions among neurons in the ANNs and could cause oscillation. In the past decade, the state estimation problem for ANNs with time-delays has received persistent research attentions and many results are available.

Among the existing state estimation approaches for the ANNs with time-delays, the robust state estimation methods have gained much research interest owing to the ability to guarantee specific robustness to uncertainties and external disturbances [160]. As a classic robust state estimation method, the l_2-l_∞ state estimation is able to constrain the effect from the energy-bounded external disturbances with unknown statistical characteristic on the estimation error. To be specific, the purpose of the l_2-l_∞ state estimation is to design a state estimator with which the l_2-l_∞ gain is under a certain level [22]. So far, there are plenty of results available on the l_2-l_∞ state estimation problem for the ANNs with time-delays. For example, the l_2-l_∞ state estimation problem has been studied in [24] for switched neural networks with discrete-time delays, in [189] for Markovian jump recurrent neural networks with time-varying delays, and in [177] for discrete-time neural networks with distributed delays.

In reality, the ANNs implemented in the industrial field are sometimes realized by the hardware devices (e.g. the electronic circuit and the field programmable gate array) due to the advantages like high computational speed and low cost [115]. For the ANNs in the industry, the remote state

DOI: 10.1201/9781032619507-7

estimation scenario commonly exists because of the large scale of the ANNs and the high computational burden. In such a scenario, the state estimator is implemented on a site that is different with the site where the ANN is implemented, and the communication channel is introduced under which the measurement outputs of the ANN are transmitted to the remote state estimator. As a widely used communication channel in the industry, the process field bus is employed to connect the industrial devices (e.g. the ANNs and the remote sate estimator) [14]. In the communication channel of the process field bus, the data transmission rate is up to 31.25 kbit/s which is much higher than the data generation rates of the sensors (e.g. 0.37kb/s for 13-b LM95172 temperature sensor and 1.2kb/s for 12-b DT138 acceleration sensor). Communication networks like process field bus are called high-rate communication networks in [215] and will give rise to a special situation that the sensor transmits the same measurement output multiple times in one sampling period. To date, despite the important engineering significance, the remote state estimation problem for the ANNs under high-rate communication channels has not been investigated yet.

Due to the large scale of the ANNs, a large number of sensors are required to fully measure the ANNs in order to accurately estimate the state. During the data transmissions through the high-rate communication channels, the simultaneous transmissions of the *numerous* sensors would most likely lead to data collisions despite the fast data transmission rate of the high-rate communication channels. As such, to alleviate the data collisions, communication protocols that are capable of scheduling the transmission orders of the sensors according to certain principles are employed in the high-rate communication channels [170]. Among various communication protocols, the RR protocol is a periodic communication protocol that regulates the transmission access of the sensors according to a fixed *circular* order and is widely utilized in industry [32]. Note that, the introduction of the RR protocol in the high-rate communication channels will further complicate the signal transmission behavior, and bring substantial difficulties to the corresponding state estimator design problems. To this end, an interesting yet challenging research problem with clear engineering insight is the state estimation for the ANNs under high-rate communication channels with RR protocol, and much effort will be devoted to investigating such a problem in this chapter.

Summarizing the above discussions, in this chapter, the l_2-l_∞ state estimation problem is studied for time-delayed ANNs under high-rate communication channels with RR protocol. The main contributions are that: 1) the state estimation problem is, for the first time, studied for the ANNs under high-rate communication channels with RR protocol; 2) the complicated transmission mechanism of the sensors, resulting from the coupling between the high-rate communication channels and the RR protocol, is modeled by a periodic sequence; and 3) sufficient conditions are provided under which the estimation error dynamics is exponentially stable and the l_2-l_∞ performance requirement is ensured.

7.1 Problem Formulation

Consider the following m-neuron ANN with time-delays:

$$\begin{cases} x_i(t_{k+1}) = a_i x_i(t_k) + \sum_{j=1}^{m} b_{ij}^1 f_j(x_j(t_k)) \\ \qquad\qquad + \sum_{j=1}^{m} b_{ij}^2 f_j(x_j(t_{k-\tau_j})) + d_i w_i(t_k), \quad i = 1, 2, \ldots, m \\ z_q(t_k) = \sum_{j=1}^{m} h_{qj} x_j(t_k), \quad q = 1, 2, \ldots, l \\ x_i(t_s) = \phi_i(t_s), \quad s \in \mathcal{T} \triangleq \left[-\max_{1 \le j \le m} \{\tau_j\}, 0 \right] \end{cases} \tag{7.1}$$

where $t_k (k \geq 0)$ represents the $(k+1)$-th sampling instant, $x_i(t_k) \in \mathbb{R}$ is the state of the i-th neuron, $z_q(t_k) \in \mathbb{R}$ is the q-th signal to be estimated, $w_i(t_k)$ is the disturbance input belonging to $l_2([0, \infty); \mathbb{R})$, and $f_j(\cdot)$ is the non-linear activation function of the j-th neuron. a_i is the state feedback coefficient. b_{ij}^1 and b_{ij}^2 are the connection weight and the delayed connection weight, respectively. d_i and h_{qj} are known real scalars, τ_j is a positive integer representing the time-delay of the j-th neuron, and $\phi_i(t_s)$ are the given initial conditions.

Denote

$$\begin{aligned} x(t_k) &\triangleq \text{col}\{x_1(t_k), x_2(t_k), \cdots, x_m(t_k)\}, \\ z(t_k) &\triangleq \text{col}\{z_1(t_k), z_2(t_k), \cdots, z_l(t_k)\}, \\ \phi(t_s) &\triangleq \text{col}\{\phi_1(t_s), \phi_2(t_s), \cdots, \phi_m(t_s)\}, \\ f(x(t_k)) &\triangleq \text{col}\{f_1(x_1(t_k)), f_2(x_2(t_k)), \cdots, f_m(x_m(t_k))\}, \\ w(t_k) &\triangleq \text{col}\{w_1(t_k), w_2(t_k), \cdots, w_m(t_k)\}, \\ A &\triangleq \text{diag}\{a_1, a_2, \cdots, a_m\}, \ D \triangleq \text{diag}\{d_1, d_2, \cdots, d_m\}, \\ B_1 &\triangleq \left[b_{ij}^1\right]_{m \times m}, \quad B_2 \triangleq \left[b_{ij}^2\right]_{m \times m}, \quad H \triangleq \left[h_{qj}\right]_{l \times m}, \\ L_i &\triangleq \text{diag}\{\underbrace{0, \cdots, 0}_{i-1}, 1, \underbrace{0, \cdots, 0}_{m-i}\}. \end{aligned}$$

The ANN (7.1) is written in a compact form as follows:

$$\begin{cases} x(t_{k+1}) = Ax(t_k) + B_1 f(x(t_k)) + \sum_{i=1}^{m} B_2 L_i f(x(t_{k-\tau_i})) + Dw(t_k), \\ z(t_k) = Hx(t_k), \\ x(t_s) = \phi(t_s), \quad s \in \mathcal{T}. \end{cases} \tag{7.2}$$

Assumption 7.1 *The non-linear function $f(\cdot)$ satisfies the following condition:*

$$\|f(a) - f(b)\| \leq \|G(a-b)\| \tag{7.3}$$

where $a, b \in \mathbb{R}^m$ are real vectors and G is a known diagonal matrix.

The ANN (7.2) is measured by n sensors which are modeled as follows:

$$y_i(t_k) = C_i x(t_k), \quad i = 1, 2, \ldots, n \tag{7.4}$$

where $y_i(t_k) \in \mathbb{R}^{n_y}$ is the measurement output of the i-th sensor before being transmitted and C_i is a known matrix with appropriate dimensions. Moreover, the sampling period of the sensors is denoted as $h_s \triangleq t_{k+1} - t_k$.

In this chapter, the measurement outputs of the sensors are transmitted to the remote state estimator through the high-rate communication channel. Since the number of deployed sensors is large, data collisions are very likely to occur when the sensors transmit their measurement outputs simultaneously. In order to alleviate data collisions, the RR protocol is employed in the high-rate communication channels to determine the transmission sequence of the sensors.

Before going further, as in [215], the following assumption on the high-rate communication channels is made.

Assumption 7.2 *The time interval between two adjacent transmissions in the high-rate communication channels is h_t which satisfies $h_t = h_s/d$ with $d \geq 1$ being a positive integer. Moreover, the first transmission occurs at the initial sampling instant t_0.*

Denoting the $(k'+1)$-th $(k' \geq 0)$ transmission time instant as $T_{k'}$, we have $T_{k'+1} = T_{k'} + h_t$. Moreover, it is easily known that the transmitted measurement output from sensor i at time instant $T_{k'}$ is

$$y_i(T_{k'}) = y_i(t_k), \quad T_{k'} \in [t_k, t_{k+1}). \tag{7.5}$$

Now, we are going to elaborate the scheduling effect of the RR protocol in the high-rate communication channels. Let $\bar{y}_i(T_{k'})$ be the actual received measurement output from sensor i by the state estimator at time instant $T_{k'}$. From the knowledge of the RR protocol [89], we have

$$\bar{y}_i(T_{k'}) = \begin{cases} y_i(T_{k'}) + E_i v_i(T_{k'}), & \text{if } \xi(T_{k'}) = i; \\ \bar{y}_i(T_{k'-1}), & \text{otherwise} \end{cases} \tag{7.6}$$

where $\xi(T_{k'}) \triangleq \text{mod}(k'-1, n) + 1$ represents the sensor which is given the transmission access at time instant $T_{k'}$, $v_i(T_{k'})$ is the network-induced noise belonging to $l_2([0,\infty); \mathbb{R}^{n_v})$, and E_i is a known matrix with appropriate dimensions. Moreover, we set $\bar{y}_i(T_{k'}) = 0$ for any $k' < 0$.

Denote

$$
\begin{aligned}
\bar{y}(T_{k'}) &\triangleq \mathrm{col}\{\bar{y}_1(T_{k'}), \bar{y}_2(T_{k'}), \cdots, \bar{y}_n(T_{k'})\},\\
y(t_k) &\triangleq \mathrm{col}\{y_1(t_k), y_2(t_k), \cdots, y_n(t_k)\},\\
v(T_{k'}) &\triangleq \mathrm{col}\{v_1(T_{k'}), v_2(T_{k'}), \cdots, v_n(T_{k'})\},\\
E &\triangleq \mathrm{diag}\{E_1, E_2, \cdots, E_n\}, \quad C \triangleq \mathrm{col}\{C_1, C_2, \cdots, C_n\},\\
\Psi_{\xi(T_{k'})} &\triangleq \mathrm{diag}\{\delta(\xi(T_{k'}), 1)I, \delta(\xi(T_{k'}), 2)I, \cdots, \delta(\xi(T_{k'}), n)I\}
\end{aligned}
$$

with $\delta(\cdot,\cdot)$ being the Kronecker delta function. From (7.5) and (7.6), one has

$$
\begin{aligned}
\bar{y}(T_{k'}) =& \Psi_{\xi(T_{k'})} y(t_k) + \Psi_{\xi(T_{k'})} E v(T_{k'})\\
&+ (I - \Psi_{\xi(T_{k'})})\bar{y}(T_{k'-1}), \quad T_{k'} \in [t_k, t_{k+1}).
\end{aligned} \tag{7.7}
$$

From Assumption 7.2, we know that there are d transmissions occurring in the time interval $(t_{k-1}, t_k]$. Denote the time instant when the ι-th transmission in the time interval $(t_{k-1}, t_k]$ occurs as $\bar{t}_k^{\iota} \triangleq t_{k-1} + \iota h_t$ ($\iota = 1, 2, \ldots, d$). From (7.7), we have

$$
\begin{aligned}
\bar{y}(t_k) =& \Psi_{\xi(t_k)} y(t_k) + \Psi_{\xi(t_k)} E v(t_k) + \big(I - \Psi_{\xi(t_k)}\big)\bar{y}(\bar{t}_k^{d-1}),\\
\bar{y}(\bar{t}_k^{d-1}) =& \Psi_{\xi(\bar{t}_k^{d-1})} y(t_{k-1}) + \Psi_{\xi(\bar{t}_k^{d-1})} E v(\bar{t}_k^{d-1}) + \big(I - \Psi_{\xi(\bar{t}_k^{d-1})}\big)\bar{y}(\bar{t}_k^{d-2}),\\
&\vdots\\
\bar{y}(\bar{t}_k^{1}) =& \Psi_{\xi(\bar{t}_k^{1})} y(t_{k-1}) + \Psi_{\xi(\bar{t}_k^{1})} E v(\bar{t}_k^{1}) + \big(I - \Psi_{\xi(\bar{t}_k^{1})}\big)\bar{y}(t_{k-1}).
\end{aligned}
$$

Iterating the above equations, we further have

$$
\begin{aligned}
\bar{y}(t_k) =& \Psi_{\xi(t_k)} y(t_k) + \Phi^1_{\zeta(\bar{t}_k^d)} y(t_{k-1})\\
&+ \Phi^2_{\zeta(\bar{t}_k^d)} \bar{y}(t_{k-1}) + \Phi^3_{\zeta(\bar{t}_k^d)} \bar{v}(t_k)
\end{aligned} \tag{7.8}
$$

where

$$
\begin{aligned}
\zeta(\bar{t}_k^d) &\triangleq (\xi(\bar{t}_k^1), \xi(\bar{t}_k^2), \ldots, \xi(\bar{t}_k^d)),\\
\Phi^1_{\zeta(\bar{t}_k^d)} &\triangleq \sum_{i=1}^{d-1} \left(\prod_{j=0}^{i-1} \Big(I - \Psi_{\xi(\bar{t}_k^{d-j})}\Big) \right) \Psi_{\xi(\bar{t}_k^{d-i})},\\
\Phi^2_{\zeta(\bar{t}_k^d)} &\triangleq \prod_{i=0}^{d-1} \Big(I - \Psi_{\xi(\bar{t}_k^{d-i})}\Big),\\
\Phi^3_{\zeta(\bar{t}_k^d)} &\triangleq \big[\Psi_{\xi(t_k)} E \quad \bar{\Psi}_{\zeta(\bar{t}_k^d)}\big],\\
\bar{\Psi}_{\zeta(\bar{t}_k^d)} &\triangleq \Big[\bar{\Psi}^1_{\zeta(\bar{t}_k^d)} \quad \bar{\Psi}^2_{\zeta(\bar{t}_k^d)} \quad \cdots \quad \bar{\Psi}^{d-1}_{\zeta(\bar{t}_k^d)}\Big],\\
\bar{\Psi}^i_{\zeta(\bar{t}_k^d)} &\triangleq \prod_{j=0}^{i-1} \Big(I - \Psi_{\xi(\bar{t}_k^{d-j})}\Big) \Psi_{\xi(\bar{t}_k^{d-i})} E,\\
\bar{v}(t_k) &\triangleq \mathrm{col}\{v(\bar{t}_k^d), v(\bar{t}_k^{d-1}), \cdots, v(\bar{t}_k^1)\}.
\end{aligned}
$$

Remark 7.1 *It can be seen that the matrix* $X_{\zeta(\bar{t}_k^d)}$ *(*X *stands for* Φ^1, Φ^2, Φ^3, $\bar{\Psi}$, *and* $\bar{\Psi}^i$*) is dependent on* $\xi(\bar{t}_k^1), \xi(\bar{t}_k^2), \ldots, \xi(\bar{t}_k^d)$ *and therefore* $X_{\zeta(\bar{t}_k^d)}$ *should be written as* $X_{\xi(\bar{t}_k^1),\xi(\bar{t}_k^2),\ldots,\xi(\bar{t}_k^d)}$*. Nevertheless, for convenience of presentation, we introduce a new notation* $\zeta(\bar{t}_k^d)$ *to represent* $\xi(\bar{t}_k^1), \xi(\bar{t}_k^2), \ldots, \xi(\bar{t}_k^d)$*, and* $X_{\xi(\bar{t}_k^1),\xi(\bar{t}_k^2),\ldots,\xi(\bar{t}_k^d)}$ *is represented by* $X_{\zeta(\bar{t}_k^d)}$*.*

Denoting $\bar{x}(t_k) \triangleq \text{col}\{x(t_k), y(t_{k-1}), \bar{y}(t_{k-1})\}$, the ANN (7.2) with the measurement outputs (7.4) is rewritten as the following form:

$$\begin{cases} \bar{x}(t_{k+1}) = \bar{A}_{\zeta(\bar{t}_k^d)}\bar{x}(t_k) + \bar{B}_1\bar{f}(\bar{x}(t_k)) \\ \qquad\qquad + \sum_{i=1}^{m} \bar{B}_2\bar{L}_i\bar{f}(\bar{x}(t_{k-\tau_i})) + \bar{D}_{\zeta(\bar{t}_k^d)}\varpi(t_k), \\ \bar{y}(t_k) = \bar{C}_{\zeta(\bar{t}_k^d)}\bar{x}(t_k) + \bar{E}_{\zeta(\bar{t}_k^d)}\varpi(t_k), \\ z(t_k) = \bar{H}\bar{x}(t_k) \end{cases} \tag{7.9}$$

where

$$\bar{A}_{\zeta(\bar{t}_k^d)} \triangleq \begin{bmatrix} A & 0 & 0 \\ C & 0 & 0 \\ \Psi_{\xi(t_k)}C & \Phi^1_{\zeta(\bar{t}_k^d)} & \Phi^2_{\zeta(\bar{t}_k^d)} \end{bmatrix},$$

$$\bar{f}(\bar{x}(t_k)) \triangleq \begin{bmatrix} f(x(t_k)) \\ 0 \\ 0 \end{bmatrix}, \quad \bar{D}_{\zeta(\bar{t}_k^d)} \triangleq \begin{bmatrix} D & 0 \\ 0 & 0 \\ 0 & \Phi^3_{\zeta(\bar{t}_k^d)} \end{bmatrix},$$

$$\bar{E}_{\zeta(\bar{t}_k^d)} \triangleq \begin{bmatrix} 0 & \Phi^3_{\zeta(\bar{t}_k^d)} \end{bmatrix}, \quad \varpi(t_k) \triangleq \begin{bmatrix} w(t_k) \\ \bar{v}(t_k) \end{bmatrix},$$

$$\bar{C}_{\zeta(\bar{t}_k^d)} \triangleq \begin{bmatrix} \Psi_{\xi(t_k)}C & \Phi^1_{\zeta(\bar{t}_k^d)} & \Phi^2_{\zeta(\bar{t}_k^d)} \end{bmatrix},$$

$$\bar{B}_1 \triangleq \text{diag}\{B_1, 0, 0\}, \quad \bar{B}_2 \triangleq \text{diag}\{B_2, 0, 0\},$$

$$\bar{L}_i \triangleq \text{diag}\{L_i, 0, 0\}, \quad \bar{H} \triangleq \begin{bmatrix} H & 0 & 0 \end{bmatrix}.$$

For convenience of later analysis, the following property of $\zeta(\bar{t}_k^d)$ is introduced.

Lemma 7.1 *For a positive integer* $p \triangleq \min\{i \in \mathbb{N}^+ | \frac{id}{n} \in \mathbb{N}^+\}$, $\zeta(\bar{t}_k^d)$ *satisfies*

$$\zeta(\bar{t}_{k+p}^d) = \zeta(\bar{t}_k^d)$$

which means that the elements in $\zeta(\bar{t}_{k+p}^d)$ *and* $\zeta(\bar{t}_k^d)$ *are the same.*

Proof *From Assumption 7.2, it is known that* $t_{k+p} = t_0 + (k+p)h_s = T_0 + (k+p)dh_t = T_{(k+p)d}$*. Then, one has*

$$\begin{aligned} \xi(t_{k+p} + ih_t) &= \xi(T_{(k+p)d+i}) \\ &= \text{mod}((k+p)d + i - 1, n) + 1 \\ &= \text{mod}(kd + i - 1, n) + 1 \\ &= \xi(T_{kd+i}) = \xi(t_k + ih_t). \end{aligned}$$

Noting the definition of $\zeta(t_k)$*, we know that*

$$\begin{aligned}\zeta(\bar{t}^d_{k+p}) &= (\xi(t_{k+p-1}+h_t), \xi(t_{k+p-1}+2h_t), \cdots, \xi(t_{k+p}))\\ &= (\xi(t_{k-1}+h_t), \xi(t_{k-1}+2h_t), \cdots, \xi(t_k))\\ &= \zeta(\bar{t}^d_k)\end{aligned}$$

which completes the proof.

Based on Lemma 7.1, we are ready to give the following mapping method which maps $\zeta(\bar{t}^d_k)$ (i.e. $\xi(\bar{t}^1_k), \xi(\bar{t}^2_k), \ldots, \xi(\bar{t}^d_k)$) to a new periodic sequence.

Lemma 7.2 $\zeta(\bar{t}^d_k)$ *can be mapped to a new periodic sequence* $r(t_k) \in \mathcal{R} \triangleq \{r_0, r_1, \ldots, r_{p-1}\}$ *that satisfies* $r(t_{k+p}) = r(t_k)$ *by the mapping* $\mathscr{M}(\cdot)$*:*

$$r(t_k) = \mathscr{M}(\zeta(\bar{t}^d_k)) = \sum_{i=1}^{d}(\xi(\bar{t}^i_k) - 1)n^i + 1$$

where

$$r_\rho \triangleq \sum_{i=1}^{d}(\xi(\bar{t}^i_\rho) - 1)n^i + 1, \quad \rho \in \mathcal{P} \triangleq \{0, 1, \ldots, p-1\}$$

with $\bar{t}^i_\rho \triangleq t_{\rho-1} + ih_t$ *for* $\rho \in \{0, 1, \ldots, p-1\}$*. Moreover, for a given* $r(t_k)$*, the value of* $\xi(\bar{t}^i_k)$ *can be calculated by*

$$\xi(\bar{t}^i_k) = \mathrm{mod}\Big(\Big\lfloor \frac{r(t_k)-1}{n^i} \Big\rfloor, n\Big) + 1, \quad i = 1, 2, \ldots, d.$$

Proof *From Lemma 7.1, we know that*

$$r(t_{k+p}) = \mathscr{M}(\zeta(\bar{t}^d_{k+p})) = \mathscr{M}(\zeta(\bar{t}^d_k)) = r(t_k)$$

which means that $r(t_k)$ *only takes values in the set* $\{r(t_0), r(t_1), \ldots, r(t_{p-1})\}$*. Noting the mapping* $\mathscr{M}(\cdot)$*, we have*

$$r(t_\rho) = \sum_{i=1}^{d}(\xi(\bar{t}^i_\rho) - 1)n^i + 1 = r_\rho, \quad \rho \in \mathcal{P}.$$

Then, it is known that $r(t_k) \in \mathcal{R}$*.*

The rest of the proof is similar to that of Proposition 1 in [214] and therefore is omitted here.

According to Lemma 7.2, the augmented system (7.9) can be rewritten as

$$\begin{cases}\bar{x}(t_{k+1}) = \bar{A}_{r(t_k)}\bar{x}(t_k) + \bar{B}_1\bar{f}(\bar{x}(t_k))\\ \qquad\qquad + \sum_{i=1}^{m}\bar{B}_2\bar{L}_i\bar{f}(\bar{x}(t_{k-\tau_i})) + \bar{D}_{r(t_k)}\varpi(t_k),\\ \bar{y}(t_k) = \bar{C}_{r(t_k)}\bar{x}(t_k) + \bar{E}_{r(t_k)}\varpi(t_k),\\ z(t_k) = \bar{H}\bar{x}(t_k)\end{cases} \tag{7.10}$$

where

$$\bar{A}_{r(t_k)} \triangleq \bar{A}_{\zeta(\bar{t}_k^d)}, \quad \bar{C}_{r(t_k)} \triangleq \bar{C}_{\zeta(\bar{t}_k^d)},$$
$$\bar{D}_{r(t_k)} \triangleq \bar{D}_{\zeta(\bar{t}_k^d)}, \quad \bar{E}_{r(t_k)} \triangleq \bar{E}_{\zeta(\bar{t}_k^d)}.$$

Remark 7.2 *The system (7.9) can be seen as a switched system with the switching signal $\zeta(\bar{t}_k^d)$ that depends on $\xi(\bar{t}_k^1), \xi(\bar{t}_k^2), \ldots, \xi(\bar{t}_k^d)$. For analysis convenience, we have proposed a mapping method in Lemma 7.2 to map $\xi(\bar{t}_k^1), \xi(\bar{t}_k^2), \ldots, \xi(\bar{t}_k^d)$ to a new sequence $r(t_k)$. From Lemma 7.2, we know that the sequence $r(t_k)$ is a periodic sequence and takes values in a finite set $\mathcal{R}$. Moreover, with the help of Lemma 7.2, the switched system (7.9) is transformed to a new switched system (7.10) with the switching signal $r(t_k)$.*

In this chapter, the state estimator of the following form is adopted:

$$\begin{cases} \hat{x}(t_{k+1}) = \bar{A}_{r(t_k)}\hat{x}(t_k) + \bar{B}_1\bar{f}(\hat{x}(t_k)) + \sum_{i=1}^{m} \bar{B}_2\bar{L}_i\bar{f}(\hat{x}(t_{k-\tau_i})) \\ \qquad + K_{r(t_k)}\left(\bar{y}(t_k) - \bar{C}_{r(t_k)}\hat{x}(t_k)\right), \\ \hat{z}(t_k) = \bar{H}\hat{x}(t_k), \\ \hat{x}(t_s) = 0, \quad s \in \mathcal{T} \end{cases} \tag{7.11}$$

where $\hat{x}(t_k)$ is the estimate of $\bar{x}(t_k)$, $\hat{z}(t_k)$ is the estimate of $z(t_k)$, and $K_{r(t_k)}$ is the state estimator gain to be designed.

Denoting the state estimation error as $e(t_k) \triangleq \bar{x}(t_k) - \hat{x}(t_k)$ and letting $\tilde{z}(t_k) \triangleq z(t_k) - \hat{z}(t_k)$, the estimation error dynamics is obtained as

$$\begin{cases} e(t_{k+1}) = \tilde{A}_{r(t_k)}e(t_k) + \bar{B}_1\tilde{f}_e(t_k) \\ \qquad + \sum_{i=1}^{m} \bar{B}_2\bar{L}_i\tilde{f}_e(t_{k-\tau_i}) + \tilde{D}_{r(t_k)}\varpi(t_k), \\ \tilde{z}(t_k) = \bar{H}e(t_k), \\ e(t_s) = \bar{\phi}(t_s), \quad s \in \mathcal{T} \end{cases} \tag{7.12}$$

where

$$\tilde{A}_{r(t_k)} \triangleq \bar{A}_{r(t_k)} - K_{r(t_k)}\bar{C}_{r(t_k)},$$
$$\tilde{D}_{r(t_k)} \triangleq \bar{D}_{r(t_k)} - K_{r(t_k)}\bar{E}_{r(t_k)},$$
$$\tilde{f}_e(t_k) \triangleq \bar{f}(\bar{x}(t_k)) - \bar{f}(\hat{x}(t_k)),$$
$$\bar{\phi}(t_s) \triangleq \begin{bmatrix}\phi^T(t_s) & 0 & 0\end{bmatrix}^T.$$

Definition 7.1 *The system (7.12) with $\varpi(t_k) = 0$ is said to be exponentially stable if there exist scalars $\mu > 0$ and $0 < \nu < 1$ such that*

$$\|e(t_k)\|^2 \leq \mu\nu^k \sup_{s\in\mathcal{T}} \|\bar{\phi}(t_s)\|^2.$$

The aim of this chapter is to design a state estimator in the form of (7.11) such that:

1. the estimation error dynamic system (7.12) with $\varpi(t_k) = 0$ is exponentially stable;
2. under the zero initial condition, for a given disturbance attenuation level $\gamma > 0$ and all non-zero $\varpi(t_k)$, the estimation error $\tilde{z}(t_k)$ satisfies the following l_2-l_∞ performance constraint:

$$\sup_k \sqrt{\tilde{z}^T(t_k)\tilde{z}(t_k)} < \gamma \sqrt{\sum_{k=0}^{\infty} \|\varpi(t_k)\|^2}. \tag{7.13}$$

7.2 l_2-l_∞ Estimator Design

In this section, the l_2-l_∞ state estimator design problem is studied for the augmented system (7.10). First, sufficient conditions are given that ensure the exponential stability of the estimation error dynamic system (7.12). Then, sufficient conditions for the existence of the l_2-l_∞ state estimator are provided under which the estimation error dynamic system is exponentially stable and the l_2-l_∞ performance constraint is ensured. Finally, the characterization of the state estimator gains is realized by solving certain LMIs.

Before presenting the main results, the following useful lemma is introduced.

Lemma 7.3 *Denote*

$$\vec{f}_e(t_k) \triangleq \text{col}\{\tilde{f}_e(t_{k-\tau_1}), \tilde{f}_e(t_{k-\tau_2}), \cdots, \tilde{f}_e(t_{k-\tau_m})\}.$$

$\tilde{f}_e(t_k)$ *and* $\vec{f}_e(t_k)$ *satisfy*

$$\|\tilde{f}_e(t_k)\| \le \|\bar{G}e(t_k)\|, \quad \|\vec{f}_e(t_k)\| \le \|\tilde{G}\bar{e}(t_k)\| \tag{7.14}$$

where

$$\begin{aligned}\bar{e}(t_k) \triangleq& \text{col}\{e(t_{k-\tau_1}), e(t_{k-\tau_2}), \cdots, e(t_{k-\tau_m})\},\\ \bar{G} \triangleq& \begin{bmatrix} G & 0 & 0 \end{bmatrix}, \quad \tilde{G} \triangleq \text{diag}\{\underbrace{\bar{G}, \bar{G}, \cdots, \bar{G}}_{m}\}.\end{aligned}$$

Proof *It is easily known from condition* (7.3) *that*

$$\|\tilde{f}_e(t_k)\| = \|\bar{f}(\bar{x}(t_k)) - \bar{f}(\hat{x}(t_k))\| \le \|\bar{G}e(t_k)\|.$$

Furthermore, we have

$$\begin{aligned}\|\vec{f}_e(t_k)\|^2 =& \sum_{i=1}^{m} \|\tilde{f}_e(t_{k-\tau_i})\|^2 \\ \leq& \sum_{i=1}^{m} \|\bar{G}e(t_{k-\tau_i})\|^2 = \|\tilde{G}\bar{e}(t_k)\|^2\end{aligned}$$

which completes the proof.

In the following theorem, the exponential stability of the system (7.12) is analyzed and sufficient conditions are derived.

Theorem 7.1 *Let the state estimator gains K_{r_i} $(i \in \mathcal{P})$ be given. The estimation error dynamic system (7.12) is exponentially stable with $\varpi(t_k) = 0$ if there exist positive-definite matrices P_{r_i} $(i \in \mathcal{P})$ and Q such that*

$$\Xi_{r_i} = \begin{bmatrix} \Xi_{r_i}^{11} & 0 & \Xi_{r_i}^{13} & \Xi_{r_i}^{14} \\ * & \Xi^{22} & 0 & 0 \\ * & * & \Xi_{r_i}^{33} & \Xi_{r_i}^{34} \\ * & * & * & \Xi_{r_i}^{44} \end{bmatrix} < 0 \quad (i \in \mathcal{P}) \tag{7.15}$$

where

$$\begin{aligned}\Xi_{r_i}^{11} \triangleq& \tilde{A}_{r_i}^T P_{r_{i+1}} \tilde{A}_{r_i} - P_{r_i} + mQ + \bar{G}^T \bar{G}, \\ \Xi^{22} \triangleq& \tilde{G}^T \tilde{G} - \bar{Q}, \quad \Xi_{r_i}^{13} \triangleq \tilde{A}_{r_i}^T P_{r_{i+1}} \bar{B}_1, \\ \Xi_{r_i}^{14} \triangleq& \tilde{A}_{r_i}^T P_{r_{i+1}} \bar{B}_2 \bar{L}, \quad \Xi_{r_i}^{33} \triangleq \bar{B}_1^T P_{r_{i+1}} \bar{B}_1 - I, \\ \Xi_{r_i}^{34} \triangleq& \bar{B}_1^T P_{r_{i+1}} \bar{B}_2 \bar{L}, \quad \Xi_{r_i}^{44} \triangleq \bar{L}^T \bar{B}_2^T P_{r_{i+1}} \bar{B}_2 \bar{L} - I, \\ \bar{L} \triangleq& \begin{bmatrix} \bar{L}_1 & \bar{L}_2 & \cdots & \bar{L}_m \end{bmatrix}, \quad \bar{Q} \triangleq \mathrm{diag}\{Q, Q, \cdots, Q\},\end{aligned}$$

and $P_{r_{p+1}} = P_{r_1}$.

Proof *Choose the following Lyapunov-Krasovskii functional*

$$V(t_k) = V_1(t_k) + V_2(t_k) \tag{7.16}$$

where

$$\begin{aligned}V_1(t_k) \triangleq& e^T(t_k) P_{r(t_k)} e(t_k), \\ V_2(t_k) \triangleq& \sum_{i=1}^{m} \sum_{s=k-\tau_i}^{k-1} e^T(t_s) Q e(t_s).\end{aligned}$$

Calculating the differences of $V_1(t_k)$ and $V_2(t_k)$ along the trajectory of system

(7.12) *with* $\varpi(t_k)=0$, *we have*

$$\begin{aligned}\Delta V_1(t_k)=&V_1(t_{k+1})-V_1(t_k)\\=&\Big(\tilde{A}_{r(t_k)}e(t_k)+\bar{B}_1\tilde{f}_e(t_k)+\bar{B}_2\bar{L}\vec{f}_e(t_k)\Big)^TP_{r(t_{k+1})}\\&\times\Big(\tilde{A}_{r(t_k)}e(t_k)+\bar{B}_1\tilde{f}_e(t_k)+\bar{B}_2\bar{L}\vec{f}_e(t_k)\Big)\\&-e^T(t_k)P_{r(t_k)}e(t_k)\\=&e^T(t_k)\big(\tilde{A}^T_{r(t_k)}P_{r(t_{k+1})}\tilde{A}_{r(t_k)}-P_{r(t_k)}\big)e(t_k)\\&+e^T(t_k)\tilde{A}^T_{r(t_k)}P_{r(t_{k+1})}\bar{B}_1\tilde{f}_e(t_k)\\&+e^T(t_k)\tilde{A}^T_{r(t_k)}P_{r(t_{k+1})}\bar{B}_2\bar{L}\vec{f}_e(t_k)\\&+\tilde{f}_e^T(t_k)\bar{B}_1^TP_{r(t_{k+1})}\tilde{A}_{r(t_k)}e(t_k)\\&+\tilde{f}_e^T(t_k)\bar{B}_1^TP_{r(t_{k+1})}\bar{B}_1\tilde{f}_e(t_k)\\&+\tilde{f}_e^T(t_k)\bar{B}_1^TP_{r(t_{k+1})}\bar{B}_2\bar{L}\vec{f}_e(t_k)\\&+\vec{f}_e^T(t_k)\bar{L}^T\bar{B}_2^TP_{r(t_{k+1})}\tilde{A}_{r(t_k)}e(t_k)\\&+\vec{f}_e^T(t_k)\bar{L}^T\bar{B}_2^TP_{r(t_{k+1})}\bar{B}_1\tilde{f}_e(t_k)\\&+\vec{f}_e^T(t_k)\bar{L}^T\bar{B}_2^TP_{r(t_{k+1})}\bar{B}_2\bar{L}\vec{f}_e(t_k)\end{aligned}$$

and

$$\begin{aligned}\Delta V_2(t_k)=&V_2(t_{k+1})-V_2(t_k)\\=&\sum_{i=1}^{m}\sum_{s=k-\tau_i+1}^{k}e^T(t_s)Qe(t_s)-\sum_{i=1}^{m}\sum_{s=k-\tau_i}^{k-1}e^T(t_s)Qe(t_s)\\=&\sum_{i=1}^{m}e^T(t_k)Qe(t_k)-\sum_{i=1}^{m}e^T(t_{k-\tau_i})Qe(t_{k-\tau_i})\\=&me^T(t_k)Qe(t_k)-\bar{e}^T(t_k)\bar{Q}\bar{e}(t_k).\end{aligned}$$

From (7.14), *we have*

$$\begin{aligned}\Delta V(t_k)=&\Delta V_1(t_k)+\Delta V_2(t_k)\\\le&e^T(t_k)\big(\tilde{A}^T_{r(t_k)}P_{r(t_{k+1})}\tilde{A}_{r(t_k)}-P_{r(t_k)}\big)e(t_k)\\&+e^T(t_k)\tilde{A}^T_{r(t_k)}P_{r(t_{k+1})}\bar{B}_1\tilde{f}_e(t_k)\\&+e^T(t_k)\tilde{A}^T_{r(t_k)}P_{r(t_{k+1})}\bar{B}_2\bar{L}\vec{f}_e(t_k)\\&+\tilde{f}_e^T(t_k)\bar{B}_1^TP_{r(t_{k+1})}\tilde{A}_{r(t_k)}e(t_k)\\&+\tilde{f}_e^T(t_k)\bar{B}_1^TP_{r(t_{k+1})}\bar{B}_1\tilde{f}_e(t_k)\\&+\tilde{f}_e^T(t_k)\bar{B}_1^TP_{r(t_{k+1})}\bar{B}_2\bar{L}\vec{f}_e(t_k)\\&+\vec{f}_e^T(t_k)\bar{L}^T\bar{B}_2^TP_{r(t_{k+1})}\tilde{A}_{r(t_k)}e(t_k)\end{aligned}$$

$$
\begin{aligned}
&+ \vec{f}_e^T(t_k)\bar{L}^T\bar{B}_2^T P_{r(t_{k+1})}\bar{B}_1\tilde{f}_e(t_k) \\
&+ \vec{f}_e^T(t_k)\bar{L}^T\bar{B}_2^T P_{r(t_{k+1})}\bar{B}_2\bar{L}\vec{f}_e(t_k) \\
&+ me^T(t_k)Qe(t_k) - \bar{e}^T(t_k)\bar{Q}\bar{e}(t_k) \\
&- \tilde{f}_e^T(t_k)\tilde{f}_e(t_k) + e^T(t_k)\bar{G}^T\bar{G}e(t_k) \\
&- \vec{f}_e^T(t_k)\vec{f}_e(t_k) + \bar{e}^T(t_k)\tilde{G}^T\tilde{G}\bar{e}(t_k) \\
=&\eta^T(t_k)\Xi_{r(t_k)}\eta(t_k)
\end{aligned}
$$

where

$$\eta(t_k) \triangleq \mathrm{col}\{e(t_k), \bar{e}(t_k), \tilde{f}_e(t_k), \vec{f}_e(t_k)\}.$$

It is known from $\Xi_{r(t_k)} < 0$ that there exists a sufficiently small scalar $\delta > 0$ such that

$$\Delta V(t_k) + \delta\|e(t_k)\|^2 < 0.$$

By following the similar analysis as in [165], we know that the system (7.12) *is exponentially stable with $\varpi(t_k) = 0$. The proof is complete.*

In Theorem 7.1, sufficient conditions for the exponential stability are given. In the following theorem, sufficient conditions are presented under which the exponential stability as well as the l_2-l_∞ performance constraint are guaranteed simultaneously.

Theorem 7.2 *Let the disturbance attenuation level $\gamma > 0$ and the state estimator gains K_{r_i} ($i \in \mathcal{P}$) be given. The estimation error dynamic system* (7.12) *is exponentially stable with $\varpi(t_k) = 0$ and satisfies the l_2-l_∞ performance constraint* (7.13) *for all non-zero $\varpi(t_k)$ if there exist positive-definite matrices P_{r_i} ($i \in \mathcal{P}$) and Q such that*

$$\bar{\Xi}_{r_i} \triangleq \begin{bmatrix} \Xi_{r_i} & \Pi_{r_i} \\ * & \bar{\Xi}_{r_i}^{22} \end{bmatrix} < 0 \quad (i \in \mathcal{P}), \tag{7.17}$$

$$\Sigma_{r_i} \triangleq \begin{bmatrix} -I & \bar{H} \\ * & -P_{r_i} \end{bmatrix} < 0 \quad (i \in \mathcal{P}) \tag{7.18}$$

where

$$
\begin{aligned}
\bar{\Xi}_{r_i}^{22} &\triangleq \tilde{D}_{r_i}^T P_{r_{i+1}}\tilde{D}_{r_i} - \gamma^2 I, \\
\Pi_{r_i} &\triangleq \begin{bmatrix} \tilde{A}_{r_i}^T P_{r_{i+1}}\tilde{D}_{r_i} \\ 0 \\ \bar{B}_1^T P_{r_{i+1}}\tilde{D}_{r_i} \\ \bar{L}^T\bar{B}_2^T P_{r_{i+1}}\tilde{D}_{r_i} \end{bmatrix}
\end{aligned}
$$

and $P_{r_{p+1}} = P_{r_1}$.

Proof *It is straightforward to see that (7.17) implies (7.15), which means that the estimation error dynamic system (7.12) is exponentially stable.*

In the following, we are going to analyze the l_2-l_∞ performance constraint. First, an index functional is introduced as:

$$\mathcal{J}(t_k) \triangleq V(t_k) - \gamma^2 \sum_{i=0}^{k-1} \varpi^T(t_i)\varpi(t_i). \tag{7.19}$$

With the zero initial condition, one has

$$\begin{aligned}
\mathcal{J}(t_k) =& \sum_{i=0}^{k-1} \Delta V(t_i) - \gamma^2 \sum_{i=0}^{k-1} \varpi^T(t_i)\varpi(t_i) + V(t_0) \\
\leq & \sum_{i=0}^{k-1} \Big(\eta^T(t_i)\Xi_{r(t_i)}\eta(t_i) \\
& + e^T(t_i)\tilde{A}^T_{r(t_i)}P_{r(t_{i+1})}\tilde{D}_{r(t_i)}\varpi(t_i) \\
& + \tilde{f}_e^T(t_i)\bar{B}_1^T P_{r(t_{i+1})}\tilde{D}_{r(t_i)}\varpi(t_i) \\
& + \vec{f}_e^T(t_i)\bar{L}^T\bar{B}_2^T P_{r(t_{i+1})}\tilde{D}_{r(t_i)}\varpi(t_i) \\
& + \varpi^T(t_i)\tilde{D}^T_{r(t_i)}P_{r(t_{i+1})}\tilde{A}_{r(t_i)}e(t_i) \\
& + \varpi^T(t_i)\tilde{D}^T_{r(t_i)}P_{r(t_{i+1})}\bar{B}_1\tilde{f}_e(t_i) \\
& + \varpi^T(t_i)\tilde{D}^T_{r(t_i)}P_{r(t_{i+1})}\bar{B}_2\bar{L}\vec{f}_e(t_i) \\
& + \varpi^T(t_i)\tilde{D}^T_{r(t_i)}P_{r(t_{i+1})}\tilde{D}_{r(t_i)}\varpi(t_i)\Big) \\
& - \gamma^2 \sum_{i=0}^{k-1} \varpi^T(t_i)\varpi(t_i) \\
=& \sum_{i=0}^{k-1} \bar{\eta}^T(t_i)\bar{\Xi}_{r(t_i)}\bar{\eta}(t_i)
\end{aligned}$$

where

$$\bar{\eta}(t_i) \triangleq \mathrm{col}\{\eta(t_i), \varpi(t_i)\}.$$

From (7.17), we know $\mathcal{J}(t_k) < 0$ which gives that

$$V_1(t_k) < V(t_k) < \gamma^2 \sum_{i=0}^{k-1} \varpi^T(t_i)\varpi(t_i). \tag{7.20}$$

Moreover, it is known from (7.18) that

$$\tilde{z}^T(t_k)\tilde{z}(t_k) = e^T(t_k)\bar{H}^T\bar{H}e(t_k) < e^T(t_k)P_{r(t_k)}e(t_k). \tag{7.21}$$

Then, from (7.20) *and* (7.21), *one has*

$$\tilde{z}^T(t_k)\tilde{z}(t_k) < \gamma^2 \sum_{i=0}^{k-1} \varpi^T(t_i)\varpi(t_i).$$

Taking the supremum of $\tilde{z}^T(t_k)\tilde{z}(t_k)$ *over* k *and the limit of* $\sum_{i=0}^{k-1} \varpi^T(t_i)\varpi(t_i)$ *with* $k \to \infty$, *we have*

$$\sup_k \tilde{z}^T(t_k)\tilde{z}(t_k) < \gamma^2 \sum_{i=0}^{\infty} \varpi^T(t_i)\varpi(t_i)$$

which means

$$\sup_k \sqrt{\tilde{z}^T(t_k)\tilde{z}(t_k)} < \sqrt{\gamma^2 \sum_{k=0}^{\infty} \|\varpi(t_k)\|^2}.$$

The proof is thus complete.

In what follows, based on the results derived in Theorem 7.2, the l_2-l_∞ state estimator design problem is solved for a class of delayed ANNs under high-rate communication channels with RR protocol.

Theorem 7.3 *Let the disturbance attenuation level* $\gamma > 0$ *be given. The estimation error dynamic system* (7.12) *is exponentially stable with* $\varpi(t_k) = 0$ *and satisfies the* l_2-l_∞ *performance constraint* (7.13) *for all non-zero* $\varpi(t_k)$ *if there exist positive-definite matrices* P_{r_i} $(i \in \mathcal{P})$, Q *and matrices* Y_{r_i} $(i \in \mathcal{P})$ *such that*

$$\tilde{\Xi}_{r_i} \triangleq \begin{bmatrix} \tilde{\Xi}_{r_i}^{11} & 0 & \tilde{\Xi}_{r_i}^{13} & \tilde{\Xi}_{r_i}^{14} & 0 & \tilde{\Xi}_{r_i}^{16} \\ * & \Xi^{22} & 0 & 0 & 0 & 0 \\ * & * & \Xi_{r_i}^{33} & \Xi_{r_i}^{34} & \tilde{\Xi}_{r_i}^{35} & 0 \\ * & * & * & \Xi_{r_i}^{44} & \tilde{\Xi}_{r_i}^{45} & 0 \\ * & * & * & * & -\gamma^2 I & \tilde{\Xi}_{r_i}^{56} \\ * & * & * & * & * & -P_{r_{i+1}} \end{bmatrix} < 0, \tag{7.22}$$

$$\Sigma_{r_i} \triangleq \begin{bmatrix} -I & \bar{H} \\ * & -P_{r_i} \end{bmatrix} < 0 \quad (i \in \mathcal{P}) \tag{7.23}$$

where

$$\begin{aligned}
\tilde{\Xi}_{r_i}^{11} &\triangleq mQ + \bar{G}^T\bar{G} - P_{r_i}, \\
\tilde{\Xi}_{r_i}^{13} &\triangleq \bar{A}_{r_i}^T P_{r_{i+1}} \bar{B}_1 - \bar{C}_{r_i}^T Y_{r_i}^T \bar{B}_1, \\
\tilde{\Xi}_{r_i}^{14} &\triangleq \bar{A}_{r_i}^T P_{r_{i+1}} \bar{B}_2 \bar{L} - \bar{C}_{r_i}^T Y_{r_i}^T \bar{B}_2 \bar{L}, \\
\tilde{\Xi}_{r_i}^{16} &\triangleq \bar{A}_{r_i}^T P_{r_{i+1}} - \bar{C}_{r_i}^T Y_{r_i}^T, \\
\tilde{\Xi}_{r_i}^{35} &\triangleq \bar{B}_1^T P_{r_{i+1}} \bar{D}_{r_i} - \bar{B}_1^T Y_{r_i} \bar{E}_{r_i}, \\
\tilde{\Xi}_{r_i}^{45} &\triangleq \bar{L}^T \bar{B}_2^T P_{r_{i+1}} \bar{D}_{r_i} - \bar{L}^T \bar{B}_2^T Y_{r_i} \bar{E}_{r_i}, \\
\tilde{\Xi}_{r_i}^{56} &\triangleq \bar{D}_{r_i}^T P_{r_{i+1}} - \bar{E}_{r_i}^T Y_{r_i}^T
\end{aligned}$$

and $P_{r_{p+1}} = P_{r_1}$. Moreover, if (7.22) and (7.23) are feasible, the state estimator gains K_{r_i} are calculated as

$$K_{r_i} = P^{-1}_{r_{i+1}} Y_{r_i} \quad (i \in \mathcal{P}). \tag{7.24}$$

Proof *The inequality (7.17) can be rewritten as*

$$\bar{\Xi}_{r(t_k)} = \Upsilon^T_{r(t_k)} P^{-1}_{r(t_{k+1})} \Upsilon_{r(t_k)} + \Gamma_{r(t_k)} < 0$$

where

$$\Upsilon_{r(t_k)} \triangleq \begin{bmatrix} P_{r(t_{k+1})} \tilde{A}_{r(t_k)} & 0 & 0 & 0 & P_{r(t_{k+1})} \tilde{D}_{r(t_k)} \end{bmatrix},$$

$$\Gamma_{r(t_k)} \triangleq \begin{bmatrix} \tilde{\Xi}^{11}_{r(t_k)} & 0 & \Xi^{13}_{r(t_k)} & \Xi^{14}_{r(t_k)} & 0 \\ * & \Xi^{22} & 0 & 0 & 0 \\ * & * & \Xi^{33}_{r(t_k)} & \Xi^{34}_{r(t_k)} & \Xi^{35}_{r(t_k)} \\ * & * & * & \Xi^{44}_{r(t_k)} & \Xi^{45}_{r(t_k)} \\ * & * & * & * & -\gamma^2 I \end{bmatrix}$$

with

$$\Xi^{35}_{r(t_k)} \triangleq \bar{B}_1^T P_{r(t_{k+1})} \tilde{D}_{r(t_k)},$$
$$\Xi^{45}_{r(t_k)} \triangleq \bar{L}^T \bar{B}_2^T P_{r(t_{k+1})} \tilde{D}_{r(t_k)}.$$

By using Lemma 7.1, $\bar{\Xi}_{r(t_k)} < 0$ is equivalent to

$$\begin{bmatrix} \Gamma_{r(t_k)} & \Upsilon^T_{r(t_k)} \\ * & -P_{r(t_{k+1})} \end{bmatrix} < 0.$$

Letting $P_{r(t_{k+1})} K_{r(t_k)} = Y_{r(t_k)}$, we arrive at (7.22) which means that the estimation error dynamic system (7.12) is exponentially stable with $\varpi(t_k) = 0$ and satisfies the l_2-l_∞ performance constraint (7.13) for all non-zero $\varpi(t_k)$.

Moreover, it is known from $P_{r(t_{k+1})} K_{r(t_k)} = Y_{r(t_k)}$ that the estimator gains $K_{r(t_k)}$ can be obtained as (7.24). The proof is complete.

Remark 7.3 *A l_2-l_∞ state estimation scheme is developed in this chapter for the delayed ANNs under high-rate communication channels with RR protocol. In Theorems 7.1–7.2, sufficient conditions are derived that guarantee the exponential stability of the estimation error dynamics and the l_2-l_∞ performance constraint. Then, in Theorem 7.3, the desired estimator gains are obtained by solving two sets of LMIs. It can be seen that (7.22) and (7.23) include r_i, which represents the transmission behavior of the sensors in the high-rate communication channels with RR protocol, and all other system information. Note that, the state estimation algorithm obtained in this chapter can be easily specialized to the state estimation problem for systems without high-rate communication channels by setting $d = 1$.*

7.3 Simulation Examples

In this section, two numerical simulations are provided to show the effectiveness of the developed l_2-l_∞ state estimation algorithm for the delayed ANNs under high-rate communication channels with RR protocol.

Consider a 3-neuron ANN described by (7.1) with the following parameters:

$$\begin{aligned} a_1 =&0.36,\ a_2 = 0.35,\ a_3 = 0.34,\\ d_1 =&0.12,\ d_2 = 0.21,\ d_3 = 0.12,\\ B_1 =& \begin{bmatrix} 0.38 & -0.11 & 0.18 \\ 0.02 & 0.55 & -0.19 \\ 0.19 & -0.19 & 0.41 \end{bmatrix},\\ B_2 =& \begin{bmatrix} -0.32 & 0.23 & 0.32 \\ 0.14 & -0.21 & 0.13 \\ 0.19 & 0.14 & -0.21 \end{bmatrix},\\ H =& \begin{bmatrix} -0.31 & -0.21 & 0.19 \\ -0.02 & 0.12 & -0.32 \end{bmatrix}, \end{aligned}$$

and the time-delays of the neurons are set as $\tau_1 = 1$, $\tau_2 = 2$, and $\tau_3 = 1$, respectively.

The non-linear activation functions $f_i(\cdot)$ $(i = 1, 2, 3)$ are respectively chosen as

$$\begin{aligned} f_1(x_1(t_k)) &= \tanh(0.11x_1(t_k)),\\ f_2(x_2(t_k)) &= \tanh(0.10x_2(t_k)),\\ f_3(x_3(t_k)) &= \tanh(0.15x_3(t_k)). \end{aligned}$$

Then, it is obvious that $f(x(t_k))$ satisfies condition (7.3) with $G = \text{diag}\{0.11, 0.10, 0.15\}$.

Two sensors are used to measure the ANN with the parameters selected as:

$$\begin{aligned} C_1 &= \begin{bmatrix} 0.28 & 0.1 & 0.2 \\ 0.2 & 0.1 & 0.2 \end{bmatrix},\ E_1 = \begin{bmatrix} 0.15 \\ 0.2 \end{bmatrix},\\ C_2 &= \begin{bmatrix} 0.29 & 0.1 & 0.2 \\ 0.2 & 0.1 & 0.2 \end{bmatrix},\ E_2 = \begin{bmatrix} 0.1 \\ 0.15 \end{bmatrix}. \end{aligned}$$

Moreover, we set $t_0 = T_0 = 0$.

Example 7.1

In this example, we set $h_s = 3$, $h_t = 1$ and $d = 3$. Then, it is obtained that $p = 2$.

Based on the above parameters, by using the Matlab LMI toolbox, we obtain the optimal value of the attenuation level $\gamma = 0.06353$ and the estimator gains as:

$$K_{r_1} = \begin{bmatrix} -0.082 & 0.032 & 4.329 & -4.466 \\ -0.009 & 0.006 & 0.034 & -0.068 \\ -0.811 & 0.415 & 0.135 & -0.006 \\ -0.514 & 0.259 & 3.342 & -3.335 \\ -0.514 & 0.259 & 2.453 & -2.446 \\ -0.514 & 0.259 & 3.453 & -3.446 \\ -0.514 & 0.259 & 2.453 & -2.446 \\ 1 & 0 & 0 & 0 \\ 0 & 1 & 0 & 0 \\ 0 & 0 & 1 & 0 \\ 0 & 0 & 0 & 1 \end{bmatrix},$$

$$K_{r_2} = \begin{bmatrix} 1.555 & -0.883 & -0.066 & -0.007 \\ 0.447 & -0.251 & 0.058 & -0.060 \\ 0.751 & -0.294 & 0.068 & -0.004 \\ 1.700 & -0.851 & 0.011 & -0.008 \\ 1.401 & -0.702 & 0.022 & -0.016 \\ 1.738 & -0.870 & 0.010 & -0.007 \\ 1.401 & -0.702 & 0.022 & -0.016 \\ 1 & 0 & 0 & 0 \\ 0 & 1 & 0 & 0 \\ 0 & 0 & 1 & 0 \\ 0 & 0 & 0 & 1 \end{bmatrix}.$$

In the simulation, the disturbance inputs $w_i(t_k)$ $(i = 1,2,3)$ and the network-induced noises $v_i(T_{k'})$ $(i = 1,2)$ are

$$w_1(t_k) = \frac{0.5\sin(t_k)}{0.52t_k + 1}, \quad w_2(t_k) = \frac{0.7\sin(t_k)}{0.35t_k + 1},$$
$$w_3(t_k) = \frac{0.6\sin(t_k)}{0.15t_k + 1}, \quad v_1(T_{k'}) = \frac{0.2\cos(T_{k'})}{0.15T_k' + 1},$$
$$v_2(T_{k'}) = \frac{0.3\cos(T_{k'})}{0.65T_k' + 1},$$

respectively. The initial conditions are taken as $\phi(t_s) = \begin{bmatrix}0.1 & 0.2 & -0.1\end{bmatrix}^T$ $(s \in \mathcal{T})$. The simulation results are shown in Figs. 7.1–7.4. The state trajectories of the neurons and their corresponding estimates are given in Figs. 7.1–7.3. Fig. 7.4 depicts the output estimation error $\tilde{z}(t_k)$. It can be seen from Figs. 7.1–7.3 that the estimates follow the states of the neurons well and from Fig. 7.4 that the estimation error approaches zero fast. Moreover, to further verify the performance of the proposed estimator, the output estimation error $\tilde{z}(t_k)$ when the initial conditions are zero is given in Fig. 7.5, which also

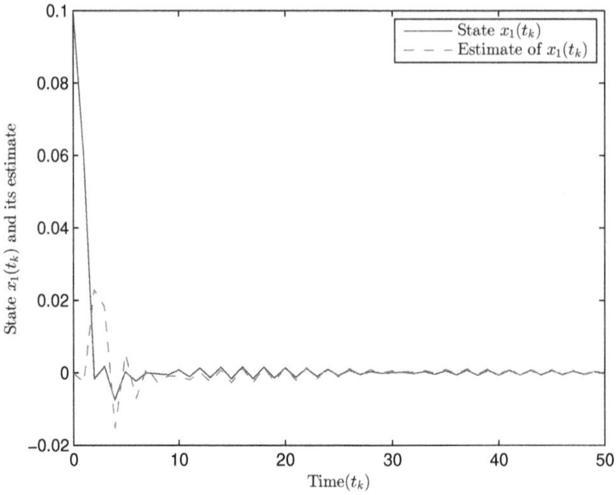

FIGURE 7.1: State $x_1(t_k)$ and its estimate when $d = 3$.

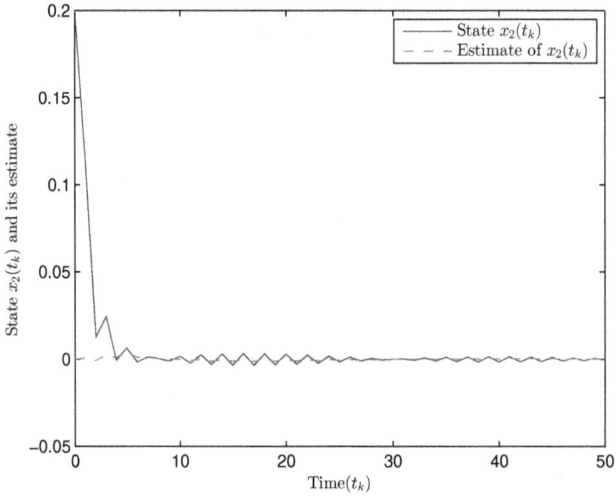

FIGURE 7.2: State $x_2(t_k)$ and its estimate when $d = 3$.

shows that the estimation error approaches zero fast. The simulation results demonstrate that the developed state estimation algorithm is effective.

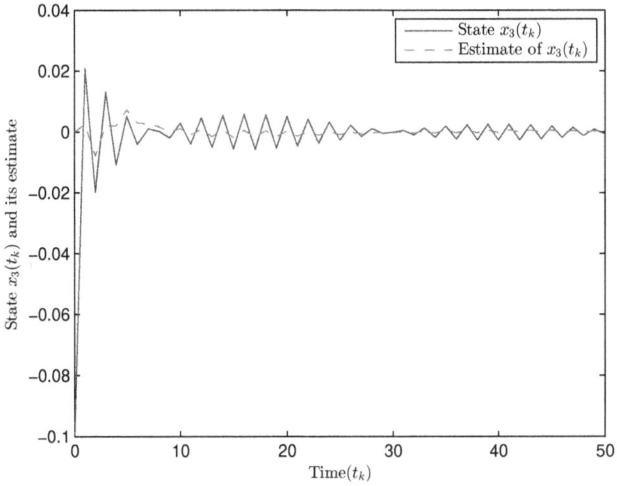

FIGURE 7.3: State $x_3(t_k)$ and its estimate when $d = 3$.

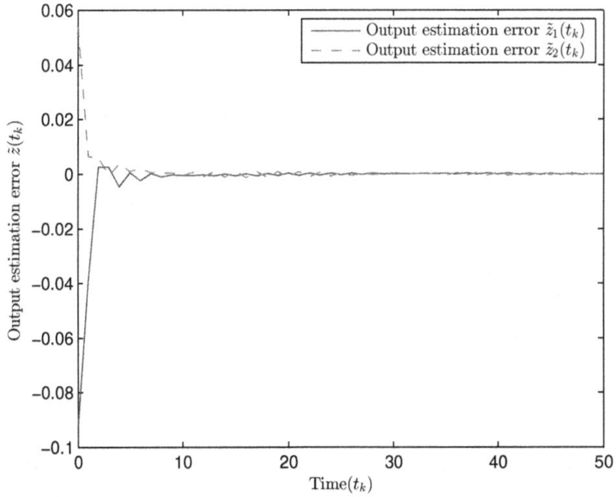

FIGURE 7.4: Output estimation error $\tilde{z}(t_k)$ when $d = 3$.

Example 7.2

To discuss the influence of h_t and d on the performance of the proposed state estimation algorithm, in this example, we set $h_s = 3$, $h_t = 3$, and $d = 1$. Then, it is obtained that $p = 2$. Other parameters are all the same with the corresponding ones in Example 7.1.

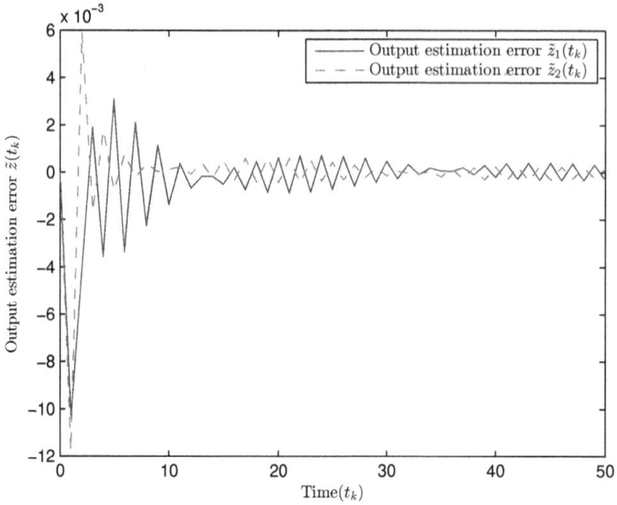

FIGURE 7.5: Output estimation error $\tilde{z}(t_k)$ when $d = 3$ with zero initial conditions.

By using the Matlab LMI toolbox, we obtain the optimal value of the attenuation level $\gamma = 0.06337$. Simulation result is given in Fig. 7.6, where only the estimate of state $x_1(t_k)$ is given for saving the space.

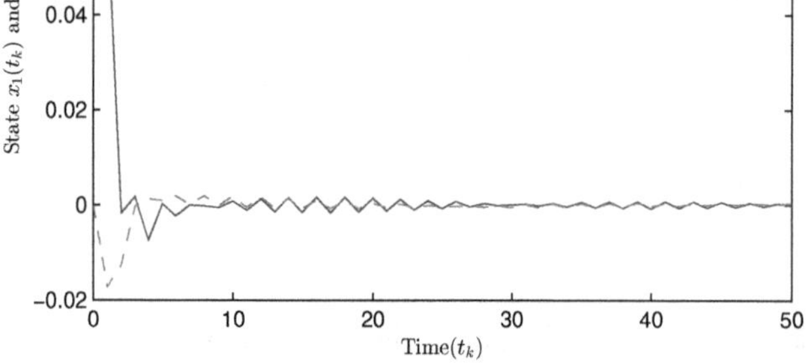

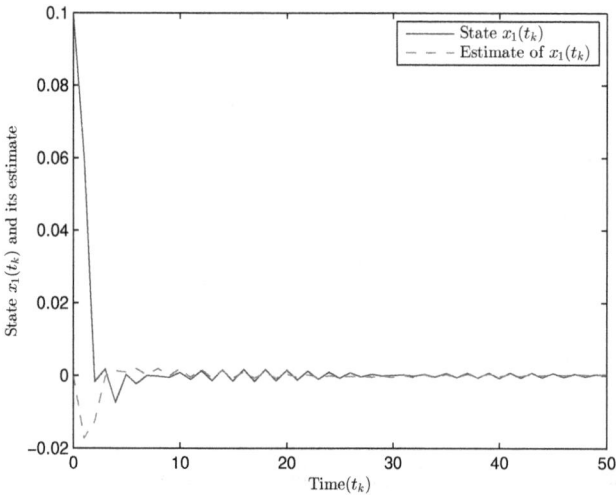

FIGURE 7.6: State $x_1(t_k)$ and its estimate when $d = 1$.

From Fig. 7.6, it can be seen that the values of h_t and d have no (or little) influence on the estimation performance. Although the number of the signal transmissions in one sampling period increases, there is no more useful information being provided.

7.4 Conclusion

In this chapter, the l_2-l_∞ state estimation problem has been studied for a class of delayed ANNs under high-rate communication channels with RR protocol. A periodic sequence has been introduced to account for the scheduling effect of the RR protocol in the high-rate communication channels. Sufficient conditions have been developed that ensure the exponential stability of the estimation error dynamics and the l_2-l_∞ disturbance attenuation capacity of the proposed state estimator. By solving two sets of LMIs, the explicit expressions of the estimator gains have been obtained. Finally, the effectiveness of the proposed l_2-l_∞ state estimation scheme has been verified by a numerical example.

8

Recursive State Estimation for Multi-Rate Systems with Distributed Time-Delays under Round-Robin Protocol

Time-delays are unavoidably encountered in many applications. Recently, distributed time-delays, which occur frequently in practical engineering, have received considerable research attention and many results have been available for single-rate systems. For example, the event-triggered H_∞ state estimation problem has been studied in [89] for single-rate systems with distributed time-delays. In [106], the robust H_∞ control problem has been investigated for uncertain non-linear single-rate systems with distributed time-delays. Nevertheless, for MRSs with distributed time-delays, such as power systems with load frequency control schemes [114], the corresponding results have been very few. Up to now, it remains an open yet challenging problem as how to characterize the impact from the distributed time-delays on the estimation/control performance of the MRSs, and this constitutes the first motivation of this chapter to look into the recursive state estimation problem for MRSs with distributed time-delays.

In networked systems, information exchanges among system components (e.g. sensors, actuators, estimators, and controllers) are executed typically through shared communication networks, and the limited-bandwidth-induced data collision is a fairly common phenomenon that is likely to occur when multiple components launch the signal transmissions simultaneously. In industry, communication protocol has proven to be an effective tool for avoiding the data collision by scheduling the sequence of signal transmissions. Under the scheduling of the communication protocol, only one component can obtain the access to transmit its information at each sampling instant. Among the periodic communication protocols, the RR protocol orchestrates the transmission sequence of the components according to a fixed circular order. Recently, the RR protocol has drawn much research attention and a large number of results have been available on the estimation/control problem for single-rate systems. Nevertheless, the recursive state estimation problem for MRSs with RR protocol has not been fully investigated due mainly to the fundamental difficulties in 1) examining the effect of the RR protocol on the estimation performance for MRSs; and 2) developing a recursive state estimation scheme guaranteeing a locally minimized upper bound on the estimation error

DOI: 10.1201/9781032619507-8

covariance. The second motivation of this chapter is to deal with these two identified difficulties.

In this chapter, the recursive state estimation problem is studied for MRSs with distributed time-delays under RR protocol. The contributions of this chapter are outlined as follows: 1) the recursive state estimation problem is, for the first time, addressed for MRSs with distributed time-delays where the RR protocol is implemented to schedule the sensors; 2) a new method is proposed to model the measurement received by the estimator from sensors scheduled by the RR protocol; and 3) a recursive state estimation scheme is designed which ensures a locally minimized upper bound for the estimation error covariance under the impacts of distributed time-delays, multi-rate sampling and RR protocol.

8.1 Problem Formulation

Consider a class of discrete time-delayed systems described by

$$\begin{aligned} x(T_{k+1}) =& A(T_k)x(T_k) + B(T_k)\sum_{j=1}^{p} \mu_j x(T_k - jh) \\ &+ f\big(x(T_k), \eta(T_k)\big) + E(T_k)w(T_k) \end{aligned} \tag{8.1}$$

where $x(T_k) \in \mathbb{R}^m$ is the system state, $w(T_k) \in \mathbb{R}^w$ is the process noise with zero mean and covariance $W(T_k) > 0$, $\eta(T_k)$ is zero-mean Gaussian noise sequence, $h \triangleq T_{k+1} - T_k$ is a positive constant that represents the state update period of the system, $A(T_k)$, $B(T_k)$, and $E(T_k)$, are known matrices with appropriate dimensions, $\mu_j \geq 0$ $(j = 1, 2, \ldots, p)$ are given positive scalars, and p is a known positive integer. The initial values $x(T_d)$ $(d = -p, -p + 1, \ldots, 0)$ are mutually independent random variables with the means $\bar{x}(T_d)$. $f(x(T_k), \eta(T_k))$ is a stochastic non-linear function that satisfies $f(0, \eta(T_k)) = 0$ with the following properties:

$$\begin{aligned} &\mathbb{E}\{f\big(x(T_k),\eta(T_k)\big)|x(T_k)\} = 0, \\ &\mathbb{E}\{f\big(x(T_k),\eta(T_k)\big)f^T\big(x(T_j), \eta(T_j)\big)|x(T_k)\} \\ &\qquad = \sum_{i=1}^{a} \Pi_i(T_k)x^T(T_k)\Phi_i(T_k)x(T_k)\delta(k - j) \end{aligned} \tag{8.2}$$

where a is a known positive integer, $\Pi_i(T_k)$ and $\Phi_i(T_k)$ $(i = 1, 2, \ldots, a)$ are known matrices with appropriate dimensions.

Remark 8.1 *In networked systems, due to randomly fluctuated network conditions, the nonlinearity may occur in a random fashion. The so-called stochastic nonlinearity can appropriately describe such a phenomenon [63].*

The first condition in (8.2) *is the first-order statistic characteristic of the stochastic nonlinearity* $f(x(T_k), \eta(T_k))$, *while the second condition in* (8.2) *is the second-order statistic characteristic of the stochastic nonlinearity* $f(x(T_k), \eta(T_k))$.

The system is measured by M sensors with a sampling period $bh \triangleq t_{k+1} - t_k$ where b is a known positive integer. For sensor i ($i = 1, 2, \ldots, M$), the measurement model is described as

$$y_i(t_k) = C_i(t_k)x(t_k) + D_i(t_k)v_i(t_k) \tag{8.3}$$

where $y_i(t_k) \in \mathbb{R}^n$ is the measurement output of the ith sensor and $v_i(t_k) \in \mathbb{R}^v$ is the measurement noise on sensor i with zero mean and covariance $V_i(t_k) > 0$. $C_i(t_k)$ and $D_i(t_k)$ are known matrices with appropriate dimensions.

Assumption 8.1 $x(T_d)$ $(d = -p, -p+1, \ldots, 0)$, $w(T_k)$, $v_i(t_k)$ $(i = 1, 2, \ldots, M)$ *and* $\eta(T_k)$ *are mutually independent random variables.*

In this chapter, the measurement outputs of the sensors are transmitted to the estimator through a shared communication network of limited bandwidth. The RR protocol is utilized to orchestrate the transmission order of the sensors for the purpose of preventing data from collision. Let $\bar{y}(t_k) = \text{col}\{\bar{y}_1(t_k), \bar{y}_2(t_k), \cdots, \bar{y}_M(t_k)\}$ denote the actually received measurement by the estimator with a zero-order holder strategy where $\bar{y}_i(t_k)$ represents the actually received measurement from sensor i. According to the RR protocol, we have

$$\bar{y}_i(t_k) = \begin{cases} y_i(t_k), & \text{if } \text{mod}(k-i, M) = 0, \\ \bar{y}_i(t_{k-1}), & \text{otherwise.} \end{cases} \tag{8.4}$$

By denoting $y(t_k) \triangleq \text{col}\{y_1(t_k), y_2(t_k), \cdots, y_M(t_k)\}$, we obtain

$$y(t_k) = \bar{C}(t_k)x(t_k) + \bar{D}(t_k)\bar{v}(t_k) \tag{8.5}$$

where

$$\begin{aligned} \bar{C}(t_k) &\triangleq \text{col}\{C_1(t_k), C_2(t_k), \cdots, C_M(t_k)\}, \\ \bar{D}(t_k) &\triangleq \text{diag}\{D_1(t_k), D_2(t_k), \cdots, D_M(t_k)\}, \\ \bar{v}(t_k) &\triangleq \text{col}\{v_1(t_k), v_2(t_k), \cdots, v_M(t_k)\}. \end{aligned}$$

According to (8.4), the measurement $\bar{y}(t_k)$ is written as

$$\bar{y}(t_k) = \sum_{l=0}^{M-1} \Gamma(t_{k-l})y(t_{k-l}) \tag{8.6}$$

where

$$\Gamma(t_k) \triangleq \text{diag}\{\delta(\zeta(t_k) - 1)I, \delta(\zeta(t_k) - 2)I, \cdots, \delta(\zeta(t_k) - M)I\}$$

with $\zeta(t_k) \triangleq \text{mod}(k-1, M) + 1$ being the sensor which gets the access to transmission at time instant t_k. Moreover, we set $\zeta(t_k) = M + k$ for $k = -M+1, -M+2, \cdots, 0$ and $y(t_k) = y(t_0)$ for $k = -M+1, -M+2, \cdots, -1$.

Remark 8.2 *Under the scheduling of the RR protocol, the transmissions of the information of the sensors are executed in a fixed circular sequence. To compensate measurement $\bar{y}_i(t_k)$, in case that the sensor i does not get the access to transmission, the zero-order holder strategy is implemented to generate measurement $\bar{y}_i(t_k)$. Under the RR protocol and the zero-order holder strategy, at time instant t_{k-l}, only the component $\bar{y}_{\zeta(t_{k-l})}(t_{k-l})$ of the actually received measurement $\bar{y}(t_{k-l})$ is updated and other components remain the same as those corresponding ones in $\bar{y}(t_{k-l-1})$. According to the fixed circular sequence of the transmission order, $\bar{y}(t_k)$ can be seen as the combination of $\bar{y}_{\zeta(t_k)}(t_k), \bar{y}_{\zeta(t_{k-1})}(t_{k-1}), \cdots, \bar{y}_{\zeta(t_{k-M-1})}(t_{k-M-1})$ that is expressed as* (8.6).

Remark 8.3 *In [156, 215], the measurement after transmitted through communication network under RR protocol is expressed as $\Gamma(t_k)y(t_k) + (I - \Gamma(t_{k-1}))\bar{y}(t_{k-1})$. With such an expression, in the recursive state estimation problem, the derivation of the upper bound for the estimation error covariance requires extra efforts for the following two reasons: 1) an augmentation of $x(t_k)$ and $\bar{y}(t_{k-1})$ is needed to avoid the tedious calculations of $\mathbb{E}\{\bar{y}(t_{k-1})\bar{y}^T(t_{k-1})\}, \mathbb{E}\{\bar{y}(t_{k-2})\bar{y}^T(t_{k-2})\}$, $\mathbb{E}\{\bar{y}(t_{k-3})\bar{y}^T(t_{k-3})\}, \cdots$, which result from the iteration of $\bar{y}(t_k)$ and 2) the introduction of an identity matrix is required to ensure the invertibility of a certain matrix which is vital in obtaining the estimator gain. In contrast to [156, 215], in this chapter, the measurement after transmission is written as* (8.6) *which avoids the iteration of $\bar{y}(t_k)$ and guarantees the invertibility of the aforementioned matrix, and the computational complexity is therefore largely reduced. Moreover, as shown later, extra efforts are needed to the calculation of the estimation error covariance with the measurement model* (8.6).

It can be seen from (8.1) and (8.3) that the state update period of the system and the sampling period of the sensors are allowed to be different, the system under consideration is actually a MRS. Now, for convenience of later analysis, let us convert the MRS to a single-rate one. By simply denoting $\sum_{j=1}^{p} \mu_j x(T_k - jh)$ as $\check{x}(T_k)$ and iteratively using (8.1), we have

$$
\begin{aligned}
x(t_{k+1}) =& \bar{A}(t_k)x(t_k) + \bar{B}(t_k)\tilde{x}(t_k) \\
&+ \vec{A}(t_k)\bar{f}(x(t_k), \eta(t_k)) + \bar{E}(t_k)\bar{w}(t_k)
\end{aligned} \tag{8.7}
$$

where

$$
\begin{aligned}
\mathscr{A}_j(t_k) \triangleq& \prod_{i=1}^{j} A(t_k + (b-i)h), \quad \bar{A}(t_k) \triangleq \mathscr{A}_b(t_k), \\
\vec{A}(t_k) \triangleq& \begin{bmatrix} \mathscr{A}_{b-1}(t_k) & \mathscr{A}_{b-2}(t_k) & \cdots & I \end{bmatrix},
\end{aligned}
$$

$$\begin{aligned}
\vec{B}(t_k) \triangleq& \mathrm{diag}\{B(t_k), B(t_k+h), \cdots, B(t_k+(b-1)h)\},\\
\vec{E}(t_k) \triangleq& \mathrm{diag}\{E(t_k), E(t_k+h), \cdots, E(t_k+(b-1)h)\},\\
\bar{B}(t_k) \triangleq& \vec{A}(t_k)\vec{B}(t_k), \quad \bar{E}(t_k) \triangleq \vec{A}(t_k)\vec{E}(t_k),\\
\tilde{x}(t_k) \triangleq& \mathrm{col}\{\breve{x}(t_k), \breve{x}(t_k+h), \cdots, \breve{x}(t_k+(b-1)h)\},\\
\bar{w}(t_k) \triangleq& \mathrm{col}\{w(t_k), w(t_k+h), \cdots, w(t_k+(b-1)h)\},\\
\bar{f}(x(t_k), \eta(t_k)) \triangleq& \mathrm{col}\{f(x(t_k), \eta(t_k)), f(x(t_k+h), \eta(t_k+h)),\\
& \cdots, f(x(t_k+(b-1)h), \eta(t_k+(b-1)h))\}.
\end{aligned}$$

Remark 8.4 *The term $\tilde{x}(t_k)$ in (8.7) results from the coupling between the distributed time-delays (shown as $\sum_{j=1}^{p} \mu_j x(T_k - jh)$) and the multi-rate sampling. As can be seen later, the calculation of $\mathbb{E}\{\tilde{x}(t_k)\tilde{x}^T(t_k)\}$, which is brought from the term $\tilde{x}(t_k)$, would involve many cross-terms that are difficult to handle. As such, there is a theoretical need to develop a novel yet effective method to overcome the identified difficulty.*

From (8.6) and (8.7), we have the following single-rate system in a compact form

$$\begin{cases}
x(t_{k+1}) =& \bar{A}(t_k)x(t_k) + \bar{B}(t_k)\tilde{x}(t_k)\\
& + \vec{A}(t_k)\bar{f}(x(t_k), \eta(t_k)) + \bar{E}(t_k)\bar{w}(t_k),\\
\bar{y}(t_k) =& \sum_{l=0}^{M-1} \Gamma(t_{k-l})\bar{C}(t_{k-l})x(t_{k-l})\\
& + \sum_{l=0}^{M-1} \Gamma(t_{k-l})\bar{D}(t_{k-l})\bar{v}(t_{k-l}).
\end{cases} \tag{8.8}$$

The estimator of the following structure is adopted to estimate the state of the system

$$\hat{x}(t_{k+1}) = \bar{A}(t_k)\hat{x}(t_k) + K(t_k)\Big(\bar{y}(t_k) - \sum_{l=0}^{M-1} \Gamma(t_{k-l})\bar{C}(t_{k-l})\hat{x}(t_{k-l})\Big) \tag{8.9}$$

where $\hat{x}(t_k)$ is the estimate of the system state with the initial values $\hat{x}(t_d) = \mathbb{E}\{x(t_d)\}$ ($d = -\lfloor \frac{p}{b} \rfloor, -\lfloor \frac{p}{b} \rfloor + 1, \cdots, 0$) and $K(t_k)$ is the estimator parameter to be determined.

From (8.8) and (8.9), the estimation error $e(t_k) \triangleq x(t_k) - \hat{x}(t_k)$ is calculated as

$$\begin{aligned}
e(t_{k+1}) =& \tilde{A}(t_k)e(t_k) + \bar{B}(t_k)\tilde{x}(t_k)\\
& + \vec{A}(t_k)\bar{f}(x(t_k), \eta(t_k)) + \bar{E}(t_k)\bar{w}(t_k)\\
& - K(t_k)\mathcal{V}(t_k) - K(t_k)\mathcal{E}(t_k)
\end{aligned} \tag{8.10}$$

where

$$\begin{aligned}\tilde{A}(t_k) \triangleq& \bar{A}(t_k) - K(t_k)\Gamma(t_k)\bar{C}(t_k),\\ \mathcal{V}(t_k) \triangleq& \sum_{l=0}^{M-1} \Gamma(t_{k-l})\bar{D}(t_{k-l})\bar{v}(t_{k-l}),\\ \mathcal{E}(t_k) \triangleq& \sum_{l=1}^{M-1} \Gamma(t_{k-l})\bar{C}(t_{k-l})e(t_{k-l}).\end{aligned}$$

The purpose of this chapter is to construct an estimator (8.9) under which an upper bound is ensured on the estimation error covariance $P(t_k) \triangleq \mathbb{E}\{e(t_k)e^T(t_k)\}$ and characterize the estimator gain $K(t_k)$ that minimizes such an upper bound.

8.2 Estimator Design

In this section, an upper bound on the state covariance $X(T_k) \triangleq \mathbb{E}\{x(T_k)x^T(T_k)\}$ is first parameterized. Then, an upper bound is given on the estimation error covariance $P(t_k)$ and the estimator gain that minimizes such an upper bound is obtained.

Noting that the estimation error dynamics (8.10) includes the state of the system, an upper bound on the state covariance is derived in the following lemma to facilitate our later analysis.

Lemma 8.1 *For a given positive scalar $\epsilon(T_k)$, the state covariance $X(T_k)$ has an upper bound $\Psi(T_k)$ with initial values $\Psi(T_d) = X(T_d)$ $(d = -p, -p+1, \cdots, 0)$, and $\Psi(T_k)$ can be obtained by solving*

$$\begin{aligned}\Psi(T_{k+1}) =& \big(1+\epsilon(T_k)\big)A(T_k)\Psi(T_k)A^T(T_k)\\ &+ \big(1+\epsilon^{-1}(T_k)\big)B(T_k)\bar{\Theta}(T_k)B^T(T_k)\\ &+ \sum_{i=1}^{a} \Pi_i(T_k)\mathrm{tr}\big(\Psi(T_k)\Phi_i(T_k)\big)\\ &+ E(T_k)W(T_k)E^T(T_k)\end{aligned} \tag{8.11}$$

where

$$\bar{\Theta}(T_k) \triangleq (p-1)\sum_{i=1}^{p} \mu_i^2\Psi(T_{k-i}).$$

Proof *The induction method is used in the proof. First, from the initial values, $X(T_d) \le \Psi(T_d)$ $(d = -p, -p+1, \cdots, 0)$ are guaranteed. Next, by assuming $X(T_k) \le \Psi(T_k)$, we are ready to prove $X(T_{k+1}) \le \Psi(T_{k+1})$.*

From (8.1), we know that

$$\begin{aligned}X(T_{k+1}) =&\mathbb{E}\{A(T_k)x(T_k)x^T(T_k)A^T(T_k)\\&+A(T_k)x(T_k)\check{x}^T(T_k)B^T(T_k)\\&+B(T_k)\check{x}(T_k)x^T(T_k)A^T(T_k)\\&+B(T_k)\check{x}(T_k)\check{x}^T(T_k)B^T(T_k)\\&+f\big(x(T_k),\eta(T_k)\big)f^T\big(x(T_k),\eta(T_k)\big)\\&+E(T_k)w(T_k)w^T(T_k)E^T(T_k)\}.\end{aligned}$$

By using the elementary inequality $ab^T + ba^T \le aa^T + bb^T$ (where a and b are vectors of appropriate dimensions) and noting that $\check{x}(T_k) = \sum_{j=1}^{p}\mu_j x(T_{k-j})$, it is derived that

$$\begin{aligned}\mathbb{E}\{\check{x}(T_k)\check{x}^T(T_k)\} =&\sum_{i=1}^{p}\sum_{j=1}^{p}\mathbb{E}\{\mu_i\mu_j x(T_{k-i})x^T(T_{k-j})\}\\=&\frac{1}{2}\sum_{i=1}^{p}\sum_{j=1}^{p}\mathbb{E}\{\mu_i\mu_j x(T_{k-i})x^T(T_{k-j})\\&+\mu_j\mu_i x(T_{k-j})x^T(T_{k-i})\}\\\le&\frac{1}{2}\sum_{i=1}^{p}\sum_{j=1}^{p}\mathbb{E}\{\mu_i^2 x(T_{k-i})x^T(T_{k-i})\\&+\mu_j^2 x(T_{k-j})x^T(T_{k-j})\} = \Theta(T_k)\end{aligned} \tag{8.12}$$

where

$$\Theta(T_k) \triangleq (p-1)\sum_{i=1}^{p}\mu_i^2 X(T_{k-i}).$$

Then, we have

$$\begin{aligned}X(T_{k+1}) \le&(1+\epsilon(T_k))A(T_k)\mathbb{E}\{x(T_k)x^T(T_k)\}A^T(T_k)\\&+\big(1+\epsilon^{-1}(T_k)\big)B(T_k)\mathbb{E}\{\check{x}(T_k)\check{x}^T(T_k)\}B^T(T_k)\\&+\mathbb{E}\{f\big(x(T_k),\eta(T_k)\big)f^T\big(x(T_k),\eta(T_k)\big)\}\\&+E(T_k)\mathbb{E}\{w(T_k)w^T(T_k)\}E^T(T_k)\\=&(1+\epsilon(T_k))A(T_k)X(T_k)A^T(T_k)\\&+\big(1+\epsilon^{-1}(T_k)\big)B(T_k)\Theta(T_k)B^T(T_k)\\&+\sum_{i=1}^{a}\Pi_i(T_k)\mathrm{tr}\big(X(T_k)\Phi_i(T_k)\big)\\&+E(T_k)W(T_k)E^T(T_k)\end{aligned}$$

$$
\begin{aligned}
&\leq \big(1+\epsilon(T_k)\big)A(T_k)\Psi(T_k)A^T(T_k) \\
&\quad + \big(1+\epsilon^{-1}(T_k)\big)B(T_k)\bar{\Theta}(T_k)B^T(T_k) \\
&\quad + \sum_{i=1}^{a}\Pi_i(T_k)\mathrm{tr}\big(\Psi(T_k)\Phi_i(T_k)\big) \\
&\quad + E(T_k)W(T_k)E^T(T_k)
\end{aligned}
$$

which means $X(T_{k+1}) \leq \Psi(T_{k+1})$. *The proof is now complete.*

The following theorem presents an upper bound on the estimation error covariance with the help of Lemma 8.1.

Theorem 8.1 *For given positive scalars* $\varepsilon_i(t_k)$ $(i = 1, 2, \ldots, 7)$, *consider the following matrix equation*

$$
\begin{aligned}
\Sigma(t_{k+1}) =&\kappa_1(t_k)\tilde{A}(t_k)\Sigma(t_k)\tilde{A}^T(t_k) + \kappa_2(t_k)\bar{B}(t_k)\bar{\Lambda}(t_k)\bar{B}^T(t_k) \\
&+ \kappa_3(t_k)\sum_{i=1}^{a}\vec{A}(t_k)\bar{\Pi}_i(t_k)\tilde{\Phi}_i(t_k)\vec{A}^T(t_k) \\
&+ \kappa_4(t_k)\bar{E}(t_k)\bar{W}(t_k)\bar{E}^T(t_k) \\
&+ \kappa_5(t_k)M\sum_{i=0}^{M-1}K(t_k)\Gamma(t_{k-i})\bar{D}(t_{k-i})\bar{V}(t_{k-i}) \\
&\quad \times \bar{D}^T(t_{k-i})\Gamma^T(t_{k-i})K^T(t_k) \\
&+ \kappa_6(t_k)(M-1)\sum_{i=1}^{M-1}K(t_k)\Gamma(t_{k-i})\bar{C}(t_{k-i})\Sigma(t_{k-i}) \\
&\quad \times \bar{C}^T(t_{k-i})\Gamma^T(t_{k-i})K^T(t_k)
\end{aligned}
\tag{8.13}
$$

where

$$
\begin{aligned}
\bar{\Lambda}(t_k) &\triangleq \sum_{i=0}^{b-1}\mathrm{tr}\big(\bar{\Theta}(t_k+ih)\big)I, \\
\bar{\Pi}_i(t_k) &\triangleq \mathrm{diag}\{\Pi_i(t_k), \Pi_i(t_k+h), \cdots, \Pi_i(t_k+(b-1)h)\}, \\
\tilde{\Phi}_i(t_k) &\triangleq \mathrm{diag}\{\mathrm{tr}\big(\Psi(t_k)\Phi_i(t_k)\big)I, \mathrm{tr}\big(\Psi(t_k+h)\Phi_i(t_k+h)\big)I, \\
&\qquad \cdots, \mathrm{tr}\big(\Psi(t_k+(b-1)h)\Phi_i(t_k+(b-1)h)\big)I\}, \\
\bar{W}(t_k) &\triangleq \mathrm{diag}\{W(t_k), W(t_k+h), \cdots, W(t_k+(b-1)h)\}, \\
\bar{V}(t_k) &\triangleq \mathrm{diag}\{V_1(t_k), V_1(t_k), \cdots, V_M(t_k)\}, \\
\kappa_1(t_k) &\triangleq 1+\varepsilon_1(t_k)+\varepsilon_2(t_k)+\varepsilon_3(t_k), \\
\kappa_2(t_k) &\triangleq 1+\varepsilon_1^{-1}(t_k)+\varepsilon_4(t_k)+\varepsilon_5(t_k)+\varepsilon_6(t_k), \\
\kappa_3(t_k) &\triangleq 1+\varepsilon_4^{-1}(t_k), \quad \kappa_4(t_k) \triangleq 1+\varepsilon_5^{-1}(t_k), \\
\kappa_5(t_k) &\triangleq 1+\varepsilon_2^{-1}(t_k)+\varepsilon_7(t_k),
\end{aligned}
$$

$$\kappa_6(t_k) \triangleq 1+\varepsilon_3^{-1}(t_k)+\varepsilon_6^{-1}(t_k)+\varepsilon_7^{-1}(t_k)$$

with initial values $\Sigma(t_d) = X(t_d)$ $(d = -\lfloor \frac{p}{b} \rfloor, -\lfloor \frac{p}{b} \rfloor + 1, \cdots, 0)$. *Then, the solution* $\Sigma(t_k)$ *to* (8.13) *is an upper bound on* $P(t_k)$, *that is,* $P(t_k) \leq \Sigma(t_k)$.

Proof *Again, we will use the induction method to carry out the proof. First, the initial values guarantee* $P(t_d) \leq \Sigma(t_d)$ $(d = -\lfloor \frac{p}{b} \rfloor, -\lfloor \frac{p}{b} \rfloor + 1, \cdots, 0)$. *Assuming that* $P(t_k) \leq \Sigma(t_k)$ *holds, what we need to do is to prove* $P(t_{k+1}) \leq \Sigma(t_{k+1})$.

It is straightforward to see from (8.10) *that*

$$\begin{aligned}
P(t_{k+1}) =& \mathbb{E}\{\tilde{A}(t_k)e(t_k)e^T(t_k)\tilde{A}^T(t_k)\} \\
&+ \mathbb{E}\{\bar{B}(t_k)\tilde{x}(t_k)\tilde{x}^T(t_k)\bar{B}^T(t_k)\} \\
&+ \mathbb{E}\{\vec{A}(t_k)\bar{f}(x(t_k),\eta(t_k))\bar{f}^T(x(t_k),\eta(t_k))\vec{A}^T(t_k)\} \\
&+ \mathbb{E}\{\bar{E}(t_k)\bar{w}(t_k)\bar{w}^T(t_k)\bar{E}^T(t_k)\} \\
&+ \mathbb{E}\{K(t_k)\mathcal{V}(t_k)\mathcal{V}^T(t_k)K^T(t_k)\} \\
&+ \mathbb{E}\{K(t_k)\mathcal{E}(t_k)\mathcal{E}^T(t_k)K^T(t_k)\} \\
&+ \Upsilon_1(t_k) + \Upsilon_1^T(t_k) - \Upsilon_2(t_k) - \Upsilon_2^T(t_k) \\
&- \Upsilon_3(t_k) - \Upsilon_3^T(t_k) + \Upsilon_4(t_k) + \Upsilon_4^T(t_k) \\
&+ \Upsilon_5(t_k) + \Upsilon_5^T(t_k) - \Upsilon_6(t_k) - \Upsilon_6^T(t_k) \\
&+ \Upsilon_7(t_k) + \Upsilon_7^T(t_k)
\end{aligned}$$

where

$$\begin{aligned}
\Upsilon_1(t_k) &\triangleq \mathbb{E}\{\tilde{A}(t_k)e(t_k)\tilde{x}^T(t_k)\bar{B}^T(t_k)\}, \\
\Upsilon_2(t_k) &\triangleq \mathbb{E}\{\tilde{A}(t_k)e(t_k)\mathcal{V}^T(t_k)K^T(t_k)\}, \\
\Upsilon_3(t_k) &\triangleq \mathbb{E}\{\tilde{A}(t_k)e(t_k)\mathcal{E}^T(t_k)K^T(t_k)\}, \\
\Upsilon_4(t_k) &\triangleq \mathbb{E}\{\bar{B}(t_k)\tilde{x}(t_k)\bar{f}^T(x(t_k),\eta(t_k))\vec{A}^T(t_k)\}, \\
\Upsilon_5(t_k) &\triangleq \mathbb{E}\{\bar{B}(t_k)\tilde{x}(t_k)\bar{w}^T(t_k)\bar{E}^T(t_k)\}, \\
\Upsilon_6(t_k) &\triangleq \mathbb{E}\{\bar{B}(t_k)\tilde{x}(t_k)\mathcal{E}^T(t_k)K^T(t_k)\}, \\
\Upsilon_7(t_k) &\triangleq \mathbb{E}\{K(t_k)\mathcal{V}(t_k)\mathcal{E}^T(t_k)K^T(t_k)\}.
\end{aligned}$$

From the properties (8.2) *of the stochastic non-linear function* $f(x(T_k), \eta(T_k))$, *one has*

$$\begin{aligned}
&\mathbb{E}\{\bar{f}(x(t_k),\eta(t_k))\bar{f}^T(x(t_k),\eta(t_k))\} \\
=&\sum_{i=1}^{a} \mathrm{diag}\{\Pi_i(t_k)\mathrm{tr}(X(t_k)\Phi_i(t_k)), \Pi_i(t_k+h)\mathrm{tr}(X(t_k+h)\Phi_i(t_k+h)), \cdots, \\
&\Pi_i(t_k+(b-1)h)\mathrm{tr}(X(t_k+(b-1)h)\Phi_i(t_k+(b-1)h))\} \\
=&\sum_{i=1}^{a} \bar{\Pi}_i(t_k)\bar{\Phi}_i(t_k)
\end{aligned}$$

where

$$\bar{\Phi}_i(t_k) \triangleq \text{diag}\{\text{tr}\big(X(t_k)\Phi_i(t_k)\big)I, \text{tr}\big(X(t_k+h)\Phi_i(t_k+h)\big)I,$$
$$\cdots, \text{tr}\big(X(t_k+(b-1)h)\Phi_i(t_k+(b-1)h)\big)I\}.$$

Moreover, it is obvious that

$$\begin{aligned}\mathbb{E}\{\bar{w}(t_k)\bar{w}^T(t_k)\} =&\text{diag}\{\mathbb{E}\{w(t_k)w^T(t_k)\}, \mathbb{E}\{w(t_k+h)w^T(t_k+h)\},\\ &\cdots, \mathbb{E}\{w(t_k+(b-1)h)w^T(t_k+(b-1)h)\}\}\\ =&\bar{W}(t_k),\\ \mathbb{E}\{\bar{v}(t_k)\bar{v}^T(t_k)\} =&\text{diag}\{\mathbb{E}\{v_1(t_k)v_1^T(t_k)\}, \mathbb{E}\{v_2(t_k)v_2^T(t_k)\},\\ &\cdots, \mathbb{E}\{v_M(t_k)v_M^T(t_k)\}\}\\ =&\bar{V}(t_k).\end{aligned}$$

Utilizing the properties of matrix operation, it is obtained that

$$\begin{aligned}\mathbb{E}\{\tilde{x}(t_k)\tilde{x}^T(t_k)\} \leq \mathbb{E}\{\tilde{x}^T(t_k)\tilde{x}(t_k)\}I &= \text{tr}\big(\mathbb{E}\{\tilde{x}(t_k)\tilde{x}^T(t_k)\}\big)I\\ &= \sum_{i=0}^{b-1}\text{tr}\big(\mathbb{E}\{\breve{x}(t_k+ih)\breve{x}^T(t_k+ih)\}\big)I = \Lambda(t_k)\end{aligned}$$

where

$$\Lambda(t_k) \triangleq \sum_{i=0}^{b-1}\text{tr}\big(\Theta(t_k+ih)\big)I.$$

Following the similar line as the derivation in (8.12), *we have*

$$\begin{aligned}&\mathbb{E}\{K(t_k)\mathcal{V}(t_k)\mathcal{V}^T(t_k)K^T(t_k)\}\\ \leq&M\sum_{i=0}^{M-1}\mathbb{E}\{K(t_k)\Gamma(t_{k-i})\bar{D}(t_{k-i})\bar{v}(t_{k-i})\\ &\qquad\times\bar{v}^T(t_{k-i})\bar{D}^T(t_{k-i})\Gamma^T(t_{k-i})K^T(t_k)\}\\ =&M\sum_{i=0}^{M-1}K(t_k)\Gamma(t_{k-i})\bar{D}(t_{k-i})\\ &\qquad\times\bar{V}(t_{k-i})\bar{D}^T(t_{k-i})\Gamma^T(t_{k-i})K^T(t_k)\end{aligned}$$

and

$$
\begin{aligned}
&\mathbb{E}\{K(t_k)\mathcal{E}(t_k)\mathcal{E}^T(t_k)K^T(t_k)\}\\
\leq&(M-1)\sum_{i=1}^{M-1}\mathbb{E}\{K(t_k)\Gamma(t_{k-i})\bar{C}(t_{k-i})e(t_{k-i})\\
&\quad\times e^T(t_{k-i})\bar{C}^T(t_{k-i})\Gamma^T(t_{k-i})K^T(t_k)\}\\
=&(M-1)\sum_{i=1}^{M-1}K(t_k)\Gamma(t_{k-i})\bar{C}(t_{k-i})\\
&\quad\times P(t_{k-i})\bar{C}^T(t_{k-i})\Gamma^T(t_{k-i})K^T(t_k).
\end{aligned}
$$

Summarizing the discussions above, it is not difficult to conclude that

$$
\begin{aligned}
P(t_{k+1})\leq&\kappa_1(t_k)\tilde{A}(t_k)\mathbb{E}\{e(t_k)e^T(t_k)\}\tilde{A}^T(t_k)\\
&+\kappa_2(t_k)\bar{B}(t_k)\mathbb{E}\{\tilde{x}(t_k)\tilde{x}^T(t_k)\}\bar{B}^T(t_k)\\
&+\kappa_3(t_k)\vec{A}(t_k)\mathbb{E}\{\bar{f}(x(t_k),\eta(t_k))\bar{f}^T(x(t_k),\eta(t_k))\}\vec{A}^T(t_k)\\
&+\kappa_4(t_k)\bar{E}(t_k)\mathbb{E}\{\bar{w}(t_k)\bar{w}^T(t_k)\}\bar{E}^T(t_k)\\
&+\kappa_5(t_k)K(t_k)\mathbb{E}\{\mathcal{V}(t_k)\mathcal{V}^T(t_k)\}K^T(t_k)\\
&+\kappa_6(t_k)K(t_k)\mathbb{E}\{\mathcal{E}(t_k)\mathcal{E}^T(t_k)\}K^T(t_k)\\
\leq&\kappa_1(t_k)\tilde{A}(t_k)P(t_k)\tilde{A}^T(t_k)+\kappa_2(t_k)\bar{B}(t_k)\Lambda(t_k)\bar{B}^T(t_k)\\
&+\kappa_3(t_k)\sum_{i=1}^{a}\vec{A}(t_k)\bar{\Pi}_i(t_k)\bar{\Phi}_i(t_k)\vec{A}^T(t_k)\\
&+\kappa_4(t_k)\bar{E}(t_k)\bar{W}(t_k)\bar{E}^T(t_k)\\
&+\kappa_5(t_k)M\sum_{i=0}^{M-1}K(t_k)\Gamma(t_{k-i})\bar{D}(t_{k-i})\bar{V}(t_{k-i})\\
&\quad\times\bar{D}^T(t_{k-i})\Gamma^T(t_{k-i})K^T(t_k)\\
&+\kappa_6(t_k)(M-1)\sum_{i=1}^{M-1}K(t_k)\Gamma(t_{k-i})\bar{C}(t_{k-i})P(t_{k-i})\\
&\quad\times\bar{C}^T(t_{k-i})\Gamma^T(t_{k-i})K^T(t_k)\\
\leq&\Sigma(t_{k+1})
\end{aligned}
$$

which completes this proof.

In Theorem 8.1, an upper bound on the estimation error covariance is derived. In the following theorem, the estimator gain that minimizes such an upper bound at each sampling instant is given.

Theorem 8.2 *The upper bound $\Sigma(t_{k+1})$ on the estimation error covariance given in* (8.13) *is minimized by the following estimator gain*

$$K(t_k) = \Omega(t_k)\Delta^{-1}(t_k) \tag{8.14}$$

where

$$\begin{aligned}\Delta(t_k) \triangleq & \kappa_1(t_k)\Gamma(t_k)\bar{C}(t_k)\Sigma(t_k)\bar{C}^T(t_k)\Gamma^T(t_k) \\ & + \kappa_5(t_k)M\sum_{i=0}^{M-1}\Gamma(t_{k-i})\bar{D}(t_{k-i})\bar{V}(t_{k-i})\bar{D}^T(t_{k-i})\Gamma^T(t_{k-i}) \\ & + \kappa_6(t_k)(M-1)\sum_{i=1}^{M-1}\Gamma(t_{k-i})\bar{C}(t_{k-i})\Sigma(t_{k-i})\bar{C}^T(t_{k-i})\Gamma^T(t_{k-i}), \\ \Omega(t_k) \triangleq & \kappa_1(t_k)\bar{A}(t_k)\Sigma(t_k)\bar{C}^T(t_k)\Gamma^T(t_k).\end{aligned}$$

Moreover, the minimal upper bound $\Sigma(t_{k+1})$ is obtained by the following recursion

$$\begin{aligned}\Sigma(t_{k+1}) = & -\Omega(t_k)\Delta^{-1}(t_k)\Omega^T(t_k) + \kappa_1(t_k)\bar{A}(t_k)\Sigma(t_k)\bar{A}^T(t_k) \\ & + \kappa_2(t_k)\bar{B}(t_k)\bar{\Lambda}(t_k)\bar{B}^T(t_k) \\ & + \kappa_4(t_k)\bar{E}(t_k)\bar{W}(t_k)\bar{E}^T(t_k) \\ & + \kappa_3(t_k)\sum_{i=1}^{a}\vec{A}(t_k)\bar{\Pi}_i(t_k)\tilde{\Phi}_i(t_k)\vec{A}^T(t_k)\end{aligned} \tag{8.15}$$

with initial values $\Sigma(t_d) = X(t_d)$ $(d = -\lfloor\frac{p}{b}\rfloor, -\lfloor\frac{p}{b}\rfloor + 1, \cdots, 0)$.

Proof *The upper bound $\Sigma(t_{k+1})$ derived in Theorem 8.1 can be expressed as*

$$\begin{aligned}\Sigma(t_{k+1}) = & K(t_k)\Delta(t_k)K^T(t_k) - \Omega(t_k)K^T(t_k) - K(t_k)\Omega^T(t_k) \\ & + \kappa_1(t_k)\bar{A}(t_k)\Sigma(t_k)\bar{A}^T(t_k) + \kappa_2(t_k)\bar{B}(t_k)\bar{\Lambda}(t_k)\bar{B}^T(t_k) \\ & + \kappa_3(t_k)\sum_{i=1}^{a}\vec{A}(t_k)\bar{\Pi}_i(t_k)\tilde{\Phi}_i(t_k)\vec{A}^T(t_k) \\ & + \kappa_4(t_k)\bar{E}(t_k)\bar{W}(t_k)\bar{E}^T(t_k) \\ = & \big(K(t_k)-\Omega(t_k)\Delta^{-1}(t_k)\big)\Delta(t_k)\big(K(t_k)-\Omega(t_k)\Delta^{-1}(t_k)\big)^T \\ & - \Omega(t_k)\Delta^{-1}(t_k)\Omega^T(t_k) + \kappa_1(t_k)\bar{A}(t_k)\Sigma(t_k)\bar{A}^T(t_k) \\ & + \kappa_2(t_k)\bar{B}(t_k)\bar{\Lambda}(t_k)\bar{B}^T(t_k) \\ & + \kappa_4(t_k)\bar{E}(t_k)\bar{W}(t_k)\bar{E}^T(t_k) \\ & + \kappa_3(t_k)\sum_{i=1}^{a}\vec{A}(t_k)\bar{\Pi}_i(t_k)\tilde{\Phi}_i(t_k)\vec{A}^T(t_k).\end{aligned}$$

Noting that $P(t_k) \geq 0$, $\Sigma(t_k) \geq P(t_k)$, $\bar{V}(t_k) > 0$ *and* $\kappa_1(t_k)$, $\kappa_5(t_k)$, $\kappa_6(t_k)$ *are positive scalars, we know* $\Delta(t_k) \geq 0$. *Then, the upper bound* $\Sigma(t_{k+1})$ *on the estimation error covariance is minimized by choosing the estimator gain as* $K(t_k) = \Omega(t_k)\Delta^{-1}(t_k)$, *and the minimized upper bound is* (8.15). *The proof is complete.*

Remark 8.5 *In this chapter, the estimation procedure is based on the calculation of* (8.9), (8.11), (8.13), *and* (8.14), *where the augmented measurement is sequentially used at each estimate update instant, and thus the proposed estimation scheme is suitable for online computation. The computational complexity can be obtained by the arithmetic operations in deriving* $\hat{x}(t_k)$, $\Psi(T_k)$, $\Sigma(t_k)$, $\Delta(t_k)$, $\Omega(t_k)$, *and* $K(t_k)$. *According to the matrix computations, at each recursion, the computational complexity is* $O(m^3b^2 + m^2Mn + m^2bw + mM^2n^2 + mb^2w^2 + mM^2nv + mM^2v^2 + M^3n^3 + M^3n^2v + M^3nv^2)$. *Compared with the estimation scheme where only one sensor is used, the computational complexity of our estimation scheme is high. Nevertheless, the centralized fusion provides more information for the estimation, and hence the estimation accuracy is improved.*

Remark 8.6 *In this chapter, the recursive state estimation problem is studied for MRSs with distributed time-delays under RR protocol. Note that, it is difficult to obtain the actual estimation error covariance because of the distributed time-delays and the RR protocol. Accordingly, an alternative method is used in our chapter. In Theorem 8.1, an upper bound on the estimation error covariance is derived recursively by solving Riccati-like equation* (8.13). *Then, the estimator gain minimizing such an upper bound at each sampling instant is designed and the corresponding minimized upper bound is given in Theorem 8.2. Unlike [156], the recursive state estimation scheme developed in this chapter can ensure the invertibility of* $\Delta(t_k)$ *without introducing an extra identity matrix. On the other hand, the conservatism of the proposed algorithm could be reduced by appropriately choosing the scalars* $\epsilon(T_k)$ *and* $\varepsilon_i(t_k)$ $(i = 1, 2, \ldots, 7)$.

8.3 Simulation Examples

In this section, two simulation examples are presented to demonstrate the validity of the proposed recursive state estimation scheme.

Example 8.1

Consider the discrete time-delayed system (8.1) with two sensors whose parameters are given as follows:

$$A(T_k) = \begin{bmatrix} 0.79 & -0.15 \\ 0.66 + 0.1\cos(T_k) & 0.77 \end{bmatrix},$$
$$B(T_k) = \begin{bmatrix} 0.08 & -0.03 - 0.01\sin(T_k) \\ 0.06 & 0.05 \end{bmatrix},$$
$$C_1(t_k) = \begin{bmatrix} 3 & 5 + \sin(t_k) \end{bmatrix}, \quad D_1(t_k) = 3,$$
$$C_2(t_k) = \begin{bmatrix} 2 + \sin(t_k) & 6 \end{bmatrix}, \quad D_2(t_k) = 2,$$
$$E(T_k) = \begin{bmatrix} 0.1 \\ -0.2 + 0.01\cos(T_k) \end{bmatrix}$$

and other parameters are chosen as $p = 2$, $\mu_1 = 0.5$, $\mu_2 = 0.4$, and $b = 2$.

The stochastic non-linear function is chosen as

$$f\big(x(T_k), \eta(T_k)\big) = \begin{bmatrix} 0.3 \\ 0.2 \end{bmatrix} \big(0.3\text{sign}(x_1(T_k))x_1(T_k)\eta_1(T_k) + 0.4\text{sign}(x_2(T_k))x_2(T_k)\eta_2(T_k)\big) \tag{8.16}$$

where $x_i(T_k)$ $(i = 1, 2)$ represents the ith element of the state $x(T_k)$, $\eta_1(T_k)$, and $\eta_2(T_k)$ are zero-mean uncorrelated Gaussian white noises with unity variance. From (8.16), we know $f\big(x(T_k), \eta(T_k)\big)$ satisfies the properties (8.2) with $a = 1$, $\Pi_1(T_k) = \begin{bmatrix} 0.3 & 0.2 \end{bmatrix}^T \begin{bmatrix} 0.3 & 0.2 \end{bmatrix}$, and $\Phi_1(T_k) = \text{diag}\{0.09, 0.16\}$. The covariances of the process noise and the measurement noises are taken as $W(T_k) = 0.01$, $V_1(t_k) = 0.01$, and $V_2(t_k) = 0.01$.

Based on the given parameters, the estimator gain $K(t_k)$ is obtained according to (8.14) and the minimal upper bound $\Sigma(t_k)$ is derived by solving (8.15). Simulation results are presented in Figs. 8.1–8.3. Figure 8.1 depicts $x_1(T_k)$ and its estimate. Figure 8.2 shows $x_2(T_k)$ and its estimate. Figure 8.3 plots the trace of $\Sigma(t_k)$ and the mean square error (MSE), which is defined by $\text{MSE}(t_k) = \frac{1}{N}\sum_{j=1}^{N}\sum_{i=1}^{2}\big(x_i^{(j)}(t_k) - \hat{x}_i^{(j)}(t_k)\big)^2$, with $N = 500$ times independent experiments. The simulation results verify that the developed recursive state estimation scheme is indeed effective.

To verify the influence of the distributed time-delays, the stochastic nonlinearities and the multi-rate sampling on the upper bound of the estimation error covariance, more simulations are conducted in which only the corresponding parameters are changed compared with the above simulation. Simulation results are shown in Figs. 8.4–8.6.

Figure 8.4 shows the trace of the minimal upper bound of the estimation error covariance and the MSE derived by the proposed recursive state estimation scheme with $b = 3$ and other parameters unchanged. It can be seen that the estimation performance when $b = 3$ is worse than that when $b = 2$. Due to the increase of b, the information available for the estimator decreases. Naturally, the estimation performance degrades.

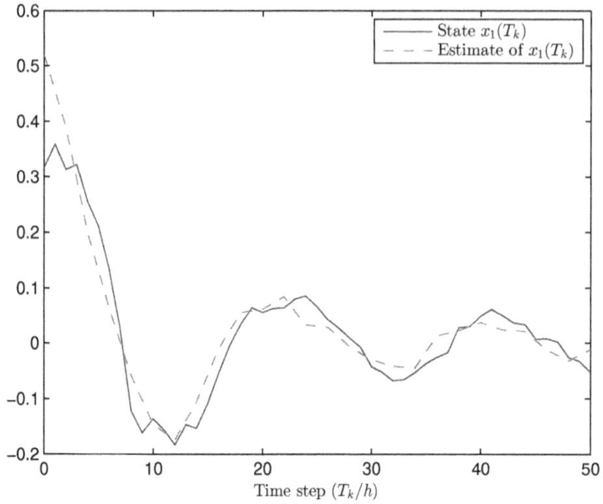

FIGURE 8.1: State $x_1(T_k)$ and the estimate of $x_1(T_k)$.

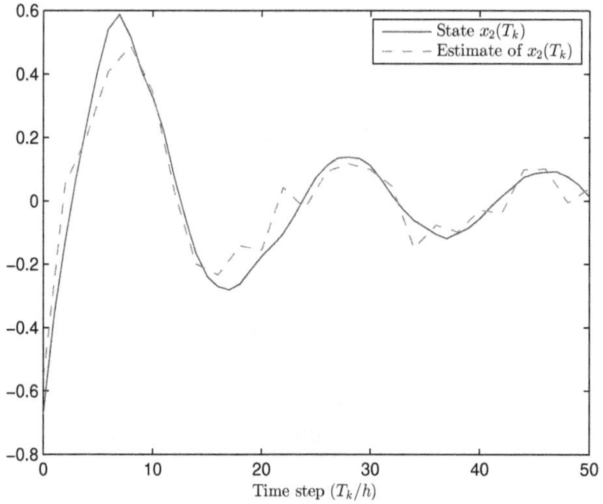

FIGURE 8.2: State $x_2(T_k)$ and the estimate of $x_2(T_k)$.

Figure 8.5 shows the trace of the minimal upper bound on the estimation error covariance and the MSE with $p = 3$, $\mu_3 = 0.3$, and other parameters unchanged. The simulation result shows that the estimation performance degrades with the increase of p. Figure 8.6 shows the trace of the minimal upper bound on the estimation error covariance and the MSE with $\Phi_1(T_k) = \text{diag}\{0.16, 0.25\}$ and other parameters unchanged.

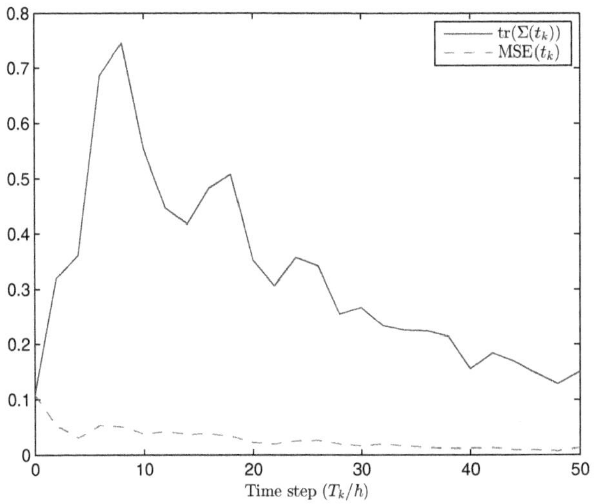

FIGURE 8.3: Trace of the minimal upper bound and the MSE.

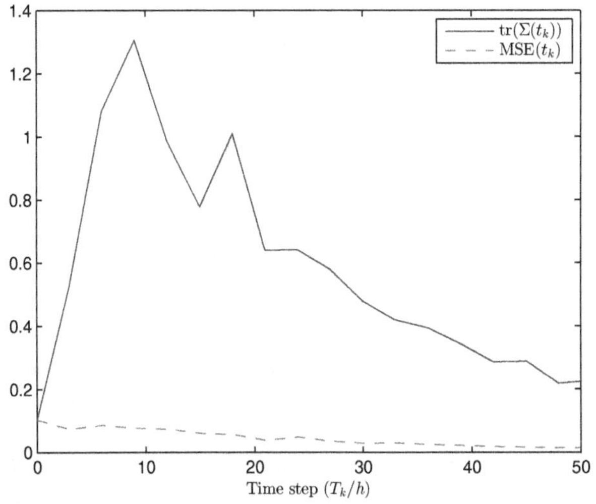

FIGURE 8.4: Trace of the minimal upper bound and the MSE with $b = 3$.

To show the influence of the scheduling effect of the RR protocol on the estimation performance, simulation without RR protocol is conducted and simulation result is given in Fig. 8.7. From Fig. 8.7, one can see that the estimation performance improves slightly. Nevertheless, the introduction of the RR protocol can effectively avoid data collisions and save network resources, and hence a slight degradation of the estimation performance is tolerable.

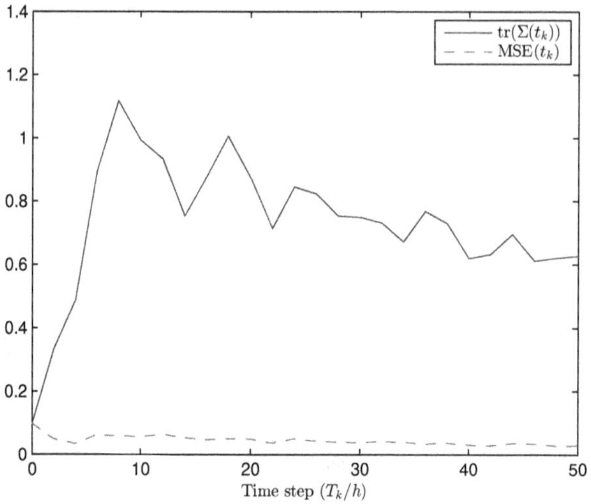

FIGURE 8.5: Trace of the minimal upper bound and the MSE with $p = 3$ and $\mu_3 = 0.3$.

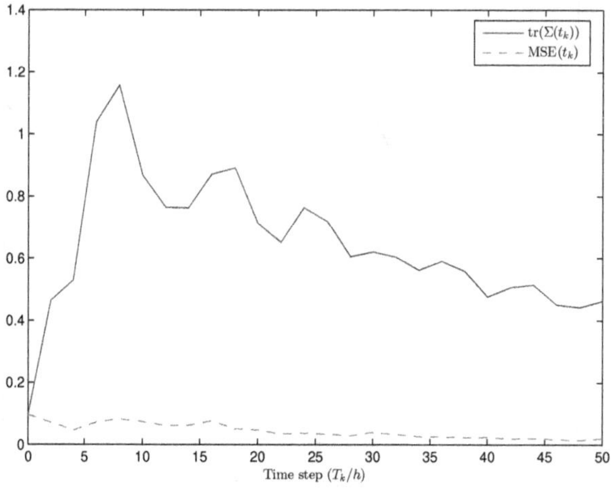

FIGURE 8.6: Trace of the minimal upper bound and the MSE with $\Phi_1(T_k) = \text{diag}\{0.16, 0.25\}$.

Example 8.2

In this example, we conduct a practical simulation to compare the performance between the proposed centralized fusion estimation scheme and the distributed fusion estimation scheme as in [150]. Firstly, along the similar line of the

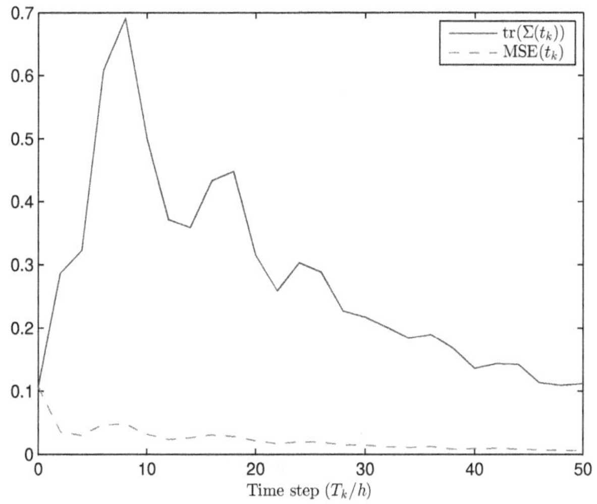

FIGURE 8.7: Trace of the minimal upper bound and the MSE without RR protocol.

analysis process in [150], we characterize the local estimators. Then, the fused estimate is obtained by fusing the estimates of the local estimators. Finally, estimation performance is compared based on a practical system.

Consider a DC servo system described by the system model (8.1) with the following parameters [213]:

$$A(T_k) = \begin{bmatrix} 1.12 + 0.3\sin(T_k) & 0.213 & -0.333 \\ 1 & 0 & 0 \\ 0 & 1 & 0 \end{bmatrix},$$

$$B(T_k) = \begin{bmatrix} -0.2193 + 0.2\cos(T_k) & 0.0219 & 0.0844 \\ 0.2177 & -0.0032 & -0.0662 \\ 0.1298 & -0.0087 & -0.0381 \end{bmatrix},$$

$$E(T_k) = \begin{bmatrix} 0.8 \\ 0 \\ 0 \end{bmatrix}, \; p = 3, \; \mu_1 = 0.01, \; \mu_2 = 0.02, \; \mu_3 = 0.01$$

and the stochastic non-linear function

$$\begin{aligned} f\big(x(T_k), \eta(T_k)\big) = \begin{bmatrix} 0.03 \\ 0.02 \\ 0.01 \end{bmatrix} \big(&0.3\text{sign}(x_1(T_k))x_1(T_k)\eta_1(T_k) \\ &+ 0.2\text{sign}(x_2(T_k))x_2(T_k)\eta_2(T_k)) \\ &+ 0.3\text{sign}(x_3(T_k))x_3(T_k)\eta_3(T_k)\big). \end{aligned}$$

The MSEs of the proposed estimator, the local state estimator developed by following [150], and the fused estimate are shown in Fig. 8.8. It can be

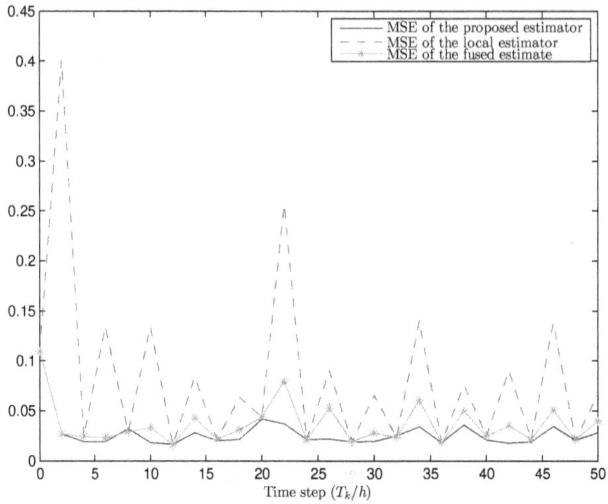

FIGURE 8.8: The MSEs of the proposed estimation scheme and the distributed fusion estimation scheme in [150].

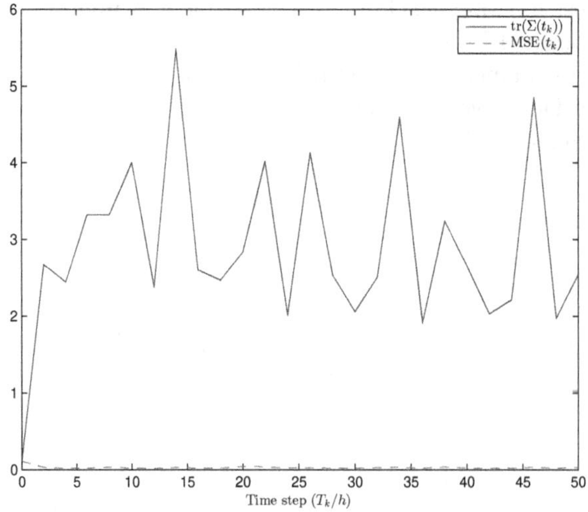

FIGURE 8.9: Trace of the minimal upper bound and the MSE of the proposed estimation algorithm.

seen that the estimate of the proposed estimator has a better accuracy than the fused estimate, and the fused estimate has a better accuracy than the estimate of the local estimator. Moreover, from Figs. 8.8–8.9, one can see that the proposed estimation scheme is indeed effective in real practice.

8.4 Conclusion

In this chapter, the recursive state estimation problem has been studied for MRSs with distributed time-delays under the RR protocol. An iterative method has been used to transform the MRS to a single-rate one. The RR protocol has been implemented to avoid the data collision. The recursive state estimation scheme proposed in this chapter has guaranteed an upper bound for the estimation error covariance, and the estimator gain has been obtained by minimizing such an upper bound. Finally, simulation examples have been provided to illustrate the effectiveness of the developed recursive state estimation scheme.

9

Fusion Estimation for Multi-Rate Linear Repetitive Processes under Weighted Try-Once-Discard Protocol

In the real world, a wide range of practical systems can be represented by the two-dimensional (2-D) model with examples including thermal processes, water steam heating, and image data processing systems [112,128]. For 2-D systems, the information propagates in two directions, and this distinct feature renders the existing theories/approaches for one-dimensional systems inapplicable to the 2-D case. Consequently, a great deal of research effort has been devoted to the study of the dynamical behaviors of 2-D systems [42, 75]. In particular, the state estimation problem, which serves as one of the fundamental topics in the area of signal processing, has aroused some initial research interest in the context of 2-D systems.

As a special case of the 2-D systems, the LRPs are dominated by a series of sweeps through a set of dynamics defined over a fixed finite duration. The sweep and the duration are called the pass and the pass length, respectively. On each pass, a pass profile is produced which acts as a forcing function on, and hence contributes to, the dynamics of the next pass profile. The main distinguishing characteristic of the LRPs is that the dynamics along the pass only evolves over a finite duration. Recently, the state estimation problem for LRPs has received considerable research attention with a large number of results available in the literature [10, 156]. It is worth pointing out that, in almost all existing literature, the underlying LRPs have been implicitly assumed to have a single sampling rate. Therefore, the primary motivation of this chapter is to look into the state estimation problem for multi-rate LRPs.

As one of the most commonly used quadratic protocols, the WTOD protocol has received consistent research attention in the areas of signal processing and control engineering [215], where most research results have been concerned with single-rate systems with synchronous sensors generating time-coincident measurements. How to design the estimator such that the estimation error covariance is minimized? These two difficulties are to be dealt with in this chapter.

The data fusion issue is central to the multi-sensor state estimation problems, whose main idea is to combine the local estimates from estimators to obtain a fused estimate that has higher accuracy than the local estimates. In the past decades, fusion estimation problem for multi-sensor systems has

DOI: 10.1201/9781032619507-9

attracted considerable research interest with fruitful results appeared on the *single-rate* cases [77], and the corresponding results for multi-rate multi-sensor systems have been really scattered. In [150], the event-triggered fusion estimation problem for multi-rate multi-sensor systems with asynchronous sensors has been studied, where the designed local filter can only generate estimates at the sampling instant of its corresponding sensor, and this makes it impossible to obtain the fused estimate at each time instant. In [193], the fusion estimation problem has been investigated for sensor networks with different measurement sampling period, measurement transmitting period and state update period, and excellent fusion results have been obtained under the assumption of *synchronous* sensors. Until now, the issue of obtaining fused estimate at each time instant for *multi-rate* multi-sensor systems with *asynchronous* sensors remains open yet challenging.

Motivated by the above discussions, in this chapter, we are concerned with the fusion estimation problem for multi-rate LRPs under WTOD protocol. The main novelties of this chapter can be highlighted as follows: 1) the fusion estimation problem is, for the first time, investigated for the multi-rate LRPs under WTOD protocol; 2) a novel selection principle of the WTOD protocol is proposed to cater for the asynchronous nature of the sensors; 3) state estimators are designed such that the estimation error covariances are minimized at each time instant; and 4) a new fusion estimation scheme is proposed to obtain the fused estimate at each time instant for multi-rate multi-sensor systems with asynchronous sensors.

9.1 Problem Formulation

Consider the discrete time-varying LRPs described by the following model with $p \geq 0$ and $0 \leq k \leq a$:

$$\begin{aligned}\breve{x}(p+1,k+1) =& \breve{A}_1(p+1,k)\breve{x}(p+1,k) + \breve{B}_1(p,k)\breve{y}(p,k) \\ &+ \breve{C}_1(p+1,k)\breve{w}(p+1,k), \qquad (9.1\text{a}) \\ \breve{y}(p+1,k) =& \breve{A}_2(p+1,k)\breve{x}(p+1,k) + \breve{B}_2(p,k)\breve{y}(p,k) \\ &+ \breve{C}_2(p+1,k)\breve{w}(p+1,k) \qquad (9.1\text{b})\end{aligned}$$

where p is the pass number, k is the along-the-pass variable, a is the pass length, $\breve{x}(p+1,k) \in \mathbb{R}^{n_x}$ is the state vector, $\breve{y}(p,k) \in \mathbb{R}^{n_y}$ is the pass profile vector, and $\breve{w}(p+1,k) \in \mathbb{R}^{n_w}$ is the zero-mean process noise with covariance $Q(p+1,k) > 0$. $\breve{A}_s(p+1,k)$, $\breve{B}_s(p,k)$ and $\breve{C}_s(p+1,k)$ $(s=1,2)$ are known matrices with appropriate dimensions. The initial values of the LRPs are specified as $\mathbb{E}\{\breve{x}(p+1,0)\} = b_{p+1}$ and $\mathbb{E}\{\breve{y}(0,k)\} = c_k$ with b_{p+1} and c_k being known vectors.

In this chapter, m sensors are employed to measure the LRPs. To cater for practical engineering, it is assumed that the sampling period of the sensor i is n_i. The measurement output of the i-th sensor is described as

$$\begin{aligned}\check{z}_i(p+1,n_ik) =&\check{D}_i(p+1,n_ik)\check{x}(p+1,n_ik)\\&+\check{E}_i(p,n_ik)\check{y}(p,n_ik)\\&+\check{F}_i(p+1,n_ik)\check{v}_i(p+1,n_ik)\end{aligned} \tag{9.2}$$

where $\check{z}_i(p+1,n_ik) \in \mathbb{R}^{n_z}$ is the measurement output of the i-th sensor and $\check{v}_i(p+1,n_ik) \in \mathbb{R}^{n_v}$ is the zero-mean measurement noise on the i-th sensor with covariance $R_i(p+1,n_ik) > 0$. $\check{D}_i(p+1,n_ik)$, $\check{E}_i(p,n_ik)$, and $\check{F}_i(p+1,n_ik)$ are known matrices with appropriate dimensions.

Remark 9.1 *In some applications such as long-wall coal cutting, the sensors are hauled back to the starting location after a pass is completed [129]. During such a process, the sensors do not generate useful measurement outputs and the time spent on such a process is much longer than the sampling periods of the sensors. Based on this fact, in this chapter, we assume that the sensors stop generating measurement outputs at the end of a pass, and start a new sampling process at the starting location of the next pass. Therefore, the measurement outputs of the sensors can be described as* (9.2)*, from which we can see that the measurement output of the ith sensor is generated at time instants $0, n_i, 2n_i, \ldots, a_in_i$ on each pass, where a_i represents the largest integer that is no larger than $\frac{a}{n_i}$.*

In networked environments, due to inherently limited bandwidth of the network, it is sometimes unrealistic for all the sensors to transmit their measurement outputs simultaneously. As such, some communication protocols are proposed to schedule the transmission sequence of the sensors. In this chapter, we consider the case that the m asynchronous sensors are orchestrated by the WTOD protocol.

To introduce the WTOD protocol on the asynchronous sensors, we let $\xi(p+1,k) \in \{0,1,2,\ldots,m\}$ be the selected sensor which gets the access to transmission at time instant $(p+1,k)$, where $\xi(p+1,k)=0$ means that there is no sensor getting the access at time instant $(p+1,k)$. Furthermore, we define

$$\tilde{z}_i(p+1,k') = \begin{cases} \check{z}_i(p+1,n_ik), & \text{if } k'=n_ik; \\ \breve{z}_i(p+1,k'), & \text{otherwise} \end{cases} \tag{9.3}$$

where $\breve{z}_i(p+1,k')$ denotes the last transmitted measurement of sensor i before time instant $(p+1,k')$.

Under the scheduling of the WTOD protocol, $\xi(p+1,k)$ can be determined by the following selection principle:

$$\xi(p+1,k) = \begin{cases} 0, & \text{if } \frac{k}{n_i} \notin \mathbb{N} \ (\forall i \in [1,m]); \\ \arg\max\limits_{1\le i\le m} \Theta_i(p+1,k), & \text{otherwise} \end{cases} \tag{9.4}$$

where $\Theta_i(p+1,k) \triangleq \big(\tilde{z}_i(p+1,k) - \breve{z}_i(p+1,k)\big)^T W_i\big(\tilde{z}_i(p+1,k) - \breve{z}_i(p+1,k)\big)$ and W_i $(i = 1, 2, \ldots, m)$ is a known positive definite matrix representing the weight matrix of the i-th sensor.

Remark 9.2 *In multi-rate multi-sensor systems, the measurement outputs of the sensors can be generated at different time instants due to possibly different sampling periods. Consequently, at certain time instants, some sensors do not have measurement outputs, and the traditional selection principle used in [215] is no longer valid since it needs the measurement outputs of all sensors to accomplish the selection. In this chapter, a novel selection principle* (9.4) *is proposed. Under such a selection principle, by employing the zero-order holder strategy, the last transmitted measurement of sensor i is used as its measurement output when the sensor i does not have measurement output, and hence the corresponding Θ_i is zero. Therefore, the sensor i will not get the access when it has no measurement. Moreover, when all the sensors do not generate measurement outputs, we have $\xi = 0$, which means that there is no sensor getting the access.*

Let $z_i(p+1, n_i k)$ be the actual information received by the i-th estimator from the i-th sensor. According to the WTOD protocol, $z_i(p+1, n_i k)$ can be written as:

$$
\begin{aligned}
z_i(p+1, n_i k) =& \delta\big(i, \xi(p+1, n_i k)\big)\breve{z}_i(p+1, n_i k) \\
&+ \Big(1 - \delta\big(i, \xi(p+1, n_i k)\big)\Big) z_i(p+1, n_i k - n_i). \qquad (9.5)
\end{aligned}
$$

where $\delta(\cdot,\cdot)$ is the Kronecker delta function. The initial values of $z_i(p+1, n_i k)$ are assumed to be $z_i(p+1, n_i j) = 0$ for $j < 0$ and $z_i(0, n_i k) = 0$.

As an effective way of handling the MRS, the lifting technique is applied in this chapter. By denoting $\bar{x}(p+1, n_i k) \triangleq \mathrm{col}\{\breve{x}(p+1, n_i k - n_i + 1), \cdots, \breve{x}(p+1, n_i k - 1), \breve{x}(p+1, n_i k)\}$ and $\bar{y}(p, n_i k) \triangleq \mathrm{col}\{\breve{y}(p, n_i k), \breve{y}(p, n_i k + 1), \cdots, \breve{y}(p, n_i k + n_i - 1)\}$, we have the following lifted single-rate system:

$$
\begin{aligned}
\bar{x}(p+1, n_i k + n_i) =& \bar{A}_1(p+1, n_i k)\bar{x}(p+1, n_i k) \\
&+ \bar{B}_1(p, n_i k)\bar{y}(p, n_i k) \\
&+ \bar{C}_1(p+1, n_i k)\bar{w}(p+1, n_i k), \qquad (9.6\text{a}) \\
\bar{y}(p+1, n_i k) =& \bar{A}_2(p+1, n_i k)\bar{x}(p+1, n_i k) \\
&+ \bar{B}_2(p, n_i k)\bar{y}(p, n_i k) \\
&+ \bar{C}_2(p+1, n_i k)\bar{w}(p+1, n_i k), \qquad (9.6\text{b}) \\
\breve{z}_i(p+1, n_i k) =& \bar{D}_i(p+1, n_i k)\bar{x}(p+1, n_i k) \\
&+ \bar{E}_i(p, n_i k)\bar{y}(p, n_i k) \\
&+ \check{F}_i(p+1, n_i k)\check{v}_i(p+1, n_i k) \qquad (9.6\text{c})
\end{aligned}
$$

where, for $s = 1, 2$,

$$
\begin{aligned}
\mathscr{F}_s^{m,n}(p+1, n_ik) &\triangleq \prod_{j=m}^{n} \check{A}_s(p+1, n_ik+j), \\
\mathscr{A}_s(p+1, n_ik) &\triangleq \mathrm{col}\{\mathscr{F}_s^{0,0}(p+1, n_ik), \mathscr{F}_s^{1,0}(p+1, n_ik), \\
&\qquad \cdots, \mathscr{F}_s^{n_i-1,0}(p+1, n_ik)\}, \\
\mathscr{L}_s^{j}(p+1, n_ik) &\triangleq \mathrm{col}\{\underbrace{0, \cdots, 0}_{j}, I, \mathscr{F}_s^{j+1,j+1}(p+1, n_ik), \\
&\qquad \cdots, \mathscr{F}_s^{n_i-1,j+1}(p+1, n_ik)\}, \\
\mathscr{B}_s(p, n_ik) &\triangleq \mathrm{diag}\{\check{B}_s(p, n_ik), \check{B}_s(p, n_ik+1), \\
&\qquad \cdots, \check{B}_s(p, n_ik+n_i-1)\}, \\
\mathscr{C}_s(p+1, n_ik) &\triangleq \mathrm{diag}\{\check{C}_s(p+1, n_ik), \check{C}_s(p+1, n_ik+1), \\
&\qquad \cdots, \check{C}_s(p+1, n_ik+n_i-1)\}, \\
\bar{\mathscr{L}}_s(p+1, n_ik) &\triangleq \begin{bmatrix} \mathscr{L}_s^{0}(p+1, n_ik) & \mathscr{L}_s^{1}(p+1, n_ik) \\ \cdots & \mathscr{L}_s^{n_i-1}(p+1, n_ik) \end{bmatrix}, \\
\bar{A}_s(p+1, n_ik) &\triangleq \begin{bmatrix} 0 & \cdots & 0 & \mathscr{A}_s(p+1, n_ik) \end{bmatrix}, \\
\bar{B}_s(p, n_ik) &\triangleq \bar{\mathscr{L}}_s(p+1, n_ik)\mathscr{B}_s(p, n_ik), \\
\bar{C}_s(p+1, n_ik) &\triangleq \bar{\mathscr{L}}_s(p+1, n_ik)\mathscr{C}_s(p+1, n_ik), \\
\bar{D}_i(p+1, n_ik) &\triangleq \begin{bmatrix} 0 & \cdots & 0 & \check{D}_i(p+1, n_ik) \end{bmatrix}, \\
\bar{E}_i(p, n_ik) &\triangleq \begin{bmatrix} \check{E}_i(p, n_ik) & 0 & \cdots & 0 \end{bmatrix}, \\
\bar{w}(p+1, n_ik) &\triangleq \mathrm{col}\{\check{w}(p+1, n_ik), \check{w}(p+1, n_ik+1), \\
&\qquad \cdots, \check{w}(p+1, n_ik+n_i-1)\}.
\end{aligned}
$$

By letting $x_i(p+1, n_ik) \triangleq \mathrm{col}\{\bar{x}(p+1, n_ik), z_i(p+1, n_ik-n_i)\}$ and $y_i(p, n_ik) \triangleq \mathrm{col}\{\bar{y}(p, n_ik), z_i(p, n_ik)\}$, we rewrite the system (9.6) in the following augmented form

$$
\begin{aligned}
x_i(p+1, n_ik+n_i) =& A_{1i}(p+1, n_ik)x_i(p+1, n_ik) \\
&+ B_{1i}(p, n_ik)y_i(p, n_ik) \\
&+ C_{1i}(p+1, n_ik)w_i(p+1, n_ik), && \text{(9.7a)} \\
y_i(p+1, n_ik) =& A_{2i}(p+1, n_ik)x_i(p+1, n_ik) \\
&+ B_{2i}(p, n_ik)y_i(p, n_ik) \\
&+ C_{2i}(p+1, n_ik)w_i(p+1, n_ik), && \text{(9.7b)} \\
z_i(p+1, n_ik) =& D_i(p+1, n_ik)x_i(p+1, n_ik) \\
&+ E_i(p, n_ik)y_i(p, n_ik) \\
&+ F_i(p+1, n_ik)w_i(p+1, n_ik) && \text{(9.7c)}
\end{aligned}
$$

where, for $s = 1, 2$,

$$A_{si}(p, n_ik) \triangleq \begin{bmatrix} \bar{A}_s(p, n_ik) & 0 \\ \delta(i, \xi(p, n_ik))\bar{D}_i(p, n_ik) & (1 - \delta(i, \xi(p, n_ik)))I \end{bmatrix},$$

$$B_{si}(p, n_ik) \triangleq \begin{bmatrix} \bar{B}_s(p, n_ik) & 0 \\ \delta(i, \xi(p+1, n_ik))\bar{E}_i(p, n_ik) & 0 \end{bmatrix},$$

$$C_{si}(p, n_ik) \triangleq \begin{bmatrix} \bar{C}_s(p, n_ik) & 0 \\ 0 & \delta(i, \xi(p, n_ik))\check{F}_i(p, n_ik) \end{bmatrix},$$

$$D_i(p, n_ik) \triangleq \begin{bmatrix} \delta(i, \xi(p, n_ik))\bar{D}_i(p, n_ik) & (1-\delta(i,\xi(p, n_ik)))I \end{bmatrix},$$

$$E_i(p, n_ik) \triangleq \begin{bmatrix} \delta(i, \xi(p+1, n_ik))\bar{E}_i(p, n_ik) & 0 \end{bmatrix},$$

$$F_i(p+1, n_ik) \triangleq \begin{bmatrix} 0 & \delta(i, \xi(p+1, n_ik))\check{F}_i(p+1, n_ik) \end{bmatrix},$$

$$w_i(p, n_ik) \triangleq \begin{bmatrix} \varpi(p, n_ik) \\ \check{v}_i(p, n_ik) \end{bmatrix}.$$

In this chapter, the i-th estimator is constructed with the following structure:

$$\begin{aligned} \hat{x}_i(p+1, n_ik + n_i) =& A_{1i}(p+1, n_ik)\hat{x}_i(p+1, n_ik) + B_{1i}(p, n_ik)\hat{y}_i(p, n_ik) \\ &+ K_{1i}(p+1, n_ik)\Big(z_i(p+1, n_ik) - E_i(p, n_ik)\hat{y}_i(p, n_ik) \\ &- D_i(p+1, n_ik)\hat{x}_i(p+1, n_ik)\Big), \end{aligned} \tag{9.8a}$$

$$\begin{aligned} \hat{y}_i(p+1, n_ik) =& A_{2i}(p+1, n_ik)\hat{x}_i(p+1, n_ik) + B_{2i}(p, n_ik)\hat{y}_i(p, n_ik) \\ &+ K_{2i}(p+1, n_ik)\Big(z_i(p+1, n_ik) - E_i(p, n_ik)\hat{y}_i(p, n_ik) \\ &- D_i(p+1, n_ik)\hat{x}_i(p+1, n_ik)\Big) \end{aligned} \tag{9.8b}$$

where $\hat{x}_i(p+1, n_ik)$ is the estimate of $x_i(p+1, n_ik)$, $\hat{y}_i(p, n_ik)$ is the estimate of $y_i(p, n_ik)$, and $K_{1i}(p+1, n_ik)$, $K_{2i}(p+1, n_ik)$ are the estimator gains to be designed. The initial values for the estimators are given as $\hat{x}_i(p+1, 0) = \mathbb{E}\{x_i(p+1, 0)\}$ and $\hat{y}_i(0, n_ik) = \mathbb{E}\{y_i(0, n_ik)\}$.

Assumption 9.1 *For $p \geq 0$, $0 \leq k \leq a$, $0 \leq n_ij \leq a$ and $i = 1, 2, \ldots, m$, the random variables $\hat{x}_i(p+1, 0)$, $\hat{y}_i(0, n_ij)$, $\breve{w}(p+1, k)$, $\check{v}_i(p+1, n_ij)$, $\breve{x}(p+1, 0)$ and $\breve{y}(0, k)$ are mutually independent.*

Define

$$\begin{aligned} \tilde{x}_i(p+1, n_ik) \triangleq& x_i(p+1, n_ik) - \hat{x}_i(p+1, n_ik), \\ \tilde{y}_i(p, n_ik) \triangleq& y_i(p, n_ik) - \hat{y}_i(p, n_ik), \\ X_i(p+1, n_ik) \triangleq& \mathbb{E}\{\tilde{x}_i(p+1, n_ik)\tilde{x}_i^T(p+1, n_ik)\}, \\ Y_i(p, n_ik) \triangleq& \mathbb{E}\{\tilde{y}_i(p, n_ik)\tilde{y}_i^T(p, n_ik)\}. \end{aligned}$$

The main aim of this chapter is to:

1. design estimators in the form of (9.8) such that there exist upper bounds for the local estimation error covariances $X_i(p+1, n_ik)$ and $Y_i(p, n_ik)$ $(i = 1, 2, \ldots, m)$ and such upper bounds are minimized by choosing appropriate estimator gains $K_{1i}(p+1, n_ik)$ and $K_{2i}(p+1, n_ik)$;
2. develop a novel fusion scheme to fuse the asynchronous local estimates generated by the estimators described by (9.8) and obtain the fused estimate at each time instant.

9.2 Estimator Design

In this section, the recursions of the local estimation error covariances $X_i(p+1, n_ik)$ and $Y_i(p, n_ik)$ are first given. Then, the upper bounds for $X_i(p+1, n_ik)$ and $Y_i(p, n_ik)$ are obtained. Moreover, the minimums of such upper bounds are derived by appropriately choosing the estimator gains $K_{1i}(p+1, n_ik)$ and $K_{2i}(p+1, n_ik)$. Finally, the fused estimate at each time instant is obtained by fusing the asynchronous local estimates in terms of the sequential covariance intersection fusion scheme.

To obtain the recursions of the local estimation error covariances $X_i(p+1, n_ik)$ and $Y_i(p, n_ik)$, we must first derive the local estimation errors $\tilde{x}_i(p+1, n_ik)$ and $\tilde{y}_i(p, n_ik)$. It follows from (9.7) and (9.8) that

$$\begin{aligned} \tilde{x}_i(p+1, n_ik+n_i) =& \mathcal{A}_{1i}(p+1, n_ik)\tilde{x}_i(p+1, n_ik) + \mathcal{B}_{1i}(p, n_ik)\tilde{y}_i(p, n_ik) \\ &+ \mathcal{C}_{1i}(p+1, n_ik)w_i(p+1, n_ik), \qquad (9.9a) \\ \tilde{y}_i(p+1, n_ik) =& \mathcal{A}_{2i}(p+1, n_ik)\tilde{x}_i(p+1, n_ik) + \mathcal{B}_{2i}(p, n_ik)\tilde{y}_i(p, n_ik) \\ &+ \mathcal{C}_{2i}(p+1, n_ik)w_i(p+1, n_ik) \qquad (9.9b) \end{aligned}$$

where, for $i = 1, 2$,

$$\begin{aligned} \mathcal{A}_{si}(p+1, n_ik) \triangleq& A_{si}(p+1, n_ik) - K_{si}(p+1, n_ik)D_i(p+1, n_ik), \\ \mathcal{B}_{si}(p, n_ik) \triangleq& B_{si}(p, n_ik) - K_{si}(p+1, n_ik)E_i(p, n_ik), \\ \mathcal{C}_{si}(p+1, n_ik) \triangleq& C_{si}(p+1, n_ik) - K_{si}(p+1, n_ik)F_i(p+1, n_ik). \end{aligned}$$

To facilitate the calculation of the cross-term $\mathbb{E}\{\tilde{x}_i(p+1, n_ik)\tilde{y}_i^T(p, n_ik)\}$, which is vital in deriving the recursions of the local estimation error covariances $X_i(p+1, n_ik)$ and $Y_i(p, n_ik)$, the following lemma is given.

Lemma 9.1 *By defining* $T_i(p+1, n_ik) \triangleq \text{col}\{\tilde{x}_i(p+1, n_ik), \tilde{y}_i(p, n_ik)\}$, *we have*

$$\begin{aligned} T_i(p+1, n_ik) = &\sum_{m=1}^{p+1} \mathcal{G}_{i,k}^{p+1-m}(p+1, n_ik) \begin{bmatrix} \tilde{x}_i(m, 0) \\ 0 \end{bmatrix} \\ &+ \sum_{n=0}^{k} \mathcal{G}_{i,k-n}^{p}(p+1, n_ik) \begin{bmatrix} 0 \\ \tilde{y}_i(0, n_in) \end{bmatrix} \\ &+ \sum_{m=1}^{p+1} \sum_{n=0}^{k} \Bigg(\mathcal{G}_{i,k-n-1}^{p+1-m}(p+1, n_ik) \begin{bmatrix} \mathcal{C}_{1i}(m, n_in) \\ 0 \end{bmatrix} \\ &+ \mathcal{G}_{i,k-n}^{p-m}(p+1, n_ik) \begin{bmatrix} 0 \\ \mathcal{C}_{2i}(m, n_in) \end{bmatrix} \Bigg) w_i(m, n_in) \end{aligned} \tag{9.10}$$

where $\mathcal{G}_{i,n}^m(p+1, n_ik) = 0$ *for* $m < 0$ *and* $n < 0$, $\mathcal{G}_{i,n}^m(p+1, n_ik) = I$ *for* $m = 0$ *or* $n = 0$, *and* $\mathcal{G}_{i,n}^m(p+1, n_ik) = \mathcal{F}_{i,1}^0(p+1, n_ik - n_i)\mathcal{G}_{i,n-1}^m(p+1, n_ik - n_i) + \mathcal{F}_{i,0}^1(p, n_ik)\mathcal{G}_{i,n}^{m-1}(p, n_ik)$ *otherwise. Moreover,*

$$\mathcal{F}_{i,1}^0(p+1, n_ik) \triangleq \begin{bmatrix} \mathcal{A}_{1i}(p+1, n_ik) & \mathcal{B}_{1i}(p, n_ik) \\ 0 & 0 \end{bmatrix},$$

$$\mathcal{F}_{i,0}^1(p+1, n_ik) \triangleq \begin{bmatrix} 0 & 0 \\ \mathcal{A}_{2i}(p+1, n_ik) & \mathcal{B}_{2i}(p, n_ik) \end{bmatrix}.$$

Proof *The induction method is used to prove this lemma. First, for* $k = 0$ *and* $p = 0$, *it is easily known that*

$$T_i(1, 0) = \begin{bmatrix} \tilde{x}_i(1, 0) \\ 0 \end{bmatrix} + \begin{bmatrix} 0 \\ \tilde{y}_i(0, 0) \end{bmatrix}$$

which satisfies (9.10).

In the following, three cases are discussed to show (9.10) *is true for all* $p \geq 0$ *and* $0 \leq k \leq a_i$.

Case 9.1 *<u>$p = 0$ and $0 \leq k \leq a_i$</u>. According to the induction method, we assume that* $T_i(p+1, n_ik)$ *satisfies* (9.10) *for* $p = 0$ *and* $0 \leq k < l$ ($l \in [1, a_i]$), *then it remains to prove that* (9.10) *is true for* $p = 0$ *and* $k = l$. *Noting the assumption and* (9.10), *for* $p = 0$ *and* $k = l - 1$, *one has*

$$\begin{aligned} T_i(1, n_il - n_i) = &\mathcal{G}_{i,l-1}^0(1, n_il - n_i) \begin{bmatrix} \tilde{x}_i(1, 0) \\ 0 \end{bmatrix} \\ &+ \sum_{n=0}^{l-1} \mathcal{G}_{i,l-n-1}^0(1, n_il - n_i) \begin{bmatrix} 0 \\ \tilde{y}_i(0, n_in) \end{bmatrix} \\ &+ \sum_{n=0}^{l-1} \mathcal{G}_{i,l-n-2}^0(1, n_il - n_i) \begin{bmatrix} \mathcal{C}_{1i}(1, n_in) \\ 0 \end{bmatrix} w_i(1, n_in). \end{aligned} \tag{9.11}$$

Moreover, from (9.7a), *it is known that*

$$T_i(1,n_il) = \begin{bmatrix} 0 \\ \tilde{y}_i(0,n_il) \end{bmatrix} + \mathcal{F}^0_{i,1}(1,n_il-n_i)T_i(1,n_il-n_i) \\ + \begin{bmatrix} \mathcal{C}_{1i}(1,n_il-n_i) \\ 0 \end{bmatrix} w_i(1,n_il-n_i). \tag{9.12}$$

Keeping $\mathcal{F}^1_{i,0}(p-1,n_ik)\mathcal{G}^{-1}_{i,n}(p-1,n_ik) = 0$ *in mind, it is quite obvious that* $\mathcal{F}^0_{i,1}(p,n_ik-n_i)\mathcal{G}^0_{i,n-1}(p,n_ik-n_i) = \mathcal{G}^0_{i,n}(p,n_ik)$. *Therefore, combining* (9.11) *and* (9.12), $T_i(1,n_il)$ *can be written as*

$$T_i(1,n_il) = \mathcal{G}^0_{i,l}(1,n_il)\begin{bmatrix} \tilde{x}_i(1,0) \\ 0 \end{bmatrix} \\ + \sum_{n=0}^{l} \mathcal{G}^0_{i,l-n}(1,n_il)\begin{bmatrix} 0 \\ \tilde{y}_i(0,n_in) \end{bmatrix} \\ + \sum_{n=0}^{l} \mathcal{G}^0_{i,l-n-1}(1,n_il)\begin{bmatrix} \mathcal{C}_{1i}(1,n_in) \\ 0 \end{bmatrix} w_i(1,n_in)$$

which satisfies (9.10). *According to the induction method and noting* (9.10) *is true for* $k=0$ *and* $p=0$, *we know that* (9.10) *is true for* $p=0$, $0 \le k \le a_i$.

Case 9.2 $p \ge 0$ *and* $k=0$. *First, assume that* $T_i(p+1,n_ik)$ *satisfies* (9.10) *for* $0 \le p < h$ ($h \in [1,+\infty)$) *and* $k=0$. *By following the similar line of the proof of the Case 9.1, we can derive that* (9.10) *is true for* $p=h$ *and* $k=0$. *Combining with the initial condition, it is known that* $T_i(p+1,n_ik)$ *satisfies* (9.10) *for* $p \ge 0$ *and* $k=0$.

Case 9.3 $p > 0$ *and* $k > 0$. *First, we also assume that* $T_i(p+1,n_ik)$ *satisfies* (9.10) *for* $(p,k) \in \{p=h, k=l-1\} \bigcup \{p=h-1, k=l\}$ ($l \in [1,a_i-1], h \in [1,+\infty)$). *Next, we are going to show that* (9.10) *is true for* $p=h$ *and* $k=l$. *It is easy to see that*

$$T_i(h,n_il) = \sum_{m=1}^{h} \mathcal{G}^{h-m}_{i,l}(h,n_il)\begin{bmatrix} \tilde{x}_i(m,0) \\ 0 \end{bmatrix} \\ + \sum_{n=0}^{l} \mathcal{G}^{h-1}_{i,l-n}(h,n_il)\begin{bmatrix} 0 \\ \tilde{y}_i(0,n_in) \end{bmatrix} \\ + \sum_{m=1}^{h}\sum_{n=0}^{l} \Bigg(\mathcal{G}^{h-m}_{i,l-n-1}(h,n_il)\begin{bmatrix} \mathcal{C}_{1i}(m,n_in) \\ 0 \end{bmatrix} \\ + \mathcal{G}^{h-m-1}_{i,l-n}(h,n_il)\begin{bmatrix} 0 \\ \mathcal{C}_{2i}(m,n_in) \end{bmatrix} \Bigg) w_i(m,n_in) \tag{9.13}$$

and

$$\begin{aligned}T_i(h+1,n_il-n_i) &= \sum_{m=1}^{h+1}\mathcal{G}_{i,l-1}^{h-m+1}(h+1,n_il-n_i)\begin{bmatrix}\tilde{x}_i(m,0)\\0\end{bmatrix}\\&\quad+\sum_{n=0}^{l-1}\mathcal{G}_{i,l-n-1}^{h}(h+1,n_il-n_i)\begin{bmatrix}0\\\tilde{y}_i(0,n_in)\end{bmatrix}\\&\quad+\sum_{m=1}^{h+1}\sum_{n=0}^{l-1}\left(\mathcal{G}_{i,l-n-2}^{h-m+1}(h+1,n_il-n_i)\begin{bmatrix}C_{1i}(m,n_in)\\0\end{bmatrix}\right.\\&\quad\left.+\mathcal{G}_{i,l-n-1}^{h-m}(h+1,n_il-n_i)\begin{bmatrix}0\\C_{2i}(m,n_in)\end{bmatrix}\right)w_i(m,n_in).\end{aligned}\tag{9.14}$$

Combining (9.13), (9.14) *and following the similar procedure in the proof of the Case 9.1, it can be obtained that*

$$\begin{aligned}T_i(h+1,n_il) &= \sum_{m=1}^{h+1}\mathcal{G}_{i,l}^{h-m+1}(h+1,n_il)\begin{bmatrix}\tilde{x}_i(m,0)\\0\end{bmatrix}\\&\quad+\sum_{n=0}^{l}\mathcal{G}_{i,l-n}^{h}(h+1,n_il)\begin{bmatrix}0\\\tilde{y}_i(0,n_in)\end{bmatrix}\\&\quad+\sum_{m=1}^{h+1}\sum_{n=0}^{l}\mathcal{G}_{i,l-n-1}^{h-m+1}(h+1,n_il)\begin{bmatrix}C_{1i}(m,n_in)\\0\end{bmatrix}w_i(m,n_in)\\&\quad+\sum_{m=1}^{h+1}\sum_{n=0}^{l}\mathcal{G}_{i,l-n}^{h-m}(h+1,n_il)\begin{bmatrix}0\\C_{2i}(m,n_in)\end{bmatrix}w_i(m,n_in)\end{aligned}$$

which shows that (9.10) *is satisfied for* $p = h$ *and* $k = l$. *Accordingly, noting Case 9.1 and Case 9.2, it can be derived that* (9.10) *is true for* $k > 0$ *and* $p > 0$ *column by column.*

By summarizing the above three cases, we conclude that (9.10) *is true for all* $p \geq 0$ *and* $0 \leq k \leq a_i$, *and the proof is thus complete.*

Now, we are ready to calculate the cross-term $\mathbb{E}\{\tilde{x}_i(p+1,n_ik)\tilde{y}_i^T(p,n_ik)\}$ by employing Lemma 9.1. Before going further, let us first give some straightforward relationships:

$$\begin{aligned}&\mathbb{E}\{\bar{w}(p+1,n_ik)\bar{w}^T(p+1,n_ik)\}\\&\quad= \text{diag}\{Q(p+1,n_ik),Q(p+1,n_ik+1),\cdots,Q(p+1,n_ik+n_i-1)\}\\&\quad:= \bar{Q}(p+1,n_ik),\\&\mathbb{E}\{w_i(p+1,n_ik)w_i^T(p+1,n_ik)\}\\&\quad= \text{diag}\{\bar{Q}(p+1,n_ik),R_i(p+1,n_ik)\} := \tilde{Q}_i(p+1,n_ik).\end{aligned}$$

Lemma 9.2 $\mathbb{E}\{\tilde{x}_i(p+1,n_ik)\tilde{y}_i^T(p,n_ik)\}$ *can be calculated by the following equation:*

$$\mathbb{E}\{\tilde{x}_i(p+1,n_ik)\tilde{y}_i^T(p,n_ik)\} = Z_iG_i(p+1,n_ik)\bar{Z}_i^T \tag{9.15}$$

where

$$\begin{aligned} Z_i &\triangleq \begin{bmatrix} I_{n_in_x+n_z} & 0_{n_in_x+n_z\times n_in_y+n_z} \end{bmatrix}, \\ \bar{Z}_i &\triangleq \begin{bmatrix} 0_{n_in_y+n_z\times n_in_x+n_z} & I_{n_in_y+n_z} \end{bmatrix}, \\ G_i(p+1,n_ik) &\triangleq f_i(X_i,Y_i,p+1,n_ik) \end{aligned}$$

with

$$\begin{aligned} &f_i(M_i,N_i,p+1,n_ik) \\ &\triangleq \sum_{m=1}^{p+1} \mathcal{G}_{i,k}^{p-m+1}(p+1,n_ik)\begin{bmatrix} M_i(m,0) & 0 \\ 0 & 0 \end{bmatrix}\left(\mathcal{G}_{i,k}^{p-m+1}(p+1,n_ik)\right)^T \\ &\quad + \sum_{n=0}^{k} \mathcal{G}_{i,k-n}^{p}(p+1,n_ik)\begin{bmatrix} 0 & 0 \\ 0 & N_i(0,n_in) \end{bmatrix}\left(\mathcal{G}_{i,k-n}^{p}(p+1,n_ik)\right)^T \\ &\quad + \sum_{m=1}^{p+1}\sum_{n=0}^{k}\Bigg(\mathcal{G}_{i,k-n-1}^{p-m+1}(p+1,n_ik)\begin{bmatrix} \mathcal{C}_{1i}(m,n_in) \\ 0 \end{bmatrix} \\ &\qquad + \mathcal{G}_{i,k-n}^{p-m}(p+1,n_ik)\begin{bmatrix} 0 \\ \mathcal{C}_{2i}(m,n_in) \end{bmatrix}\Bigg)\tilde{Q}_i(m,n_in) \\ &\quad \times \Bigg(\mathcal{G}_{i,k-n-1}^{p-m+1}(p+1,n_ik)\begin{bmatrix} \mathcal{C}_{1i}(m,n_in) \\ 0 \end{bmatrix} \\ &\qquad + \mathcal{G}_{i,k-n}^{p-m}(p+1,n_ik)\begin{bmatrix} 0 \\ \mathcal{C}_{2i}(m,n_in) \end{bmatrix}\Bigg)^T. \end{aligned} \tag{9.16}$$

Proof *It is obvious that* $\mathbb{E}\{\tilde{x}_i(p+1,n_ik)\tilde{y}_i^T(p,n_ik)\} = Z_i\mathbb{E}\{T_i(p+1,n_ik)T_i^T(p+1,n_ik)\}\bar{Z}_i^T$*, and what we need to do is to calculate* $\mathbb{E}\{T_i(p+1,n_ik)T_i^T(p+1,n_ik)\}$*. It can be seen from Assumption 9.1 that the following are true:*

$$\begin{aligned} \mathbb{E}\{\tilde{x}_i(l,0)w_i^T(m,n_in)\} &= 0, \\ \mathbb{E}\{\tilde{x}_i(l,0)\tilde{y}_i^T(0,n_in)\} &= 0, \\ \mathbb{E}\{\tilde{y}_i(0,n_ih)w_i^T(m,n_in)\} &= 0, \\ \mathbb{E}\{w_i(l,n_ih)w_i^T(m,n_in)\} &= \tilde{Q}_i(l,n_ih)\delta(l,m)\delta(h,n), \\ \mathbb{E}\{\tilde{x}_i(l,0)\tilde{x}_i^T(m,0)\} &= X_i(l,0)\delta(l,m), \\ \mathbb{E}\{\tilde{y}_i(0,n_ih)\tilde{y}_i^T(0,n_in)\} &= Y_i(0,n_ih)\delta(h,n) \end{aligned}$$

for $l, m \geq 1$ and $0 \leq h, n \leq a_i$. Consequently, it follows from (9.10) that

$$\mathbb{E}\{T_i(p+1, n_ik)T_i^T(p+1, n_ik)\} = G_i(p+1, n_ik).$$

Then, we have

$$\mathbb{E}\{\tilde{x}_i(p+1, n_ik)\tilde{y}_i^T(p, n_ik)\} = Z_iG_i(p+1, n_ik)\bar{Z}_i^T$$

which completes the proof.

Remark 9.3 *It can be seen from (9.9) that the calculation of $\mathbb{E}\{\tilde{x}_i(p+1, n_ik)\tilde{y}_i^T(p, n_ik)\}$ is one of the key steps in deriving the recursions of the local estimation error covariances $X_i(p+1, n_ik)$ and $Y_i(p, n_ik)$. Due to the complexity of the LRPs, the exact form of $\mathbb{E}\{\tilde{x}_i(p+1, n_ik)\tilde{y}_i^T(p, n_ik)\}$ is difficult to be derived, and hence some alternative methods are exploited in the literature. For example, in [156], by applying an elementary matrix inequality, the calculation of this term has been avoided, that is, the exact form of $\mathbb{E}\{\tilde{x}_i(p+1, n_ik)\tilde{y}_i^T(p, n_ik)\}$ has not been given. In [162], $\mathbb{E}\{\tilde{x}_i(p+1, n_ik)\tilde{y}_i^T(p, n_ik)\}$ has been calculated by iteratively substituting (9.9), and this leads to great computation complexity. In this chapter, a novel method is proposed to derive the exact form of $\mathbb{E}\{\tilde{x}_i(p+1, n_ik)\tilde{y}_i^T(p, n_ik)\}$, and our proposed method has less computational complexity than that in [162].*

By using Lemma 9.2, in the following lemma, the recursions of the local estimation error covariances $X_i(p+1, n_ik)$ and $Y_i(p, n_ik)$ are presented.

Lemma 9.3 *The local estimation error covariances $X_i(p+1, n_ik+n_i)$ and $Y_i(p+1, n_ik)$ can be obtained by the following recursive equations:*

$$\begin{aligned} X_i(p+1, n_ik+n_i) = {} & \mathcal{A}_{1i}(p+1, n_ik)X_i(p+1, n_ik)\mathcal{A}_{1i}^T(p+1, n_ik) \\ & + \mathcal{B}_{1i}(p, n_ik)Y_i(p, n_ik)\mathcal{B}_{1i}^T(p, n_ik) \\ & + \mathcal{C}_{1i}(p+1, n_ik)\tilde{Q}_i(p+1, n_ik)\mathcal{C}_{1i}^T(p+1, n_ik) \\ & + \mathcal{A}_{1i}(p+1, n_ik)Z_iG_i(p+1, n_ik)\bar{Z}_i^T\mathcal{B}_{1i}^T(p, n_ik) \\ & + \mathcal{B}_{1i}(p, n_ik)\bar{Z}_iG_i(p+1, n_ik)Z_i^T\mathcal{A}_{1i}^T(p+1, n_ik), \end{aligned} \tag{9.17a}$$

$$\begin{aligned} Y_i(p+1, n_ik) = {} & \mathcal{A}_{2i}(p+1, n_ik)X_i(p+1, n_ik)\mathcal{A}_{2i}^T(p+1, n_ik) \\ & + \mathcal{B}_{2i}(p, n_ik)Y_i(p, n_ik)\mathcal{B}_{2i}^T(p, n_ik) \\ & + \mathcal{C}_{2i}(p+1, n_ik)\tilde{Q}_i(p+1, n_ik)\mathcal{C}_{2i}^T(p+1, n_ik) \\ & + \mathcal{A}_{2i}(p+1, n_ik)Z_iG_i(p+1, n_ik)\bar{Z}_i^T\mathcal{B}_{2i}^T(p, n_ik) \\ & + \mathcal{B}_{2i}(p, n_ik)\bar{Z}_iG_i(p+1, n_ik)Z_i^T\mathcal{A}_{2i}^T(p+1, n_ik). \end{aligned} \tag{9.17b}$$

Proof *From the estimation error dynamics (9.9), one has*

$$
\begin{aligned}
X_i(p+1, n_i k + n_i) = &\, \mathcal{A}_{1i}(p+1, n_i k) X_i(p+1, n_i k) \mathcal{A}_{1i}^T(p+1, n_i k) \\
&+ \mathcal{B}_{1i}(p, n_i k) Y_i(p, n_i k) \mathcal{B}_{1i}^T(p, n_i k) \\
&+ \mathcal{C}_{1i}(p+1, n_i k) \tilde{Q}_i(p+1, n_i k) \mathcal{C}_{1i}^T(p+1, n_i k) \\
&+ \mathfrak{A}_{1i}(p+1, n_i k) + \mathfrak{A}_{1i}^T(p+1, n_i k) \\
&+ \mathfrak{B}_{1i}(p+1, n_i k) + \mathfrak{B}_{1i}^T(p+1, n_i k) \\
&+ \mathfrak{C}_{1i}(p+1, n_i k) + \mathfrak{C}_{1i}^T(p+1, n_i k), \\
Y_i(p+1, n_i k) = &\, \mathcal{A}_{2i}(p+1, n_i k) X_i(p+1, n_i k) \mathcal{A}_{2i}^T(p+1, n_i k) \\
&+ \mathcal{B}_{2i}(p, n_i k) Y_i(p, n_i k) \mathcal{B}_{2i}^T(p, n_i k) \\
&+ \mathcal{C}_{2i}(p+1, n_i k) \tilde{Q}_i(p+1, n_i k) \mathcal{C}_{2i}^T(p+1, n_i k) \\
&+ \mathfrak{A}_{2i}(p+1, n_i k) + \mathfrak{A}_{2i}^T(p+1, n_i k) \\
&+ \mathfrak{B}_{2i}(p+1, n_i k) + \mathfrak{B}_{2i}^T(p+1, n_i k) \\
&+ \mathfrak{C}_{2i}(p+1, n_i k) + \mathfrak{C}_{2i}^T(p+1, n_i k)
\end{aligned}
$$

where, for $s = 1, 2,$

$$
\begin{aligned}
\mathfrak{A}_{si}(p+1, n_i k) &\triangleq \mathcal{A}_{si}(p+1, n_i k) \mathbb{E}\{\tilde{x}_i(p+1, n_i k) \\
&\qquad \times \tilde{y}_i^T(p, n_i k)\} \mathcal{B}_{si}^T(p, n_i k), \\
\mathfrak{B}_{si}(p+1, n_i k) &\triangleq \mathcal{A}_{si}(p+1, n_i k) \mathbb{E}\{\tilde{x}_i(p+1, n_i k) \\
&\qquad \times w_i^T(p+1, n_i k)\} \mathcal{C}_{si}^T(p+1, n_i k), \\
\mathfrak{C}_{si}(p+1, n_i k) &\triangleq \mathcal{B}_{si}(p, n_i k) \mathbb{E}\{\tilde{y}_i(p, n_i k) \\
&\qquad \times w_i^T(p+1, n_i k)\} \mathcal{C}_{si}^T(p+1, n_i k).
\end{aligned}
$$

From Lemma 9.1, it is known that

$$
\begin{aligned}
\mathbb{E}\{\tilde{x}_i(p+1, n_i k) w_i^T(p+1, n_i k)\} &= \mathbb{E}\{Z_i T_i(p+1, n_i k) w_i^T(p+1, n_i k)\} = 0, \\
\mathbb{E}\{\tilde{y}_i(p, n_i k) w_i^T(p+1, n_i k)\} &= \mathbb{E}\{\bar{Z}_i T_i(p+1, n_i k) w_i^T(p+1, n_i k)\} - 0.
\end{aligned}
$$

Moreover, it follows from Lemma 9.2 that

$$
\mathfrak{A}_{si}(p+1, n_i k) = \mathcal{A}_{si}(p+1, n_i k) Z_i G_i(p+1, n_i k) \bar{Z}_i^T \mathcal{B}_{si}^T(p, n_i k), \quad (s = 1, 2).
$$

Summarizing the above equations, (9.17) can be obtained, which completes the proof.

Following Lemma 9.3, the local estimation error covariances $X_i(p+1, n_i k + n_i)$ and $Y_i(p+1, n_i k)$ can be minimized by choosing the estimator gains as

$$
\begin{aligned}
K_{1i}(p+1, n_i k) &= \Sigma_{1i}(p+1, n_i k) \Pi_i^{-1}(p+1, n_i k), \\
K_{2i}(p+1, n_i k) &= \Sigma_{2i}(p+1, n_i k) \Pi_i^{-1}(p+1, n_i k)
\end{aligned}
$$

where

$$\begin{aligned}\Pi_i(p+1,n_ik) \triangleq\, & D_i(p+1,n_ik)X_i(p+1,n_ik)D_i^T(p+1,n_ik)\\ & + E_i(p,n_ik)Y_i(p,n_ik)E_i^T(p,n_ik)\\ & + F_i(p+1,n_ik)\tilde{Q}_i(p+1,n_ik)F_i^T(p+1,n_ik)\\ & + D_i(p+1,n_ik)Z_iG_i(p+1,n_ik)\bar{Z}_i^TE_i^T(p,n_ik)\\ & + E_i(p,n_ik)\bar{Z}_iG_i(p+1,n_ik)Z_i^TD_i^T(p+1,n_ik).\end{aligned}$$

However, due to the scheduling effect of the WTOD protocol, the matrix $\Pi_i(p+1,n_ik)$ might be uninvertible when $\delta\big(i,\xi(p+1,n_ik)\big)=0$. Therefore, it is impossible to obtain the globally minimal local estimation error covariances $X_i(p+1,n_ik)$ and $Y_i(p,n_ik)$. In the following theorem, upper bounds for the local estimation error covariances are derived and the estimator gains which minimize such upper bounds are also obtained.

Theorem 9.1 *For given positive scalars ϑ_1 and ϑ_2, the local estimation error covariances $X_i(p+1,n_ik+n_i)$ and $Y_i(p+1,n_ik)$ have upper bounds $\bar{X}_i(p+1,n_ik+n_i)$ and $\bar{Y}_i(p+1,n_ik)$ which can be calculated by the following recursions:*

$$\begin{aligned}\bar{X}_i(p+1,n_ik+n_i) =\, & \mathcal{A}_{1i}(p+1,n_ik)\bar{X}_i(p+1,n_ik)\mathcal{A}_{1i}^T(p+1,n_ik)\\ & + \mathcal{B}_{1i}(p,n_ik)\bar{Y}_i(p,n_ik)\mathcal{B}_{1i}^T(p,n_ik)\\ & + \mathcal{C}_{1i}(p+1,n_ik)\tilde{Q}_i(p+1,n_ik)\mathcal{C}_{1i}^T(p+1,n_ik)\\ & + \vartheta_1K_{1i}(p+1,n_ik)K_{1i}^T(p+1,n_ik)\\ & + \mathcal{A}_{1i}(p+1,n_ik)Z_i\bar{G}_i(p+1,n_ik)\bar{Z}_i^T\mathcal{B}_{1i}^T(p,n_ik)\\ & + \mathcal{B}_{1i}(p,n_ik)\bar{Z}_i\bar{G}_i(p+1,n_ik)Z_i^T\mathcal{A}_{1i}^T(p+1,n_ik),\end{aligned} \tag{9.18a}$$

$$\begin{aligned}\bar{Y}_i(p+1,n_ik) =\, & \mathcal{A}_{2i}(p+1,n_ik)\bar{X}_i(p+1,n_ik)\mathcal{A}_{2i}^T(p+1,n_ik)\\ & + \mathcal{B}_{2i}(p,n_ik)\bar{Y}_i(p,n_ik)\mathcal{B}_{2i}^T(p,n_ik)\\ & + \mathcal{C}_{2i}(p+1,n_ik)\tilde{Q}_i(p+1,n_ik)\mathcal{C}_{2i}^T(p+1,n_ik)\\ & + \vartheta_2K_{2i}(p+1,n_ik)K_{2i}^T(p+1,n_ik)\\ & + \mathcal{A}_{2i}(p+1,n_ik)Z_i\bar{G}_i(p+1,n_ik)\bar{Z}_i^T\mathcal{B}_{2i}^T(p,n_ik)\\ & + \mathcal{B}_{2i}(p,n_ik)\bar{Z}_i\bar{G}_i(p+1,n_ik)Z_i^T\mathcal{A}_{2i}^T(p+1,n_ik)\end{aligned} \tag{9.18b}$$

with initial values $\bar{X}_i(p+1,0)=X_i(p+1,0)$ and $\bar{Y}_i(0,n_ik)=Y_i(0,n_ik)$, where

$$\bar{G}_i(p+1,n_ik) \triangleq f_i(\bar{X}_i,\bar{Y}_i,p+1,n_ik).$$

Moreover, the upper bounds $\bar{X}_i(p+1, n_ik+n_i)$ *and* $\bar{Y}_i(p+1, n_ik)$ *can be minimized by the following estimator gains*

$$K_{1i}(p+1, n_ik) = \Omega_{1i}(p+1, n_ik)\Phi_{1i}^{-1}(p+1, n_ik), \tag{9.19a}$$

$$K_{2i}(p+1, n_ik) = \Omega_{2i}(p+1, n_ik)\Phi_{2i}^{-1}(p+1, n_ik) \tag{9.19b}$$

where the minimums of $\bar{X}_i(p+1, n_ik+n_i)$ *and* $\bar{Y}_i(p+1, n_ik)$ *can be respectively calculated by*

$$\begin{aligned}
\bar{X}_i(p+1, n_ik+n_i) = &- \Omega_{1i}(p+1, n_ik)\Phi_{1i}^{-1}(p+1, n_ik)\Omega_{1i}^T(p+1, n_ik) \\
&+ A_{1i}(p+1, n_ik)\bar{X}_i(p+1, n_ik)A_{1i}^T(p+1, n_ik) \\
&+ B_{1i}(p, n_ik)\bar{Y}_i(p, n_ik)B_{1i}^T(p, n_ik) \\
&+ C_{1i}(p+1, n_ik)\tilde{Q}_i(p+1, n_ik)C_{1i}^T(p+1, n_ik) \\
&+ A_{1i}(p+1, n_ik)Z_i\bar{G}_i(p+1, n_ik)\bar{Z}_i^T B_{1i}^T(p, n_ik) \\
&+ B_{1i}(p, n_ik)\bar{Z}_i\bar{G}_i(p+1, n_ik)Z_i^T A_{1i}^T(p+1, n_ik), \\
\bar{Y}_i(p+1, n_ik) = &- \Omega_{2i}(p+1, n_ik)\Phi_{2i}^{-1}(p+1, n_ik)\Omega_{2i}^T(p+1, n_ik) \\
&+ A_{2i}(p+1, n_ik)\bar{X}_i(p+1, n_ik)A_{2i}^T(p+1, n_ik) \\
&+ B_{2i}(p, n_ik)\bar{Y}_i(p, n_ik)B_{2i}^T(p, n_ik) \\
&+ C_{2i}(p+1, n_ik)\tilde{Q}_i(p+1, n_ik)C_{2i}^T(p+1, n_ik) \\
&+ A_{2i}(p+1, n_ik)Z_i\bar{G}_i(p+1, n_ik)\bar{Z}_i^T B_{2i}^T(p, n_ik) \\
&+ B_{2i}(p, n_ik)\bar{Z}_i\bar{G}_i(p+1, n_ik)Z_i^T A_{2i}^T(p+1, n_ik).
\end{aligned}$$

Here, for $s = 1, 2$,

$$\begin{aligned}
\Phi_{si}(p+1, n_ik) \triangleq\, & D_i(p+1, n_ik)\bar{X}_i(p+1, n_ik)D_i^T(p+1, n_ik) \\
&+ E_i(p, n_ik)\bar{Y}_i(p, n_ik)E_i^T(p, n_ik) \\
&+ F_i(p+1, n_ik)\tilde{Q}_i(p+1, n_ik)F_i^T(p+1, n_ik) \\
&+ D_i(p+1, n_ik)Z_i\bar{G}_i(p+1, n_ik)\bar{Z}_i^T E_i^T(p, n_ik) \\
&+ E_i(p, n_ik)\bar{Z}_i\bar{G}_i(p+1, n_ik)Z_i^T D_i^T(p+1, n_ik) + \vartheta_s I, \\
\Omega_{si}(p+1, n_ik) \triangleq\, & A_{si}(p+1, n_ik)\bar{X}_i(p+1, n_ik)D_i^T(p+1, n_ik) \\
&+ B_{si}(p, n_ik)\bar{Y}_i(p, n_ik)E_i^T(p, n_ik) \\
&+ C_{si}(p+1, n_ik)\tilde{Q}_i(p+1, n_ik)F_i^T(p+1, n_ik) \\
&+ A_{si}(p+1, n_ik)Z_i\bar{G}_i(p+1, n_ik)\bar{Z}_i^T E_i^T(p, n_ik) \\
&+ B_{si}(p, n_ik)\bar{Z}_i\bar{G}_i(p+1, n_ik)Z_i^T D_i^T(p+1, n_ik).
\end{aligned}$$

Proof *It follows from* (9.16) *and* (9.18) *that*

$$\begin{aligned}
&\bar{X}_i(p+1, n_ik+n_i) - X_i(p+1, n_ik+n_i) \\
&\qquad = \mathcal{A}_{1i}(p+1, n_ik)\big(\bar{X}_i(p+1, n_ik)
\end{aligned}$$

$$
\begin{aligned}
&\quad - X_i(p+1,n_ik)\big)\mathcal{A}_{1i}^T(p+1,n_ik)\\
&\quad + \mathcal{B}_{1i}(p,n_ik)\big(\bar{Y}_i(p,n_ik)\\
&\quad - Y_i(p,n_ik)\big)\mathcal{B}_{1i}^T(p,n_ik)\\
&\quad + \vartheta_1 K_{1i}(p+1,n_ik)K_{1i}^T(p+1,n_ik),\\
\bar{Y}_i(p+1,n_ik) - Y_i(p+1,n_ik)&\\
&= \mathcal{A}_{2i}(p+1,n_ik)\big(\bar{X}_i(p+1,n_ik)\\
&\quad - X_i(p+1,n_ik)\big)\mathcal{A}_{2i}^T(p+1,n_ik)\\
&\quad + \mathcal{B}_{2i}(p,n_ik)\big(\bar{Y}_i(p,n_ik)\\
&\quad - Y_i(p,n_ik)\big)\mathcal{B}_{2i}^T(p,n_ik)\\
&\quad + \vartheta_2 K_{2i}(p+1,n_ik)K_{2i}^T(p+1,n_ik).
\end{aligned}
\tag{9.20}
$$

First, let us prove that $X_i(p+1,n_ik) \le \bar{X}_i(p+1,n_ik)$ *is valid for all* $p \ge 0$ *and* $0 \le k \le a_i$ *column by column.*

Assume that $X_i(p+1,n_in) \le \bar{X}_i(p+1,n_in)$ *is valid for all* $p \ge 0$ *and a given integer* $n \in [0, a_i - 1]$. *Then, according to the induction method and the initial conditions* $X_i(p+1,0) \le \bar{X}_i(p+1,0)$, *what we need to do is to prove that* $X_i(p+1,n_in+n_i) \le \bar{X}_i(p+1,n_in+n_i)$ *is valid for all* $p \ge 0$.

From the relationship (9.20), *the validity of* $X_i(p+1,n_in+n_i) \le \bar{X}_i(p+1,n_in+n_i)$ *is achieved if* $Y_i(p,n_in) \le \bar{Y}_i(p,n_in)$ *is valid.*

With the assumption $X_i(1,n_in) \le \bar{X}_i(1,n_in)$ *and initial condition* $Y_i(0,n_in) \le \bar{Y}_i(0,n_in)$, *we have* $Y_i(1,n_in) \le \bar{Y}_i(1,n_in)$ *by using the relationship* (9.20). *Next, by assuming that* $Y_i(m,n_in) \le \bar{Y}_i(m,n_in)$ *is true for* $m \ge 1$, *it is easy to derive from* $X_i(m+1,n_in) \le \bar{X}_i(m+1,n_in)$ *and the relationship* (9.20) *that* $Y_i(m+1,n_in) \le \bar{Y}_i(m+1,n_in)$, *which means that* $Y_i(p,n_in) \le \bar{Y}_i(p,n_in)$ *is valid for all* $p \ge 0$ *according to the induction method.*

Based on the assumptions $X_i(p+1,n_in) \le \bar{X}_i(p+1,n_in)$ *and* $Y_i(p,n_in) \le \bar{Y}_i(p,n_in)$, *we have* $X_i(p+1,n_in+n_i) \le \bar{X}_i(p+1,n_in+n_i)$ *from the relationship* (9.20). *Consequently,* $X_i(p+1,n_ik) \le \bar{X}_i(p+1,n_ik)$ *is valid for all* $p \ge 0$ *and* $0 \le k \le a_i$.

The validity of $Y_i(p,n_ik) \le \bar{Y}_i(p,n_ik)$ *for all* $p \ge 0$ *and* $0 \le k \le a_i$ *can be guaranteed by using the similar method line by line and is therefore omitted here.*

In the following, we will prove that the upper bounds $\bar{X}_i(p+1,n_ik+n_i)$ *and* $\bar{Y}_i(p+1,n_ik)$ *can be minimized by choosing the estimator gains as* (9.19). *For this purpose, the upper bounds are rewritten as*

$$
\begin{aligned}
\bar{X}_i(p+1,n_ik+n_i) = {}& \Upsilon_{1i}(p+1,n_ik)\Phi_{1i}(p+1,n_ik)\Upsilon_{1i}^T(p+1,n_ik)\\
& - \Omega_{1i}(p+1,n_ik)\Phi_{1i}^{-1}(p+1,n_ik)\Omega_{1i}^T(p+1,n_ik)\\
& + A_{1i}(p+1,n_ik)\bar{X}_i(p+1,n_ik)A_{1i}^T(p+1,n_ik)\\
& + B_{1i}(p,n_ik)\bar{Y}_i(p,n_ik)B_{1i}^T(p,n_ik)\\
& + C_{1i}(p+1,n_ik)\tilde{Q}_i(p+1,n_ik)C_{1i}^T(p+1,n_ik)
\end{aligned}
$$

$$
\begin{aligned}
&+ A_{1i}(p+1,n_ik)Z_i\bar{G}_i(p+1,n_ik)\bar{Z}_i^T B_{1i}^T(p,n_ik) \\
&+ B_{1i}(p,n_ik)\bar{Z}_i\bar{G}_i(p+1,n_ik)Z_i^T A_{1i}^T(p+1,n_ik), \\
\bar{Y}_i(p+1,n_ik) =& \Upsilon_{2i}(p+1,n_ik)\Phi_{2i}(p+1,n_ik)\Upsilon_{2i}^T(p+1,n_ik) \\
&- \Omega_{2i}(p+1,n_ik)\Phi_{2i}^{-1}(p+1,n_ik)\Omega_{2i}^T(p+1,n_ik) \\
&+ A_{2i}(p+1,n_ik)\bar{X}_i(p+1,n_ik)A_{2i}^T(p+1,n_ik) \\
&+ B_{2i}(p,n_ik)\bar{Y}_i(p,n_ik)B_{2i}^T(p,n_ik) \\
&+ C_{2i}(p+1,n_ik)\tilde{Q}_i(p+1,n_ik)C_{2i}^T(p+1,n_ik) \\
&+ A_{2i}(p+1,n_ik)Z_i\bar{G}_i(p+1,n_ik)\bar{Z}_i^T B_{2i}^T(p,n_ik) \\
&+ B_{2i}(p,n_ik)\bar{Z}_i\bar{G}_i(p+1,n_ik)Z_i^T A_{2i}^T(p+1,n_ik)
\end{aligned}
$$

where

$$
\begin{aligned}
\Upsilon_{1i}(p+1,n_ik) \triangleq& K_{1i}(p+1,n_ik) - \Omega_{1i}(p+1,n_ik)\Phi_{1i}^{-1}(p+1,n_ik), \\
\Upsilon_{2i}(p+1,n_ik) \triangleq& K_{2i}(p+1,n_ik) - \Omega_{2i}(p+1,n_ik)\Phi_{2i}^{-1}(p+1,n_ik).
\end{aligned}
$$

Therefore, the upper bounds $\bar{X}_i(p+1,n_ik+n_i)$ and $\bar{Y}_i(p+1,n_ik)$ are minimized when the estimator gains are chosen as (9.19). The proof is complete.

Remark 9.4 *Under the scheduling effect of the WTOD protocol, the actual minimal local estimation error covariances are impossible to be calculated since the matrix $\Pi_i(p+1,n_ik)$ may be uninvertible. Therefore, in Theorem 9.1, upper bounds for the local estimation error covariances are derived, where positive scalars ϑ_1 and ϑ_2 are introduced to ensure the invertibility of the matrix $\Phi_{1i}(p+1,n_ik)$ and $\Phi_{2i}(p+1,n_ik)$, respectively. It can be seen that smaller ϑ_1 and ϑ_2 would lead to tighter minimal upper bounds.*

In Theorem 9.1, the estimator gains that minimize the upper bounds for the local estimation error covariances are derived, and the minimal upper bounds are given. In the following, a fusion scheme is proposed that fuses the local estimates and the minimal estimation error covariances.

To obtain the fused estimate at each time instant, we shall first derive the local estimates and the minimal estimation error covariances at each time instant. From (9.8) and the definition of $\bar{x}(p+1,n_ik)$, it is easily known that the local estimates $\vec{x}_i(p+1,k)$, $\vec{y}_i(p,k)$ and the estimation errors $\breve{x}_i(p+1,k)$, $\breve{y}_i(p,k)$ at time instant $(p+1,k)$ can be obtained by

$$
\begin{aligned}
\vec{x}_i(p+1,k) =& I_{i,k}\hat{x}_i(p+1,n_il_{i,k}), \\
\breve{x}_i(p+1,k) =& I_{i,k}\tilde{x}_i(p+1,n_il_{i,k}),
\end{aligned}
\tag{9.21}
$$

$$
\begin{aligned}
\vec{y}_i(p,k) =& \bar{I}_{i,k}\hat{y}_i(p,n_i\bar{l}_{i,k}), \\
\breve{y}_i(p,k) =& \bar{I}_{i,k}\tilde{y}_i(p,n_i\bar{l}_{i,k})
\end{aligned}
\tag{9.22}
$$

where $l_{i,k} \triangleq \hat{k}_i + 1$ and $\bar{l}_{i,k} \triangleq \check{k}_i$ with $\check{k}_i$ representing the largest integer that is no larger than $\frac{k}{n_i}$ and $\hat{k}_i$ representing the largest integer that is smaller than $\frac{k}{n_i}$, and

$$I_{i,k} \triangleq \Big[\underbrace{0 \ \cdots \ 0}_{k-\hat{k}_i n_i - 1} \ I \ \underbrace{0 \ \cdots \ 0}_{n_i - k + \hat{k}_i n_i + 1}\Big],$$

$$\bar{I}_{i,k} \triangleq \Big[\underbrace{0 \ \cdots \ 0}_{k-\bar{l}_{i,k} n_i} \ I \ \underbrace{0 \ \cdots \ 0}_{n_i - k + \bar{l}_{i,k} n_i}\Big].$$

Based on (9.21) and (9.22), the local estimation error covariances $\check{X}_i(p+1,k) \triangleq \mathbb{E}\{\check{x}_i(p+1,k)\check{x}_i^T(p+1,k)\}$ and $\check{Y}_i(p,k) \triangleq \mathbb{E}\{\check{y}_i(p,k)\check{y}_i^T(p,k)\}$ can be written as

$$\begin{aligned}
\check{X}_i(p+1,k) &= I_{i,k}\mathbb{E}\{\tilde{x}_i(p+1,n_i l_{i,k})\tilde{x}_i^T(p+1,n_i l_{i,k})\}I_{i,k}^T \\
&= I_{i,k}X_i(p+1,n_i l_{i,k})I_{i,k}^T, \\
\check{Y}_i(p,k) &= \bar{I}_{i,k}\mathbb{E}\{\tilde{y}_i(p,n_i\bar{l}_{i,k})\tilde{y}_i^T(p,n_i\bar{l}_{i,k})\}\bar{I}_{i,k}^T \\
&= \bar{I}_{i,k}Y_i(p,n_i\bar{l}_{i,k})\bar{I}_{i,k}^T.
\end{aligned}$$

In the following theorem, the well-known sequential covariance intersection fusion scheme (see e.g. [30] and the references therein) is exploited to derive the fused estimates and the covariances of the corresponding estimation errors. The proof is fairly straightforward and is therefore omitted for conciseness.

Theorem 9.2 *The fused estimates $\vec{x}^{m-1}(p+1,k)$, $\vec{y}^{m-1}(p,k)$ and the covariances $\check{X}^{m-1}(p+1,k)$, $\check{Y}^{m-1}(p,k)$ of the corresponding estimation errors can be derived recursively by the following equations:*

$$\begin{aligned}
\check{X}^{i-1}(p+1,k) =& \big[\omega_{i-1}^1(\check{X}^{i-2}(p+1,k))^{-1} + (1-\omega_{i-1}^1)\check{X}_i^{-1}(p+1,k)\big]^{-1}, \\
\vec{x}^{i-1}(p+1,k) =& \check{X}^{i-1}(p+1,k)\big[\omega_{i-1}^1(\check{X}^{i-2}(p+1,k))^{-1}\vec{x}^{i-2}(p+1,k) \\
&+ (1-\omega_{i-1}^1)\check{X}_i^{-1}(p+1,k)\vec{x}_i(p+1,k)\big], \\
\check{Y}^{i-1}(p,k) =& \big[\omega_{i-1}^2(\check{Y}^{i-2}(p,k))^{-1} + (1-\omega_{i-1}^2)\check{Y}_i^{-1}(p,k)\big]^{-1}, \\
\vec{y}^{i-1}(p,k) =& \check{Y}^{i-1}(p,k)\big[\omega_{i-1}^2(\check{Y}^{i-2}(p,k))^{-1}\vec{y}^{i-2}(p,k) \\
&+ (1-\omega_{i-1}^2)\check{Y}_i^{-1}(p,k)\vec{y}_i(p,k)\big]
\end{aligned}$$

for $i = 2,\ldots,m$, with the initial conditions $\vec{x}^0(p+1,k) = \vec{x}_1(p+1,k)$, $\check{X}^0(p+1,k) = \check{X}_1(p+1,k)$, $\vec{y}^0(p,k) = \vec{y}_1(p,k)$ and $\check{Y}^0(p,k) = \check{Y}_1(p,k)$. Furthermore, the optimal weighting coefficients ω_{i-1}^1 and ω_{i-1}^2 are obtained by *solving following optimization problems*

$$\min_{\omega_{i-1}^1 \in [0,1]} \{\mathrm{tr}(\breve{X}^{i-1}(p+1,k))\},$$

$$\min_{\omega_{i-1}^2 \in [0,1]} \{\mathrm{tr}(\breve{Y}^{i-1}(p,k))\}, \quad i = 2, \ldots, m.$$

Moreover, for $i = 1, 2, \cdots, m$*, we have*

$$\breve{X}^{m-1}(p+1,k) \leq \breve{X}_i(p+1,k), \ \breve{Y}^{m-1}(p,k) \leq \breve{Y}_i(p,k).$$

Remark 9.5 *From (9.21) and (9.22), it can be seen that the cross-covariances* $\breve{X}_{ij}(p+1,k+1) = \mathbb{E}\{\breve{x}_i(p+1,k+1)\breve{x}_j^T(p+1,k+1)\}$ *and* $\breve{Y}_{ij}(p+1,k) = \mathbb{E}\{\breve{y}_i(p+1,k)\breve{y}_j^T(p+1,k)\}$ *between the ith and the jth estimator contain, respectively, the terms* $\mathbb{E}\{\tilde{x}_i(p+1, n_i l_{i,k+1} - n_i)\tilde{y}_j(p, n_j l_{j,k+1} - n_j)\}$ *and* $\mathbb{E}\{\tilde{x}_i(p+1, n_i \bar{l}_{i,k})\tilde{y}_j(p, n_j \bar{l}_{j,k})\}$*. Due to the coupling between the LRPs and the multi-rate sampling, the terms* $\mathbb{E}\{\tilde{x}_i(p+1, n_i l_{i,k+1} - n_i)\tilde{y}_j(p, n_j l_{j,k+1} - n_j)\}$ *and* $\mathbb{E}\{\tilde{x}_i(p+1, n_i \bar{l}_{i,k})\tilde{y}_j(p, n_j \bar{l}_{j,k})\}$ *are difficult to be calculated. Therefore, in this chapter, the sequential covariance intersection fusion scheme is employed to fuse the local estimates.*

Remark 9.6 *Until now, the fusion estimation problem for multi-rate LRPs under WTOD protocol has been studied. In Theorem 9.1, the upper bounds for the local estimation error covariances have been first derived and the estimator gains which minimize the obtained upper bounds have been given. In Theorem 9.2, based on the estimates from the designed local estimators, a sequential covariance intersection fusion scheme has been proposed to obtain the fused estimate. In the next section, an illustrative example will be given to show the effectiveness of the proposed fusion estimation scheme.*

9.3 A Simulation Example

In this section, a simulation example is given to show the performance of the proposed estimation scheme. Moreover, the effectiveness of the developed fusion scheme is also verified.

Consider a discrete time-varying linear repetitive process with path length as $a = 20$ and the following system matrices:

$$\breve{A}_1(p+1,k) = \begin{bmatrix} 0.76 + 0.01\sin(p) & 0.05 \\ 0.04 & 0.54 + 0.01\cos(k) \end{bmatrix},$$

$$\breve{A}_2(p+1,k) = \begin{bmatrix} 0.47 + 0.02\sin(p) & 0.11 \\ 0.06 & -0.49 + 0.01\cos(k) \end{bmatrix},$$

$$\breve{B}_1(p+1,k) = \begin{bmatrix} 0.32 + 0.01\cos(p) & 0.11 \\ 0.25 & 0.74 \end{bmatrix},$$

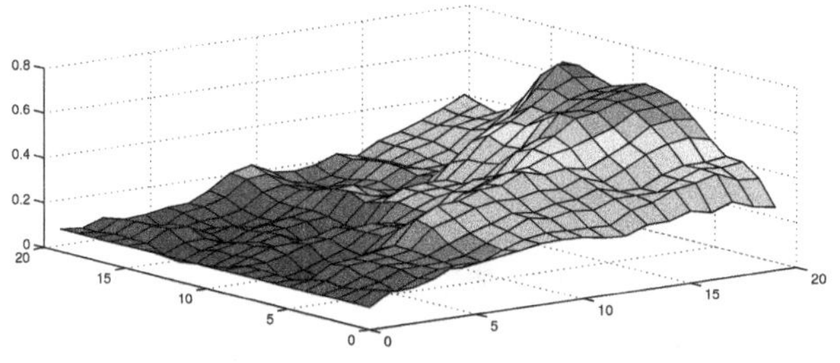

FIGURE 9.1: The state $\breve{x}(p+1,k)$.

$$\breve{B}_2(p+1,k) = \begin{bmatrix} 0.42 + 0.02\sin(2p) & 0.14 \\ 0.22 + 0.01\cos(k) & 0.71 \end{bmatrix},$$

$$\breve{C}_1(p+1,k) = \begin{bmatrix} 0.11 \\ 0.22 \end{bmatrix}, \ \breve{C}_2(p+1,k) = \begin{bmatrix} 0.31 \\ 0.23 \end{bmatrix}.$$

The linear repetitive process is measured by two sensors with $n_1 = 2$, $n_2 = 1$ and the following parameters:

$$\breve{D}_1(p+1,2k) = \begin{bmatrix} 7\sin(2p) \\ 9\cos(2k) \end{bmatrix}, \ \breve{D}_2(p+1,k) = \begin{bmatrix} 5\sin(2p) \\ 8\cos(k) \end{bmatrix},$$

$$\breve{E}_1(p,2k) = \begin{bmatrix} 5\sin(2p) \\ 6\cos(2k) \end{bmatrix}, \ \breve{E}_2(p,k) = \begin{bmatrix} 4\sin(2p) \\ 5\cos(k) \end{bmatrix},$$

$$\breve{F}_1(p+1,2k) = 2\sin(p), \ \breve{F}_2(p+1,k) = 3\sin(p).$$

The initial values of the linear repetitive process are chosen as $b_{p+1} = \begin{bmatrix} 0.1 & 0.15 \end{bmatrix}^T$ and $c_k = \begin{bmatrix} 0.1 & 0.15 \end{bmatrix}^T$. The covariances of the noises are set as $Q(p+1,k) = 0.02$, $R_1(p+1,k) = 0.01$, and $R_2(p+1,k) = 0.02$. The simulation results are shown in Figs. 9.1–9.10, where only the first element of the states and estimates are presented for saving the space.

The performance of the proposed estimation scheme is verified in Figs. 9.1–9.8. The state $\breve{x}(p+1,k)$ and its estimates from 1th and 2th estimators are given in Figs. 9.1–9.3, respectively. The state $\breve{y}(p,k)$ and its estimates from 1th and 2th estimators are given in Fig. 9.4–9.6, respectively. In Fig. 9.7, the actual estimation error covariance $X_2(p+1,k)$ is depicted based on 100 independent experiments, and the upper bound $\bar{X}_2(p+1,k)$ for the estimation error covariance is shown in Fig. 9.8. It can be seen from Figs. 9.1–9.8 that the developed estimation scheme is effective. The effectiveness of the proposed fusion scheme is examined in Figs. 9.9–9.10 where the fused estimates are given. It can be seen from Figs. 9.9–9.10 that the estimation accuracy of the proposed fusion scheme outperform the local estimators.

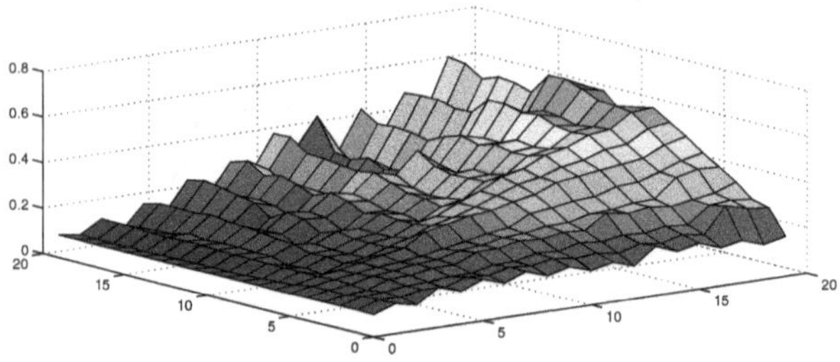

FIGURE 9.2: The estimate of $\breve{x}(p+1,k)$ from the 1th estimator.

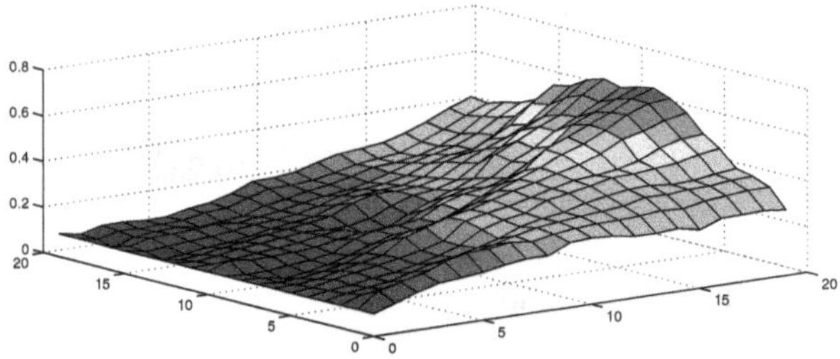

FIGURE 9.3: The estimate of $\breve{x}(p+1,k)$ from the 2th estimator.

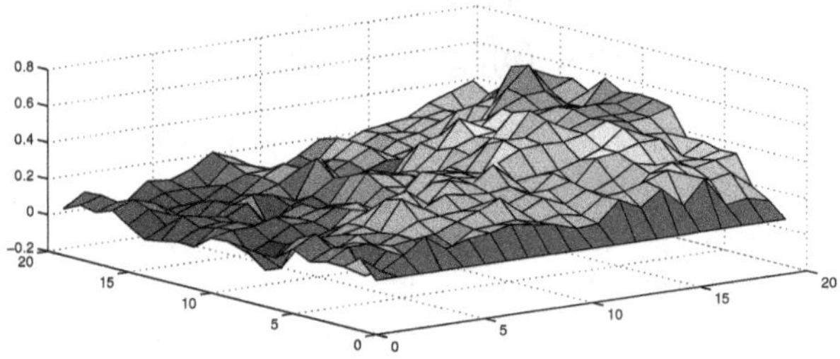

FIGURE 9.4: The state $\breve{y}(p,k)$.

9.4 Conclusion

In this chapter, the fusion estimation problem has been investigated for a class of discrete time-varying multi-rate LRPs under WTOD protocol. The lifting technique has been applied to convert the multi-rate LRPs to single-rate ones, and a novel selection principle of the WTOD protocol has been proposed to cater for sensors with different sampling periods. A set of recursive

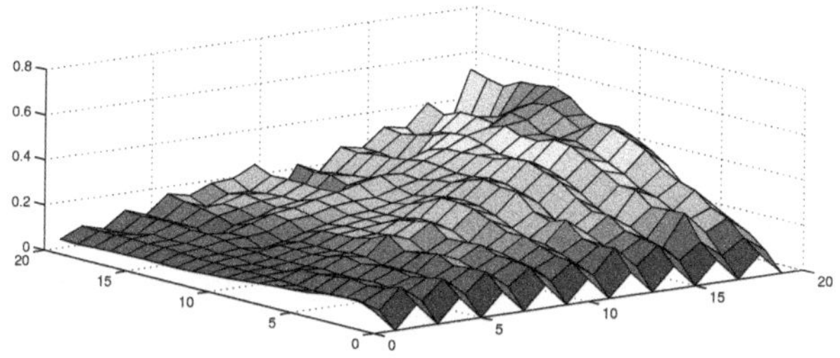

FIGURE 9.5: The estimate of $\check{y}(p,k)$ from the 1th estimator.

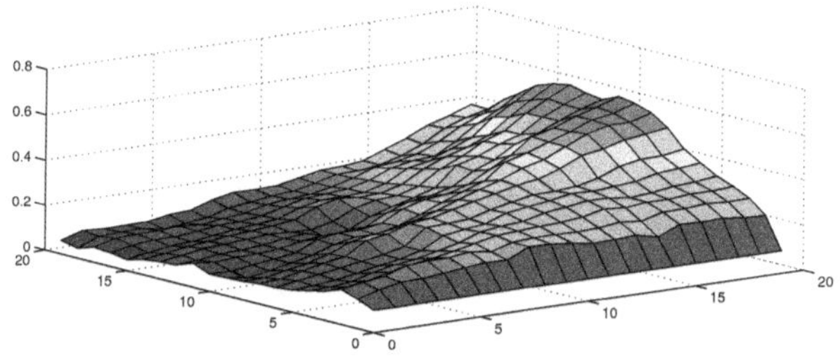

FIGURE 9.6: The estimate of $\check{y}(p,k)$ from the 2th estimator.

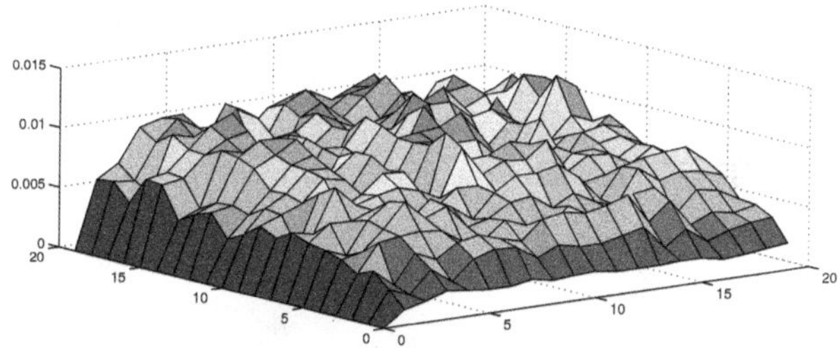

FIGURE 9.7: The actual estimation error covariance $X_2(p+1,k)$.

estimators has been constructed to estimate the states of the LRPs. Based on the proposed estimation scheme, the upper bounds for the estimation error covariances have been derived. Then, such upper bounds have been minimized at each time instant by appropriately choosing the estimator gains. Furthermore, the local estimates have been fused by recurring to the sequential covariance intersection fusion method. Finally, a numerical simulation example has been given to show the effectiveness of the overall fusion estimation scheme.

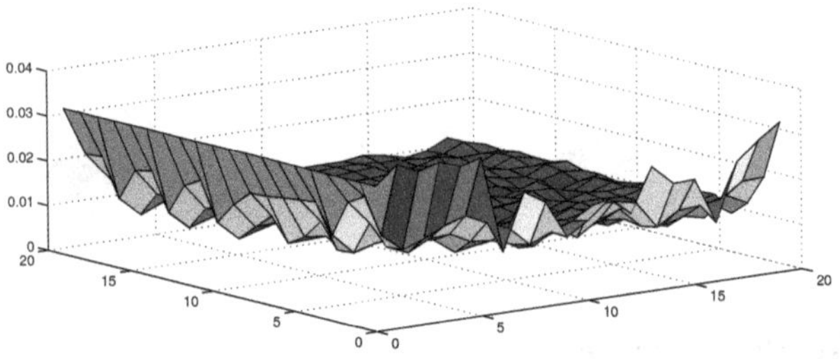

FIGURE 9.8: The upper bound $\bar{X}_2(p+1,k)$ for the estimation error covariance.

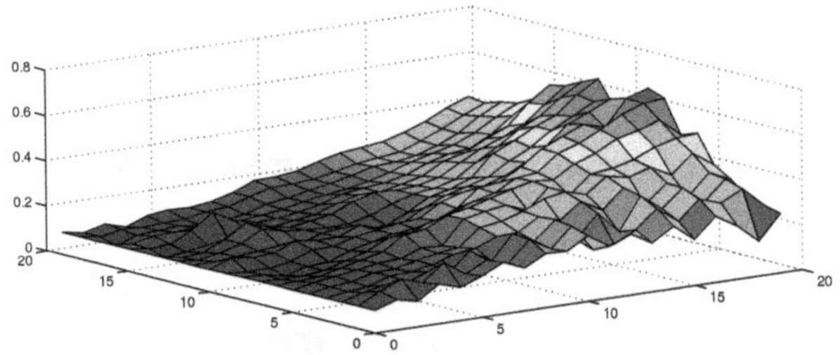

FIGURE 9.9: The fused estimate of $\breve{x}(p+1,k)$.

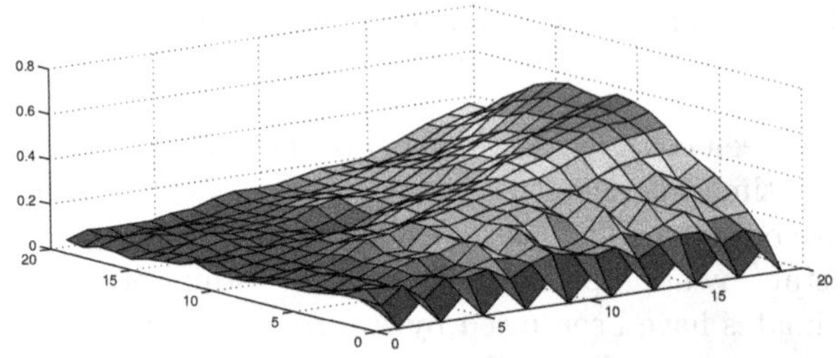

FIGURE 9.10: The fused estimate of $\breve{y}(p,k)$.

10

Outlier-Resistant Recursive Filtering for Multi-Rate Systems under Weighted Try-Once-Discard Protocol

Measurement outliers have long been a central topic in the areas of statistics and data analytics [5]. The measurement outliers are generally defined as contaminated measurements which deviate significantly from the normal measurements [153]. In networked systems, during the signal transmission along communication channel of limited bandwidth, the measurement outputs may undergo abrupt yet large disturbances due probably to unpredictable sensor failures, sudden environmental changes or even intended cyber-attacks, thereby resulting in the occurrence of the measurement outliers. Clearly, traditional filtering methods are no longer applicable as the measurement outliers may lead to abnormal magnitude changes with the innovation and, in turn, deteriorate or even destabilize the filter. Therefore, it is vitally important to design filters whose performance indices are insensitive to the appearance of measurement outliers and, accordingly, much research has been done along this line, see e.g. [1,121] for outlier resistance, [158] for outlier detection and [6] for outlier rejection.

Despite the recurring research interest in outlier-resistant filter/observer design problems, the corresponding results on recursive filtering problem for time-varying systems with error-variance constraints have been scattered, which is mainly due to the following three identified challenges: 1) how to design a suitable filter structure that can constrain/compensate the adverse effect of the measurement outliers while facilitating the recursive nature of the filter algorithm? 2) how to ensure an upper-bounded filtering error covariance which can then be minimized at each time step against the measurement outliers? and 3) how to guarantee the uniform convergence of the filtering error dynamics by resorting to the exponential boundedness regardless the measurement outliers? As such, we will make a dedicated effort in this chapter to deal with the listed three challenges coupled by the complexities resulting from both the multi-rate sampling and the WTOD protocols.

Motivated by the above discussions, in this chapter, we launch a major study on the multi-sensor MRSs with measurement outliers under the WTOD protocol. The contributions of this chapter are summarized as follows: 1) we, as the first of few attempts, investigate the recursive filtering problem for MRSs under the WTOD protocol; 2) an outlier-resistant filter is proposed

DOI: 10.1201/9781032619507-10

that maintains a satisfactory filtering performance in case of the measurement outliers; 3) an upper bound is ensured on the filtering error covariance by resorting to the solution to a matrix difference equation and such an upper bound is later minimized; and 4) a sufficient condition is given that ensures the exponential boundedness of the filtering error dynamics.

10.1 Problem Formulation

Consider the following class of discrete time-varying systems:

$$x(s_{k+1}) = A(s_k)x(s_k) + B(s_k)w(s_k) \tag{10.1}$$

where $x(s_k) \in \mathbb{R}^{n_x}$ is the system state and $w(s_k) \in \mathbb{R}^{n_w}$ is the zero-mean process noise with covariance $Q(s_k) > 0$. $A(s_k)$ and $B(s_k)$ are known time-varying matrices with appropriate dimensions. The initial value $x(s_0)$ is a random variable with the mean $d(s_0)$ and the covariance $X(s_0)$.

Let $h \triangleq s_{k+1} - s_k$ be the state update period of the system where h is a known positive constant. The system is measured by n sensors with a sampling period $bh \triangleq q_{k+1} - q_k$ where $b \geq 2$ is a known positive integer. For sensor i $(i = 1, 2, \ldots, n)$, the measurement model is described as follows:

$$y_i(q_k) = C_i(q_k)x(q_k) + D_i(q_k)v_i(q_k) \tag{10.2}$$

where $y_i(q_k) \in \mathbb{R}^{n_y}$ is the measurement output of sensor i and $v_i(q_k) \in \mathbb{R}^{n_v}$ is the zero-mean measurement noise for sensor i with covariance $R_i(q_k) > 0$. $C_i(q_k)$ and $D_i(q_k)$ are known time-varying matrices with appropriate dimensions.

Assumption 10.1 *$x(s_0)$, $w(s_k)$ $(k \geq 0)$, and $v_i(q_k)$ $(1 \leq i \leq n, k \geq 0)$ are mutually independent random variables.*

In this chapter, the measurement outputs of the sensors are transmitted to the filter through a shared communication network with limited bandwidth. To avoid data collisions, as with the case in industry, certain network protocol is utilized whose main idea is to ensure that only one sensor is allowed to transmit its measurement output at each sampling instant. In this chapter, due to its merits in dynamically granting accesses according to the individual needs from the sensors, the WTOD protocol is chosen to schedule the transmission order of the sensors in the sensor-to-filter channel.

Let $\xi(q_k)$ denote the selected sensor that is granted the privilege to transmit signals at time instant q_k. According to the WTOD-based selection principle, $\xi(q_k)$ can be determined as

$$\xi(q_k) = \arg\max_{1 \leq i \leq n} (y_i(q_k) - y_i^*(q_k))^T \Omega_i (y_i(q_k) - y_i^*(q_k))$$

where $y_i^*(q_k)$ represents the last transmitted signal from sensor i before time instant q_k and Ω_i $(i = 1, 2, \ldots, n)$ is the known positive definite weight matrix of sensor i.

Taking the WTOD protocol into account, the *actual* measurement $\breve{y}_i(q_k)$ received by the filter with a zero-order holder from sensor i is formulated as

$$\breve{y}_i(q_k) = \delta(\xi(q_k), i) y_i(q_k) + (1 - \delta(\xi(q_k), i))\, \breve{y}_i(q_{k-1})$$

where $\delta(\cdot, \cdot)$ is the Kronecker delta function. Moreover, it is assumed that $\breve{y}_i(q_k) = 0$ for any $k < 0$.

By denoting $\bar{y}(q_k) \triangleq \mathrm{col}\{\breve{y}_1(q_k), \breve{y}_2(q_k), \cdots, \breve{y}_n(q_k)\}$, we derive an augmented system with the following compact form:

$$\begin{cases} x(s_{k+1}) = A(s_k)x(s_k) + B(s_k)w(s_k) \\ \quad \bar{y}(q_k) = \Phi(q_k)\vec{C}(q_k)x(q_k) + \Phi(q_k)\vec{D}(q_k)\vec{v}(q_k) \\ \qquad\qquad + (I - \Phi(q_k))\bar{y}(q_{k-1}) \end{cases} \tag{10.3}$$

where

$$\begin{aligned} \Phi(q_k) &\triangleq \mathrm{diag}\{\delta(\xi(q_k), 1)I, \delta(\xi(q_k), 2)I, \cdots, \delta(\xi(q_k), n)I\}, \\ \vec{D}(q_k) &\triangleq \mathrm{diag}\{D_1(q_k), D_2(q_k), \cdots, D_n(q_k)\}, \\ \vec{C}(q_k) &\triangleq \mathrm{col}\{C_1(q_k), C_2(q_k), \cdots, C_n(q_k)\}, \\ \vec{v}(q_k) &\triangleq \mathrm{col}\{v_1(q_k), v_2(q_k), \cdots, v_n(q_k)\}. \end{aligned}$$

It is easily known from (10.3) that the system under consideration is essentially a MRS. To facilitate the filter design, we like to convert the MRS to a single-rate one by employing the lifting technique. In doing so, by denoting $\tilde{x}(q_k) \triangleq \mathrm{col}\{x(q_{k-1} + h), \cdots, x(q_k - h), x(q_k)\}$, we have the following single-rate system:

$$\begin{cases} \tilde{x}(q_{k+1}) = \tilde{A}(q_k)\tilde{x}(q_k) + \tilde{B}(q_k)\tilde{w}(q_k) \\ \quad \bar{y}(q_k) = \Phi(q_k)\tilde{C}(q_k)\tilde{x}(q_k) + \Phi(q_k)\vec{D}(q_k)\vec{v}(q_k) \\ \qquad\qquad + (I - \Phi(q_k))\bar{y}(q_{k-1}) \end{cases} \tag{10.4}$$

where

$$\begin{aligned} \tilde{w}(q_k) &\triangleq \mathrm{col}\{w(q_k), w(q_k + h), \cdots, w(q_{k+1} - h)\}, \\ \tilde{A}(q_k) &\triangleq \Big[\underbrace{0_{bn_x \times n_x} \quad \cdots \quad 0_{bn_x \times n_x}}_{b-1} \quad \mathscr{A}(q_k) \Big], \\ \tilde{B}(q_k) &\triangleq \bar{\mathscr{A}}(q_k)\mathscr{B}(q_k), \ \ \mathscr{F}_t^s(q_k) \triangleq \prod_{i=t}^{s} A(q_{k+1} - ih), \\ \mathscr{A}(q_k) &\triangleq \mathrm{col}\{\mathscr{F}_b^b(q_k), \mathscr{F}_{b-1}^b(q_k), \cdots, \mathscr{F}_1^b(q_k)\}, \end{aligned}$$

$$\tilde{\mathscr{A}}(q_k) \triangleq \begin{bmatrix} I & 0 & \cdots & 0 \\ \mathscr{F}_{b-1}^{b-1}(q_k) & I & \cdots & 0 \\ \mathscr{F}_{b-2}^{b-1}(q_k) & \mathscr{F}_{b-2}^{b-2}(q_k) & \cdots & 0 \\ \vdots & \vdots & \cdots & 0 \\ \mathscr{F}_{1}^{b-1}(q_k) & \mathscr{F}_{1}^{b-2}(q_k) & \cdots & I \end{bmatrix},$$

$$\mathscr{B}(q_k) \triangleq \mathrm{diag}\{B(q_k), B(q_k+h), \cdots, B(q_{k+1}-h)\},$$

$$\tilde{C}(q_k) \triangleq \Big[\underbrace{0_{nn_y\times n_x} \quad \cdots \quad 0_{nn_y\times n_x}}_{b-1} \quad \vec{C}(q_k)\Big].$$

Defining $\bar{x}(q_k) \triangleq \begin{bmatrix}\tilde{x}^T(q_k) & \bar{y}^T(q_{k-1})\end{bmatrix}^T$, we have

$$\begin{cases} \bar{x}(q_{k+1}) = \bar{A}(q_k)\bar{x}(q_k) + \bar{B}(q_k)\omega(q_k) \\ \bar{y}(q_k) = \bar{C}(q_k)\bar{x}(q_k) + \bar{D}(q_k)\omega(q_k) \end{cases} \tag{10.5}$$

where

$$\bar{A}(q_k) \triangleq \begin{bmatrix} \tilde{A}(q_k) & 0 \\ \Phi(q_k)\tilde{C}(q_k) & I-\Phi(q_k) \end{bmatrix},$$

$$\bar{B}(q_k) \triangleq \begin{bmatrix} \tilde{B}(q_k) & 0 \\ 0 & \Phi(q_k)\vec{D}(q_k) \end{bmatrix}, \quad \omega(q_k) \triangleq \begin{bmatrix} \tilde{w}(q_k) \\ \vec{v}(q_k) \end{bmatrix},$$

$$\bar{C}(q_k) \triangleq \begin{bmatrix} \Phi(q_k)\tilde{C}(q_k) & I-\Phi(q_k) \end{bmatrix},$$

$$\bar{D}(q_k) \triangleq \begin{bmatrix} 0 & \Phi(q_k)\vec{D}(q_k) \end{bmatrix}.$$

As discussed in the introduction, measurement outliers are a prevalent phenomenon in engineering practice which, if not adequately handled, would be likely to cause abnormal innovation in the filtering process and even render the filtering error divergent. To restrain the negative impact from the measurement outliers, in this chapter, a saturation function is introduced in the filter whose saturation level can be determined according to the prior knowledge on the range of the innovation. To be specific, the filter is constructed as follows:

$$\hat{x}(q_{k+1}) = \bar{A}(q_k)\hat{x}(q_k) + K(q_k)\sigma\big(\bar{y}(q_k) - \bar{C}(q_k)\hat{x}(q_k)\big) \tag{10.6}$$

where $\hat{x}(q_k)$ is the estimate of $\bar{x}(q_k)$ and $K(q_k)$ is the filter gain to be designed. The initial value of the filter is $\hat{x}(q_0) = \mathbb{E}\{\bar{x}(q_0)\}$. Here, the purposely introduced saturation function $\sigma(\cdot) : \mathbb{R}^{nn_y} \to \mathbb{R}^{nn_y}$ is defined as follows:

$$\sigma(l) \triangleq \begin{bmatrix}\sigma_1(l_1) & \sigma_2(l_2) & \cdots & \sigma_{nn_y}(l_{nn_y})\end{bmatrix}^T$$

with

$$\sigma_i(l_i) \triangleq \mathrm{sign}(l_i)\min\{l_{i,\max}, |l_i|\} \tag{10.7}$$

where $\mathrm{sign}(\cdot)$ is the signum function and $l_{i,\max}$, i.e. saturation level, is the i-th element of the vector $l_{\max}$.

Remark 10.1 *During the signal transmission from the sensors to the filter, the measurement outliers may occur due to a variety of reasons such as intermittent component failures, cyber attacks or sudden environmental changes. In general, measurement outliers are contaminated measurements that deviate significantly from normal measurements. It is obvious that the deviation of the measurements leads directly to the deviation of the innovations which, in turn, has an adverse effect on the filtering performance. As such, in this chapter, we develop an outlier-resistant filter in the form of* (10.6) *to maintain a satisfactory filtering performance in case of measurement outliers.*

Remark 10.2 *The measurement outliers, if directly introduced into the innovation* $\bar{y}(q_k)-\bar{C}(q_k)\hat{x}(q_k)$*, would result in abnormal deviation of the innovation from its usual pattern, thereby deteriorating the filtering performance. In this chapter, an outlier-resistant filter of the structure* (10.6) *is proposed where a specific saturation function* $\sigma(\cdot)$ *is put forward. It is observed from* (10.7) *that the innovation is now constrained within a given range that can be determined a priori according to engineering practice. The superiority of the proposed filter is its capability of restraining the innovation deviation within a predefined range and, consequently, the influence from the measurement outliers to the filtering performance is mitigated. It is worth mentioning that the designed filter specializes to the standard one (see, e.g. [136]) when the saturation levels* $l_{i,\max}$ *approach infinity.*

By denoting the filtering error as $e(q_k) \triangleq \bar{x}(q_k) - \hat{x}(q_k)$, we have

$$
\begin{aligned}
e(q_{k+1}) =&\bar{A}(q_k)e(q_k) + \bar{B}(q_k)\omega(q_k)\\
&- K(q_k)\sigma\big(\mathcal{C}(q_k)\big)
\end{aligned}
\tag{10.8}
$$

where

$$\mathcal{C}(q_k) \triangleq \bar{C}(q_k)e(q_k) + \bar{D}(q_k)\omega(q_k).$$

The purpose of this chapter is to develop a filter (10.6) such that 1) an upper bound $\Xi(q_k)$ is ensured for the filtering error covariance $P(q_k) \triangleq \mathbb{E}\{e(q_k)e^T(q_k)\}$, and 2) an appropriate filter gain $K(q_k)$ is designed to minimize the obtain $\Xi(q_k)$ at each sampling instant.

10.2 Outlier-Resistant Filter Design

In this section, an upper bound $\Xi(q_k)$ is first derived by resorting to the solution to a matrix difference equation. Then, the filter gain minimizing such an upper bound $\Xi(q_k)$ is obtained with explicit expression of the resulting minimal upper bound.

The following lemma is useful in obtaining the main results of this chapter.

Lemma 10.1 *[136] For a matrix $A = A^T > 0$ and a function $\phi_k(A) = \phi_k^T(A)$ $(k \geq 0)$. If*

$$\phi_k(A) \leq \phi_k(B), \quad \forall A \leq B,$$

then the solutions W_k and V_k to

$$W_{k+1} = \phi_k(W_k), \quad V_{k+1} \leq \phi_k(V_k), \quad W_0 = V_0$$

satisfy

$$V_{k+1} \leq W_{k+1}.$$

Before processing further, the saturation function $\sigma\big(\mathcal{C}(q_k)\big)$ is rewritten as follows to facilitate later analysis. First, we rewrite $\mathcal{C}(q_k)$ as

$$\mathcal{C}(q_k) = \text{col}\{\mathcal{C}_1(q_k), \mathcal{C}_2(q_k), \cdots, \mathcal{C}_{nn_y}(q_k)\}$$

where $\mathcal{C}_i(q_k)$ is the i-th element of $\mathcal{C}(q_k)$. From (10.7), we know that

$$\sigma_i(\mathcal{C}_i(q_k)) = \begin{cases} \mathcal{C}_i(q_k), & \text{if } |\mathcal{C}_i(q_k)| \leq l_{i,\max}; \\ \text{sign}(\mathcal{C}_i(q_k)) l_{i,\max}, & \text{otherwise.} \end{cases}$$

Next, we introduce a function $\lambda(\cdot, \cdot)$ satisfying

$$\lambda(a, b) = \begin{cases} 0, & \text{if } a \leq b; \\ 1, & \text{otherwise,} \end{cases}$$

and $\sigma_i(\mathcal{C}_i(q_k))$ can then be formulated as

$$\begin{aligned}\sigma_i(\mathcal{C}_i(q_k)) =& (1 - \lambda(|\mathcal{C}_i(q_k)|, l_{i,\max}))\mathcal{C}_i(q_k) \\ &+ \lambda(|\mathcal{C}_i(q_k)|, l_{i,\max})\text{sign}(\mathcal{C}_i(q_k)) l_{i,\max}.\end{aligned}$$

Next, the saturation function $\sigma\big(\mathcal{C}(q_k)\big)$ can be written as

$$\sigma\big(\mathcal{C}(q_k)\big) = (I - \Lambda(q_k))\mathcal{C}(q_k) + \Lambda(q_k) L(q_k) \tag{10.9}$$

where

$$\begin{aligned}\Lambda(q_k) \triangleq& \text{diag}\{\lambda(|\mathcal{C}_1(q_k)|, l_{1,\max}), \lambda(|\mathcal{C}_2(q_k)|, l_{2,\max}), \\ &\cdots, \lambda(|\mathcal{C}_{nn_y}(q_k)|, l_{nn_y,\max})\}, \\ L(q_k) \triangleq& \text{col}\{\text{sign}(\mathcal{C}_1(q_k)) l_{1,\max}, \text{sign}(\mathcal{C}_2(q_k)) l_{2,\max}, \\ &\cdots, \text{sign}(\mathcal{C}_{nn_y}(q_k)) l_{nn_y,\max}\}.\end{aligned}$$

Noting (10.9), the filtering error dynamics (10.8) can be rewritten as

$$\begin{aligned}e(q_{k+1}) =& \mathcal{A}(q_k)e(q_k) + \mathcal{B}(q_k)\omega(q_k) \\ &+ K(q_k)\Lambda(q_k)\bar{C}(q_k)e(q_k) \\ &+ K(q_k)\Lambda(q_k)\bar{D}(q_k)\omega(q_k) \\ &- K(q_k)\Lambda(q_k)L(q_k)\end{aligned} \tag{10.10}$$

where

$$\begin{aligned}\mathcal{A}(q_k) &\triangleq \bar{A}(q_k) - K(q_k)\bar{C}(q_k),\\ \mathcal{B}(q_k) &\triangleq \bar{B}(q_k) - K(q_k)\bar{D}(q_k).\end{aligned}$$

In light of (10.10), an upper bound $\Xi(q_k)$ is constructed in the following theorem.

Theorem 10.1 *Let positive scalars $\varepsilon_i(q_k)$ $(i = 1, 2, \ldots, 9)$ be given. Consider the following recursion for $\Xi(q_{k+1})$:*

$$\begin{aligned}\Xi(q_{k+1}) =& \zeta_1(q_k)\mathcal{A}(q_k)\Xi(q_k)\mathcal{A}^T(q_k)\\ &+ \zeta_2(q_k)\mathcal{B}(q_k)W(q_k)\mathcal{B}^T(q_k)\\ &+ \zeta_3(q_k)K(q_k)\mathrm{tr}\{\bar{C}(q_k)\Xi(q_k)\bar{C}^T(q_k)\}IK^T(q_k)\\ &+ \zeta_4(q_k)K(q_k)\mathrm{tr}\{\bar{D}(q_k)W(q_k)\bar{D}^T(q_k)\}IK^T(q_k)\\ &+ \zeta_5(q_k)nn_y\bar{\sigma}K(q_k)K^T(q_k)\end{aligned} \tag{10.11}$$

where

$$\begin{aligned}W(q_k) &\triangleq \mathrm{diag}\{\tilde{Q}(q_k), \tilde{R}(q_k)\},\ \bar{\sigma} \triangleq \sum_{i=1}^{nn_y} l_{i,\max}^2,\\ \tilde{Q}(q_k) &\triangleq \mathrm{diag}\{Q(q_k), Q(q_k + h), \cdots, Q(q_{k+1} - h)\},\\ \tilde{R}(q_k) &\triangleq \mathrm{diag}\{R_1(q_k), R_2(q_k), \cdots, R_n(q_k)\},\\ \zeta_1(q_k) &\triangleq 1 + \varepsilon_1(q_k) + \varepsilon_2(q_k) + \varepsilon_3(q_k),\\ \zeta_2(q_k) &\triangleq 1 + \varepsilon_4(q_k) + \varepsilon_5(q_k) + \varepsilon_6(q_k),\\ \zeta_3(q_k) &\triangleq 1 + \varepsilon_1^{-1}(q_k) + \varepsilon_4^{-1}(q_k) + \varepsilon_7(q_k) + \varepsilon_8(q_k),\\ \zeta_4(q_k) &\triangleq 1 + \varepsilon_2^{-1}(q_k) + \varepsilon_5^{-1}(q_k) + \varepsilon_7^{-1}(q_k) + \varepsilon_9(q_k),\\ \zeta_5(q_k) &\triangleq 1 + \varepsilon_3^{-1}(q_k) + \varepsilon_6^{-1}(q_k) + \varepsilon_8^{-1}(q_k) + \varepsilon_9^{-1}(q_k)\end{aligned}$$

with initial value $\Xi(q_0) = P(q_0)$. Then, $\Xi(q_{k+1})$ is an upper bound on the filtering error covariance $P(q_{k+1})$.

Proof *From* (10.10), *it is clear that*

$$\begin{aligned}P(q_{k+1}) =& \mathcal{A}(q_k)P(q_k)\mathcal{A}^T(q_k)\\ &+ \mathcal{B}(q_k)\mathbb{E}\{\omega(q_k)\omega^T(q_k)\}\mathcal{B}^T(q_k)\\ &+ \mathbb{E}\{K(q_k)\Lambda(q_k)\bar{C}(q_k)e(q_k)e^T(q_k)\bar{C}^T(q_k)\Lambda^T(q_k)K^T(q_k)\}\\ &+ \mathbb{E}\{K(q_k)\Lambda(q_k)\bar{D}(q_k)\omega(q_k)\omega^T(q_k)\bar{D}^T(q_k)\Lambda^T(q_k)K^T(q_k)\}\\ &+ \mathbb{E}\{K(q_k)\Lambda(q_k)L(q_k)L^T(q_k)\Lambda^T(q_k)K^T(q_k)\}\\ &+ \Upsilon_1(q_k) + \Upsilon_1^T(q_k) + \Upsilon_2(q_k) + \Upsilon_2^T(q_k)\end{aligned}$$

$$\begin{aligned}&- \Upsilon_3(q_k) - \Upsilon_3^T(q_k) + \Upsilon_4(q_k) + \Upsilon_4^T(q_k)\\&+ \Upsilon_5(q_k) + \Upsilon_5^T(q_k) - \Upsilon_6(q_k) - \Upsilon_6^T(q_k)\\&+ \Upsilon_7(q_k) + \Upsilon_7^T(q_k) - \Upsilon_8(q_k) - \Upsilon_8^T(q_k)\\&- \Upsilon_9(q_k) - \Upsilon_9^T(q_k)\end{aligned} \tag{10.12}$$

where

$$\begin{aligned}\Upsilon_1(q_k) &\triangleq \mathbb{E}\{\mathcal{A}(q_k)e(q_k)e^T(q_k)\bar{C}^T(q_k)\Lambda^T(q_k)K^T(q_k)\},\\\Upsilon_2(q_k) &\triangleq \mathbb{E}\{\mathcal{A}(q_k)e(q_k)\omega^T(q_k)\bar{D}^T(q_k)\Lambda^T(q_k)K^T(q_k)\},\\\Upsilon_3(q_k) &\triangleq \mathbb{E}\{\mathcal{A}(q_k)e(q_k)L^T(q_k)\Lambda^T(q_k)K^T(q_k)\},\\\Upsilon_4(q_k) &\triangleq \mathbb{E}\{\mathcal{B}(q_k)\omega(q_k)e^T(q_k)\bar{C}^T(q_k)\Lambda^T(q_k)K^T(q_k)\},\\\Upsilon_5(q_k) &\triangleq \mathbb{E}\{\mathcal{B}(q_k)\omega(q_k)\omega^T(q_k)\bar{D}^T(q_k)\Lambda^T(q_k)K^T(q_k)\},\\\Upsilon_6(q_k) &\triangleq \mathbb{E}\{\mathcal{B}(q_k)\omega(q_k)L^T(q_k)\Lambda^T(q_k)K^T(q_k)\},\\\Upsilon_7(q_k) &\triangleq \mathbb{E}\{K(q_k)\Lambda(q_k)\bar{C}(q_k)e(q_k)\omega^T(q_k)\bar{D}^T(q_k)\Lambda^T(q_k)K^T(q_k)\},\\\Upsilon_8(q_k) &\triangleq \mathbb{E}\{K(q_k)\Lambda(q_k)\bar{C}(q_k)e(q_k)L^T(q_k)\Lambda^T(q_k)K^T(q_k)\},\\\Upsilon_9(q_k) &\triangleq \mathbb{E}\{K(q_k)\Lambda(q_k)\bar{D}(q_k)\omega(q_k)L^T(q_k)\Lambda^T(q_k)K^T(q_k)\}.\end{aligned}$$

With the help of the elementary inequality $ab^T + ba^T \leq \varepsilon aa^T + \varepsilon^{-1}bb^T$ *(where* a *and* b *are vectors of appropriate dimensions), it is straightforward to see that*

$$\begin{aligned}\Upsilon_1(q_k) + \Upsilon_1^T(q_k) \leq{}& \varepsilon_1(q_k)\mathcal{A}(q_k)P(q_k)\mathcal{A}^T(q_k)\\&+ \varepsilon_1^{-1}(q_k)\mathbb{E}\{K(q_k)\Lambda(q_k)\bar{C}(q_k)e(q_k)\\&\quad\times e^T(q_k)\bar{C}^T(q_k)\Lambda^T(q_k)K^T(q_k)\},\\\Upsilon_2(q_k) + \Upsilon_2^T(q_k) \leq{}& \varepsilon_2(q_k)\mathcal{A}(q_k)P(q_k)\mathcal{A}^T(q_k)\\&+ \varepsilon_2^{-1}(q_k)\mathbb{E}\{K(q_k)\Lambda(q_k)\bar{D}(q_k)\omega(q_k)\\&\quad\times \omega^T(q_k)\bar{D}^T(q_k)\Lambda^T(q_k)K^T(q_k)\},\\-\Upsilon_3(q_k) - \Upsilon_3^T(q_k) \leq{}& \varepsilon_3(q_k)\mathcal{A}(q_k)P(q_k)\mathcal{A}^T(q_k)\\&+ \varepsilon_3^{-1}(q_k)\mathbb{E}\{K(q_k)\Lambda(q_k)L(q_k)\\&\quad\times L^T(q_k)\Lambda^T(q_k)K^T(q_k)\},\\\Upsilon_4(q_k) + \Upsilon_4^T(q_k) \leq{}& \varepsilon_4(q_k)\mathcal{B}(q_k)\mathbb{E}\{\omega(q_k)\omega^T(q_k)\}\mathcal{B}^T(q_k)\\&+ \varepsilon_4^{-1}(q_k)\mathbb{E}\{K(q_k)\Lambda(q_k)\bar{C}(q_k)e(q_k)\\&\quad\times e^T(q_k)\bar{C}^T(q_k)\Lambda^T(q_k)K^T(q_k)\},\\\Upsilon_5(q_k) + \Upsilon_5^T(q_k) \leq{}& \varepsilon_5(q_k)\mathcal{B}(q_k)\mathbb{E}\{\omega(q_k)\omega^T(q_k)\}\mathcal{B}^T(q_k)\\&+ \varepsilon_5^{-1}(q_k)\mathbb{E}\{K(q_k)\Lambda(q_k)\bar{D}(q_k)\omega(q_k)\\&\quad\times \omega^T(q_k)\bar{D}^T(q_k)\Lambda^T(q_k)K^T(q_k)\},\\-\Upsilon_6(q_k) - \Upsilon_6^T(q_k) \leq{}& \varepsilon_6(q_k)\mathcal{B}(q_k)\mathbb{E}\{\omega(q_k)\omega^T(q_k)\}\mathcal{B}^T(q_k)\end{aligned}$$

$$
\begin{aligned}
&+ \varepsilon_6^{-1}(q_k)\mathbb{E}\{K(q_k)\Lambda(q_k)L(q_k) \\
&\quad \times L^T(q_k)\Lambda^T(q_k)K^T(q_k)\}, \\
\Upsilon_7(q_k) + \Upsilon_7^T(q_k) \leq\, &\varepsilon_7(q_k)\mathbb{E}\{K(q_k)\Lambda(q_k)\bar{C}(q_k)e(q_k) \\
&\quad \times e^T(q_k)\bar{C}^T(q_k)\Lambda^T(q_k)K^T(q_k)\} \\
&+ \varepsilon_7^{-1}(q_k)\mathbb{E}\{K(q_k)\Lambda(q_k)\bar{D}(q_k)\omega(q_k) \\
&\quad \times \omega^T(q_k)\bar{D}^T(q_k)\Lambda^T(q_k)K^T(q_k)\}, \\
\Upsilon_8(q_k) + \Upsilon_8^T(q_k) \leq\, &\varepsilon_8(q_k)\mathbb{E}\{K(q_k)\Lambda(q_k)\bar{C}(q_k)e(q_k) \\
&\quad \times e^T(q_k)\bar{C}^T(q_k)\Lambda^T(q_k)K^T(q_k)\} \\
&+ \varepsilon_8^{-1}(q_k)\mathbb{E}\{K(q_k)\Lambda(q_k)L(q_k) \\
&\quad \times L^T(q_k)\Lambda^T(q_k)K^T(q_k)\}, \\
\Upsilon_9(q_k) + \Upsilon_9^T(q_k) \leq\, &\varepsilon_9(q_k)\mathbb{E}\{K(q_k)\Lambda(q_k)\bar{D}(q_k)\omega(q_k) \\
&\quad \times \omega^T(q_k)\bar{D}^T(q_k)\Lambda^T(q_k)K^T(q_k)\} \\
&+ \varepsilon_9^{-1}(q_k)\mathbb{E}\{K(q_k)\Lambda(q_k)L(q_k) \\
&\quad \times L^T(q_k)\Lambda^T(q_k)K^T(q_k)\}.
\end{aligned}
$$

From Assumption 10.1, it is obtained that

$$
\begin{aligned}
\mathbb{E}\{\tilde{w}(q_k)\tilde{w}^T(q_k)\} &= \mathrm{diag}\{Q(q_k), Q(q_k+h), \cdots, Q(q_{k+1}-h)\} \\
&= \tilde{Q}(q_k), \\
\mathbb{E}\{\vec{v}(q_k)\vec{v}^T(q_k)\} &= \mathrm{diag}\{R_1(q_k), R_2(q_k), \cdots, R_n(q_k)\} \\
&= \tilde{R}(q_k), \\
\mathbb{E}\{\tilde{w}(q_k)\vec{v}^T(q_k)\} &= \mathbb{E}\{\vec{v}(q_k)\tilde{w}^T(q_k)\} = 0.
\end{aligned}
$$

Then, we have

$$
\begin{aligned}
\mathbb{E}\{\omega(q_k)\omega^T(q_k)\} &= \mathbb{E}\left\{\begin{bmatrix} \tilde{w}(q_k)\tilde{w}^T(q_k) & \tilde{w}(q_k)\vec{v}^T(q_k) \\ \vec{v}(q_k)\tilde{w}^T(q_k) & \vec{v}(q_k)\vec{v}^T(q_k) \end{bmatrix}\right\} \\
&= \mathrm{diag}\{\tilde{Q}(q_k), \tilde{R}(q_k)\} = W(q_k).
\end{aligned}
$$

By applying the matrix operation and using the properties of the trace of matrix, one has

$$
\begin{aligned}
&\mathbb{E}\{K(q_k)\Lambda(q_k)\bar{C}(q_k)e(q_k)e^T(q_k)\bar{C}^T(q_k)\Lambda^T(q_k)K^T(q_k)\} \\
&\quad \leq K(q_k)\mathbb{E}\{\mathrm{tr}\{\Lambda(q_k)\bar{C}(q_k)e(q_k)e^T(q_k)\bar{C}^T(q_k)\Lambda^T(q_k)\}I\}K^T(q_k) \\
&\quad \leq K(q_k)\mathrm{tr}\{\mathbb{E}\{\bar{C}(q_k)e(q_k)e^T(q_k)\bar{C}^T(q_k)\}\}IK^T(q_k) \\
&\quad = K(q_k)\mathrm{tr}\{\bar{C}(q_k)P(q_k)\bar{C}^T(q_k)\}IK^T(q_k).
\end{aligned}
$$

Similarly, it can also be derived that

$$\begin{aligned}
&\mathbb{E}\{K(q_k)\Lambda(q_k)\bar{D}(q_k)\omega(q_k)\omega^T(q_k)\bar{D}^T(q_k)\Lambda^T(q_k)K^T(q_k)\} \\
&\quad\leq K(q_k)\mathbb{E}\{\mathrm{tr}\{\Lambda(q_k)\bar{D}(q_k)\omega(q_k)\omega^T(q_k)\bar{D}^T(q_k)\Lambda^T(q_k)\}I\}K^T(q_k) \\
&\quad\leq K(q_k)\mathrm{tr}\{\mathbb{E}\{\bar{D}(q_k)\omega(q_k)\omega^T(q_k)\bar{D}^T(q_k)\}\}IK^T(q_k) \\
&\quad= K(q_k)\mathrm{tr}\{\bar{D}(q_k)W(q_k)\bar{D}^T(q_k)\}IK^T(q_k)
\end{aligned}$$

and

$$\begin{aligned}
&\mathbb{E}\{K(q_k)\Lambda(q_k)L(q_k)L^T(q_k)\Lambda^T(q_k)K^T(q_k)\} \\
&\quad\leq K(q_k)\mathbb{E}\{\mathrm{tr}\{\Lambda(q_k)L(q_k)L^T(q_k)\Lambda^T(q_k)\}I\}K^T(q_k) \\
&\quad\leq K(q_k)\mathrm{tr}\{\mathbb{E}\{L(q_k)L^T(q_k)\}\}IK^T(q_k).
\end{aligned}$$

It follows from the definition of $L(q_k)$ *that*

$$\mathbb{E}\{L(q_k)L^T(q_k)\} \leq \mathbb{E}\big\{\mathrm{tr}\{L^T(q_k)L(q_k)\}I\big\} = \bar{\sigma}I_{nn_y}$$

and, accordingly, we have

$$\begin{aligned}
&\mathbb{E}\{K(q_k)\Lambda(q_k)L(q_k)L^T(q_k)\Lambda^T(q_k)K^T(q_k)\} \\
&\quad\leq K(q_k)\mathrm{tr}\{\mathbb{E}\{L(q_k)L^T(q_k)\}\}IK^T(q_k) \\
&\quad\leq K(q_k)nn_y\bar{\sigma}IK^T(q_k) = nn_y\bar{\sigma}K(q_k)K^T(q_k).
\end{aligned}$$

Summarizing the above discussions, we have

$$\begin{aligned}
P(q_{k+1}) \leq\, &\zeta_1(q_k)\mathcal{A}(q_k)P(q_k)\mathcal{A}^T(q_k) \\
&+ \zeta_2(q_k)\mathcal{B}(q_k)W(q_k)\mathcal{B}^T(q_k) \\
&+ \zeta_3(q_k)K(q_k)\mathrm{tr}\{\bar{C}(q_k)P(q_k)\bar{C}^T(q_k)\}IK^T(q_k) \\
&+ \zeta_4(q_k)K(q_k)\mathrm{tr}\{\bar{D}(q_k)W(q_k)\bar{D}^T(q_k)\}IK^T(q_k) \\
&+ \zeta_5(q_k)nn_y\bar{\sigma}K(q_k)K^T(q_k). \qquad (10.13)
\end{aligned}$$

Noting (10.11)–(10.13) *and the initial condition* $\Xi(q_0) = P(q_0)$, *it is known from Lemma 10.1 that* $\Xi(q_{k+1}) \leq P(q_{k+1})$.

Remark 10.3 *Due to the appearance of the saturation function, the filtering error dynamics is expressed as* (10.8) *and, subsequently, the filter gain is derived by applying the completing the square method to obtain the upper bound on the filtering error covariance. Technically, it is difficult to design the desired filter directly from the filtering error dynamics* (10.8) *because of the lack of the cross-terms* $-K(q_k)\Gamma^T(q_k)$ *and* $-\Gamma(q_k)K^T(q_k)$ *(*$\Gamma(q_k)$ *is an arbitrary matrix with appropriate dimensions), which are required to derive the upper bound on the filtering error covariance. As such, we reformulate the saturation function* $\sigma\big(\mathcal{C}(q_k)\big)$ *as* (10.9) *to bring in the technically required cross-terms so as to facilitate the filter design. Such a reformulation might lead to certain conservatism, which could be reduced by appropriately choosing the scalars* $\varepsilon_i(q_k)$ $(i = 1, 2, \ldots, 9)$.

In Theorem 10.1, an upper bound is derived for the filtering error covariance by resorting to a matrix difference equation. In the following theorem, we are going to obtain the filter gain by minimizing the acquired upper bound.

Theorem 10.2 *For given positive scalars $\varepsilon_i(q_k)$ $(i = 1, 2, \dots, 9)$, the upper bound $\Xi(q_{k+1})$ for the filtering error covariance is minimized by designing the filter gain $K(q_k)$ as*

$$K(q_k) = \Theta(q_k)\Pi^{-1}(q_k) \tag{10.14}$$

where

$$\begin{aligned}\Pi(q_k) \triangleq &\zeta_1(q_k)\bar{C}(q_k)\Xi(q_k)\bar{C}^T(q_k)\\ &+ \zeta_2(q_k)\bar{D}(q_k)W(q_k)\bar{D}^T(q_k)\\ &+ \zeta_3(q_k)\mathrm{tr}\{\bar{C}(q_k)\Xi(q_k)\bar{C}^T(q_k)\}I\\ &+ \zeta_4(q_k)\mathrm{tr}\{\bar{D}(q_k)W(q_k)\bar{D}^T(q_k)\}I\\ &+ \zeta_5(q_k)nn_y\bar{\sigma}I,\\ \Theta(q_k) \triangleq &\zeta_1(q_k)\bar{A}(q_k)\Xi(q_k)\bar{C}^T(q_k)\\ &+ \zeta_2(q_k)\bar{B}(q_k)W(q_k)\bar{D}^T(q_k)\end{aligned}$$

and $\zeta_i(q_k)$ $(i = 1, 2, \dots, 5)$ are defined in Theorem 10.1. Moreover, the minimal upper bound for the filtering error covariance can be obtained as

$$\begin{aligned}\Xi(q_{k+1}) \triangleq &-\Theta(q_k)\Pi^{-1}(q_k)\Theta^T(q_k)\\ &+ \zeta_1(q_k)\bar{A}(q_k)\Xi(q_k)\bar{A}^T(q_k)\\ &+ \zeta_2(q_k)\bar{B}(q_k)W(q_k)\bar{B}^T(q_k).\end{aligned} \tag{10.15}$$

Proof *From* (10.11)*, the upper bound $\Xi(q_{k+1})$ can be rewritten as*

$$\begin{aligned}\Xi(q_{k+1}) = &\zeta_1(q_k)\bar{A}(q_k)\Xi(q_k)\bar{A}^T(q_k)\\ &+ \zeta_1(q_k)K(q_k)\bar{C}(q_k)\Xi(q_k)\bar{C}^T(q_k)K^T(q_k)\\ &- \zeta_1(q_k)\bar{A}(q_k)\Xi(q_k)\bar{C}^T(q_k)K^T(q_k)\\ &- \zeta_1(q_k)K(q_k)\bar{C}(q_k)\Xi(q_k)\bar{A}^T(q_k)\\ &+ \zeta_2(q_k)\bar{B}(q_k)W(q_k)\bar{B}^T(q_k)\\ &+ \zeta_2(q_k)K(q_k)\bar{D}(q_k)W(q_k)\bar{D}^T(q_k)K^T(q_k)\\ &- \zeta_2(q_k)\bar{B}(q_k)W(q_k)\bar{D}^T(q_k)K^T(q_k)\\ &- \zeta_2(q_k)K(q_k)\bar{D}(q_k)W(q_k)\bar{B}^T(q_k)\\ &+ \zeta_3(q_k)K(q_k)\mathrm{tr}\{\bar{C}(q_k)\Xi(q_k)\bar{C}^T(q_k)\}IK^T(q_k)\\ &+ \zeta_4(q_k)K(q_k)\mathrm{tr}\{\bar{D}(q_k)W(q_k)\bar{D}^T(q_k)\}IK^T(q_k)\\ &+ \zeta_5(q_k)nn_y\bar{\sigma}K(q_k)K^T(q_k)\end{aligned}$$

$$
\begin{aligned}
&= K(q_k)\Pi(q_k)K^T(q_k) - \Theta(q_k)K^T(q_k) \\
&\quad - K(q_k)\Theta^T(q_k) + \zeta_1(q_k)\bar{A}(q_k)\Xi(q_k)\bar{A}^T(q_k) \\
&\quad + \zeta_2(q_k)\bar{B}(q_k)W(q_k)\bar{B}^T(q_k) \\
&= (K(q_k) - \Theta(q_k)\Pi^{-1}(q_k))\Pi(q_k) \\
&\qquad\quad \times (K(q_k) - \Theta(q_k)\Pi^{-1}(q_k))^T \\
&\quad - \Theta(q_k)\Pi^{-1}(q_k)\Theta^T(q_k) \\
&\quad + \zeta_1(q_k)\bar{A}(q_k)\Xi(q_k)\bar{A}^T(q_k) \\
&\quad + \zeta_2(q_k)\bar{B}(q_k)W(q_k)\bar{B}^T(q_k).
\end{aligned}
$$

Then, from $\Pi(q_k) > 0$*, we know that the minimum of* $\Xi(q_{k+1})$ *is obtained when the filter gain equals* $\Theta(q_k)\Pi^{-1}(q_k)$*. Moreover, it is easy to see that the minimum of* $\Xi(q_{k+1})$ *is* (10.15).

10.3 Boundedness Analysis

The boundedness of the filtering error dynamics is discussed in this section. The following definition is first introduced for discussion convenience.

Definition 10.1 *For real numbers* $\phi > 0$*,* $\varphi > 0$*, and* $0 < \nu < 1$*, if*

$$\mathbb{E}\{\|\eta(q_k)\|^2\} \leq \phi\mathbb{E}\{\|\eta(q_0)\|^2\}\nu^k + \varphi$$

holds for all $k \geq 0$*, then the stochastic process* $\eta(q_k)$ *is exponentially bounded in the mean square sense.*

The following assumption is essential in deriving our results.

Assumption 10.2 *The following conditions are satisfied for all* $k \geq 0$*:*

$$
\begin{aligned}
\|\bar{A}(q_k)\| &\leq \bar{a}, \; \underline{c}^2 I \leq \bar{C}(q_k)\bar{C}^T(q_k) \leq \bar{c}^2 I, \\
\underline{b}^2 I &\leq \bar{B}(q_k)\bar{B}^T(q_k) \leq \bar{b}^2 I, \\
\underline{d}^2 I &\leq \bar{D}(q_k)\bar{D}^T(q_k) \leq \bar{d}^2 I, \\
\underline{w} I &\leq W(q_k) \leq \bar{w} I, \; \underline{p} I \leq \Xi(q_k) \leq \bar{p} I
\end{aligned}
$$

where $\bar{a} < 1$*,* $\bar{b}$*,* $\underline{b}$*,* $\bar{c}$*,* $\underline{c}$*,* $\bar{d}$*,* $\underline{d}$*,* $\bar{w}$*,* $\underline{w}$*,* $\bar{p}$*, and* $\underline{p}$ *are positive real numbers.*

Theorem 10.3 *Consider the discrete multi-sensor MRSs* (10.1)–(10.2) *with the filter in the form of* (10.6) *whose gain is in* (10.14)*. Under Assumption 10.2, the filtering error dynamics is exponentially bounded in mean square sense.*

Proof *First, the filtering error dynamics* (10.8) *is rewritten as*

$$e(q_{k+1}) = \bar{A}(q_k)e(q_k) + u(q_k) + q(q_k)$$

where

$$\begin{aligned} q(q_k) &\triangleq \bar{B}(q_k)\omega(q_k), \\ u(q_k) &\triangleq -K(q_k)\sigma(\mathcal{C}(q_k)). \end{aligned}$$

From Assumption 10.2 and noting (10.14), *it is derived that*

$$\begin{aligned} \|K(q_k)\| \leq & \left\|\zeta_1(q_k)\bar{A}(q_k)\Xi(q_k)\bar{C}^T(q_k)\right\| \\ & \times \left\|\left(\zeta_1(q_k)\bar{C}(q_k)\Xi(q_k)\bar{C}^T(q_k)\right)^{-1}\right\| \\ & + \left\|\zeta_2(q_k)\bar{B}(q_k)W(q_k)\bar{D}^T(q_k)\right\| \\ & \times \left\|\left(\zeta_2(q_k)\bar{D}(q_k)W(q_k)\bar{D}^T(q_k)\right)^{-1}\right\| \\ \leq & \frac{\bar{a}\bar{p}\bar{c}}{\underline{p}\underline{c}^2} + \frac{\bar{b}\bar{w}\bar{d}}{\underline{w}\underline{d}^2} := \bar{k}. \end{aligned}$$

Noting the definition of the saturation function $\sigma(\cdot)$, *one has*

$$\begin{aligned} & \mathbb{E}\{\sigma(\mathcal{C}(q_k))\sigma^T(\mathcal{C}(q_k))\} \\ & \leq \mathbb{E}\Big\{\mathrm{tr}\Big(\sigma(\mathcal{C}(q_k))\sigma^T(\mathcal{C}(q_k))\Big)\Big\}I \\ & \leq \mathbb{E}\{\begin{bmatrix} l_{1,\max} & l_{2,\max} & \cdots & l_{nn_y,\max} \end{bmatrix} \\ & \qquad \times \begin{bmatrix} l_{1,\max} & l_{2,\max} & \cdots & l_{nn_y,\max} \end{bmatrix}^T\}I \\ & = \bar{\sigma}I. \end{aligned}$$

Then, it follows from the properties of the matrix trace that

$$\begin{aligned} & \mathbb{E}\{u^T(q_k)u(q_k)\} \\ & = \mathbb{E}\Big\{\sigma^T(\mathcal{C}(q_k))K^T(q_k)K(q_k)\sigma(\mathcal{C}(q_k))\Big\} \\ & = \mathrm{tr}\Big\{K(q_k)\mathbb{E}\{\sigma(\mathcal{C}(q_k))\sigma^T(\mathcal{C}(q_k))\}K^T(q_k)\Big\} \\ & \leq nn_y\bar{\sigma}\bar{k}^2 := \bar{u} \end{aligned}$$

and

$$\begin{aligned} & \mathbb{E}\{q^T(q_k)q(q_k)\} \\ & = \mathbb{E}\{\omega^T(q_k)\bar{B}^T(q_k)\bar{B}(q_k)\omega(q_k)\} \\ & = \mathrm{tr}\Big\{\bar{B}(q_k)\mathbb{E}\{\omega(q_k)\omega^T(q_k)\}\bar{B}^T(q_k)\Big\} \\ & \leq (bn_x + nn_y)\bar{w}\bar{b}^2 := \bar{q}. \end{aligned}$$

Subsequently, consider the following iterative matrix equation with respect to $\Psi(q_k)$*:*

$$\Psi(q_{k+1}) = \bar{A}(q_k)\Psi(q_k)\bar{A}^T(q_k) + \bar{B}(q_k)W(q_k)\bar{B}^T(q_k) + \kappa I \tag{10.16}$$

with the initial value being

$$\Psi(q_0) = \bar{B}(q_0)W(q_0)\bar{B}^T(q_0) + \kappa I$$

and $\kappa > 0$ *being a scalar. Then, it is easy to know that*

$$\begin{aligned}\|\Psi(q_{k+1})\| \leq& \|\bar{A}(q_k)\|^2\|\Psi(q_k)\| \\ &+ \|\bar{B}(q_k)W(q_k)\bar{B}^T(q_k)\| + \|\kappa I\| \\ \leq& \bar{a}^2\|\Psi(q_k)\| + \bar{b}^2\bar{w} + \kappa\end{aligned}$$

which, by iteration, leads to

$$\|\Psi(q_k)\| \leq \bar{a}^{2k}\|\Psi(q_0)\| + (\bar{b}^2\bar{w} + \kappa)\sum_{i=0}^{k-1}\bar{a}^{2i}.$$

From Assumption 10.2, we know $0 < \bar{a}^2 < 1$*, and therefore*

$$\begin{aligned}\|\Psi(q_k)\| \leq& \|\Psi(q_0)\| + (\bar{b}^2\bar{w} + \kappa)\sum_{i=0}^{\infty}\bar{a}^{2i} \\ =& \|\Psi(q_0)\| + \frac{\bar{b}^2\bar{w} + \kappa}{1 - \bar{a}^2}.\end{aligned} \tag{10.17}$$

Furthermore, it is clear from (10.16) *that*

$$\Psi(q_k) \geq \kappa I. \tag{10.18}$$

In view of (10.17) *and* (10.18)*, there exists a positive scalar* $\bar{\psi}$ *such that* $\kappa I \leq \Psi(q_k) \leq \bar{\psi} I$ *for* $k \geq 0$*.*

Denote $V_k(e(q_k)) \triangleq e^T(q_k)\Psi^{-1}(q_k)e(q_k)$*. For an arbitrary positive scalar* ρ*, we have*

$$\begin{aligned}&\mathbb{E}\{V_{k+1}(e(q_{k+1}))|e(q_k)\} - (1+\rho)V_k(e(q_k)) \\ =& \mathbb{E}\Big\{\big(\bar{A}(q_k)e(q_k) + u(q_k) + q(q_k)\big)^T\Psi^{-1}(q_{k+1}) \\ &\times \big(\bar{A}(q_k)e(q_k) + u(q_k) + q(q_k)\big)\Big\} \\ &- (1+\rho)e^T(q_k)\Psi^{-1}(q_k)e(q_k) \\ \leq& (1+\rho)\mathbb{E}\Big\{e^T(q_k)\big(\bar{A}^T(q_k)\Psi^{-1}(q_{k+1})\bar{A}(q_k) - \Psi^{-1}(q_k)\big)e(q_k)\Big\} \\ &+ (1+\rho^{-1})\mathbb{E}\{u^T(q_k)\Psi^{-1}(q_{k+1})u(q_k)\} \\ &+ \mathbb{E}\{q^T(q_k)\Psi^{-1}(q_{k+1})q(q_k)\}.\end{aligned} \tag{10.19}$$

By employing the matrix inversion lemma, it follows that

$$\begin{aligned}
&\bar{A}^T(q_k)\Psi^{-1}(q_{k+1})\bar{A}(q_k) - \Psi^{-1}(q_k)\\
&\quad=\bar{A}^T(q_k)\Big(\bar{A}(q_k)\Psi(q_k)\bar{A}^T(q_k)\\
&\qquad+\bar{B}(q_k)W(q_k)\bar{B}^T(q_k)+\kappa I\Big)^{-1}\bar{A}(q_k)-\Psi^{-1}(q_k)\\
&\quad=-\Big(\Psi(q_k)+\Psi(q_k)\bar{A}^T(q_k)\\
&\qquad\times\big(\bar{B}(q_k)W(q_k)\bar{B}^T(q_k)+\kappa I\big)^{-1}\bar{A}(q_k)\Psi(q_k)\Big)^{-1}\\
&\quad=-\Big(I+\bar{A}^T(q_k)\big(\bar{B}(q_k)W(q_k)\bar{B}^T(q_k)+\kappa I\big)^{-1}\\
&\qquad\times\bar{A}(q_k)\Psi(q_k)\Big)^{-1}\Psi^{-1}(q_k)\\
&\quad\le-\left(1+\frac{\bar{a}^2\bar{\psi}}{\underline{b}^2\underline{w}}\right)^{-1}\Psi^{-1}(q_k).
\end{aligned}\tag{10.20}$$

Combining the inequalities (10.19) *and* (10.20)*, it is derived that*

$$\begin{aligned}
&\mathbb{E}\{V_{k+1}(e(q_{k+1}))|e(q_k)\}-(1+\rho)V_k(e(q_k))\\
&\quad\le-(1+\rho)\left(1+\frac{\bar{a}^2\bar{\psi}}{\underline{b}^2\underline{w}}\right)^{-1}V_k(e(q_k))+(1+\rho^{-1})\frac{\bar{u}}{\kappa}+\frac{\bar{q}}{\kappa}
\end{aligned}$$

which can be rearranged as

$$\mathbb{E}\{V_{k+1}(e(q_{k+1}))|e(q_k)\}\le\chi V_k(e(q_k))+\varsigma \tag{10.21}$$

where

$$\begin{aligned}
\chi&\triangleq(1+\rho)\left(1-\left(1+\frac{\bar{a}^2\bar{\psi}}{\underline{b}^2\underline{w}}\right)^{-1}\right),\\
\varsigma&\triangleq(1+\rho^{-1})\frac{\bar{u}}{\kappa}+\frac{\bar{q}}{\kappa}.
\end{aligned}$$

It is obvious that there exist some $\rho>0$ *which guarantees* $\chi\in(0,1)$*. Accordingly, one has from* (10.21) *that:*

$$\begin{aligned}
\mathbb{E}\{\|e(q_{k+1})\|^2\}&\le\frac{\bar{\psi}}{\kappa}\mathbb{E}\{\|e(q_0)\|^2\}\chi^{k+1}+\varsigma\bar{\psi}\sum_{i=0}^{k}\chi^i\\
&\le\frac{\bar{\psi}}{\kappa}\mathbb{E}\{\|e(q_0)\|^2\}\chi^{k+1}+\varsigma\bar{\psi}\sum_{i=0}^{\infty}\chi^i\\
&=\frac{\bar{\psi}}{\kappa}\mathbb{E}\{\|e(q_0)\|^2\}\chi^{k+1}+\frac{\varsigma\bar{\psi}}{1-\chi}.
\end{aligned}$$

According to Definition 10.1, we know that the filtering error $e(q_k)$ *is exponentially bounded in mean square sense.*

Remark 10.4 *In Theorem 10.3, a sufficient condition is given to ensure the exponential boundedness in the mean square sense for the filtering error dynamics $e(q_k)$. Note that, in engineering practice, the spectral norm of matrix $\bar{A}(q_k)$, the variances of matrices $\bar{B}(q_k)$, $\bar{C}(q_k)$, and $\bar{D}(q_k)$, and the variance $W(q_k)$ of the noises are all bounded due to energy constraints. As such, Assumption 10.2 is reasonable.*

Remark 10.5 *Until now, the outlier-resistant recursive filtering problem has been solved for a class of MRSs under WTOD protocol. In Theorem 10.1, an upper bound for the filtering error covariance is derived. The filter gain is acquired to minimize the upper bound, and the minimal upper bound is obtained in Theorem 10.2. The derived minimal upper bound reflects all the information about the multi-rate sampling, the WTOD protocol, and the saturation function. Furthermore, Theorem 10.3 gives a sufficient condition ensuring the exponential boundedness in mean square sense on the filtering error dynamics.*

10.4 Illustrate Examples

In this section, two illustrate examples are presented to show the validity of the proposed outlier-resistant recursive filtering scheme.

Example 10.1

In this example, we consider a discrete time-varying systems (10.1) with parameters shown as follows:

$$A(s_k) = \begin{bmatrix} 0.97 + 0.01\cos(s_k) & 0.53 \\ -0.08 & 0.97 \end{bmatrix}, \quad B(s_k) = \begin{bmatrix} 0.08 \\ 0.06 \end{bmatrix}.$$

The system output is measured by three sensors with a sampling period $2h$ and the following parameters:

$$\begin{aligned} C_1(q_k) &= \begin{bmatrix} 7\sin(q_k) & 9\cos(q_k) \end{bmatrix}, \ D_1(q_k) = 3\sin(q_k), \\ C_2(q_k) &= \begin{bmatrix} 9\sin(q_k) & 7\cos(q_k) \end{bmatrix}, \ D_2(q_k) = 5\sin(q_k), \\ C_3(q_k) &= \begin{bmatrix} 6\sin(q_k) & 5\cos(q_k) \end{bmatrix}, \ D_3(q_k) = 3\sin(q_k). \end{aligned}$$

Other parameters are chosen as $\Omega_1 = 0.7$, $\Omega_2 = 0.8$, $\Omega_3 = 0.5$, and $l_{\max} = \begin{bmatrix} 20 & 26 & 20 \end{bmatrix}^T$. The covariances of the process noise and the measurement noises are taken as $Q(s_k) = 0.05$, $R_1(q_k) = 0.03$, $R_2(q_k) = 0.05$, and $R_3(q_k) = 0.04$, respectively.

For the purpose of testifying the robustness of the proposed filtering scheme to the measurement outliers, comparison of the filtering performance in presence of measurement outliers is implemented between the designed filter (10.6) and the following standard filter

$$\hat{x}(q_{k+1}) = \bar{A}(q_k)\hat{x}(q_k) + K_1(q_k)\big(\bar{y}(q_k) - \bar{C}(q_k)\hat{x}(q_k)\big) \qquad (10.22)$$

where $K_1(q_k)$ is the filter gain.

The measurement outliers are characterized by a zero-mean Gaussian random variable with covariance 2000. Moreover, the occurrence of the measurement outliers is assumed to be periodic with a period $4h$. The time instant of the first-time occurrence of the measurement outlier is 0.

Let $\mathrm{MSE}(q_k)$ denote the mean square error of the estimate which is defined by $\mathrm{MSE}(q_k) = \frac{1}{N}\sum_{j=1}^{N}\sum_{i=1}^{2}\left(\bar{x}_i^{(j)}(q_k) - \hat{x}_i^{(j)}(q_k)\right)^2$ with $N = 500$, that is, 500 independent experiments are conducted.

Simulation results are presented in Figs. 10.1–10.6. Figure 10.1 depicts the state $x_1(s_k)$ and the estimate of $x_1(s_k)$ from both filters, and Fig. 10.2 depicts the state $x_2(s_k)$ and the estimate of $x_2(s_k)$ from both filters. $x_i(s_k)$ $(i = 1, 2)$ represents the ith element of the state $x(s_k)$. Figures 10.3 and 10.4 plot the filtering error $e(s_k)$ of both filters. In Fig. 10.5, the trace of the minimal upper bound for the filtering error covariance and the MSE of the proposed filtering scheme are shown. As a comparison, the trace of the minimal upper bound and the MSE of the standard filtering scheme with filter (10.22) are shown in Fig. 10.6.

From the simulation results, we know that the proposed recursive filtering scheme performs well with the existence of measurement outliers, and the influence of the measurement outliers is perfectly restrained. As for the standard

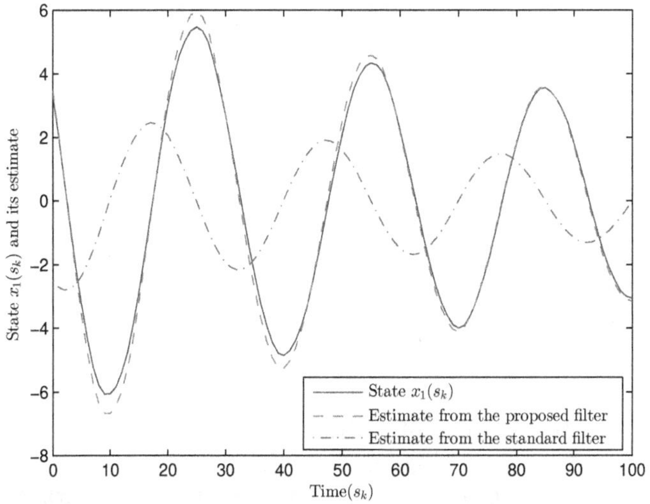

FIGURE 10.1: State $x_1(s_k)$ and its estimate.

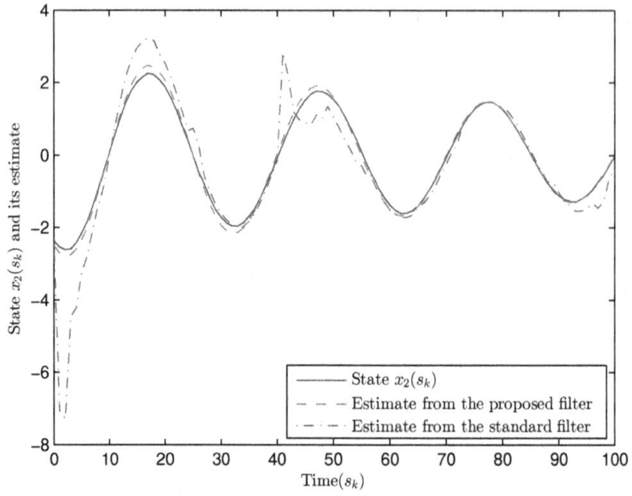

FIGURE 10.2: State $x_2(s_k)$ and its estimate.

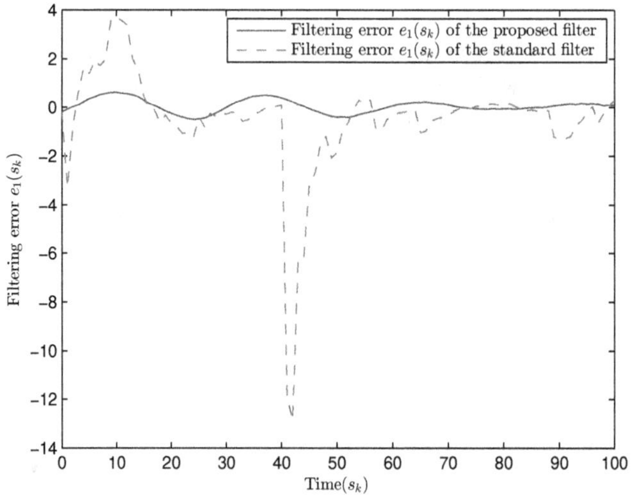

FIGURE 10.3: The filtering error $e(s_k)$.

filtering scheme, the estimate of the state deviates remarkably from the system state when the measurement outlier occurs. Consequently, the robustness of the proposed filtering scheme against the measurement outliers is indeed effective. Moreover, comparing Fig. 10.5 with Fig. 10.6, we know that the minimal upper bound tightens when the saturation levels approach infinity.

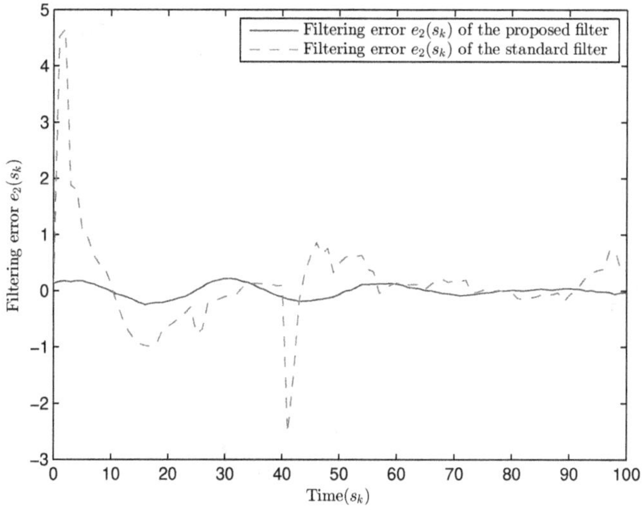

FIGURE 10.4: The filtering error $e(s_k)$.

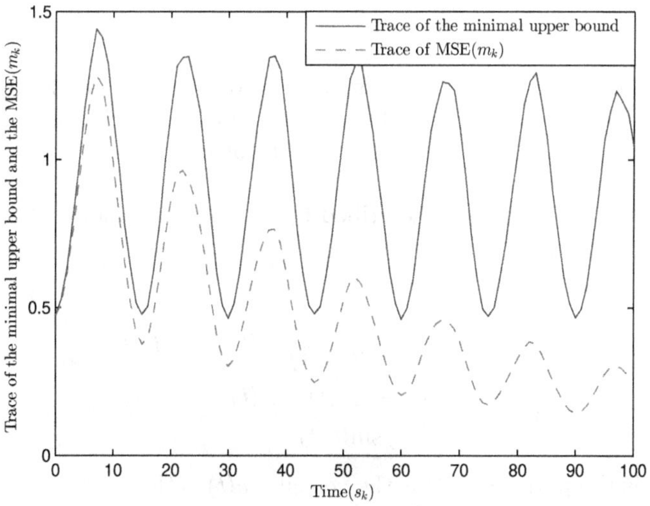

FIGURE 10.5: The trace of the minimal upper bound and the MSE with the proposed filtering scheme.

However, due to the measurement outliers, such a minimal upper bound is no longer an upper bound for the MSE.

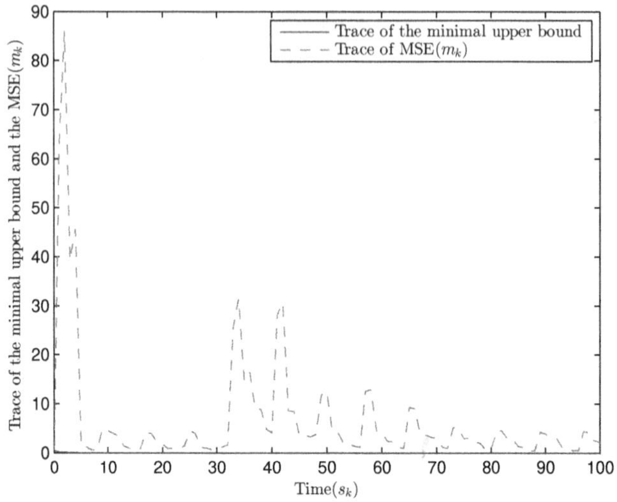

FIGURE 10.6: The trace of the minimal upper bound and the MSE with the standard filtering scheme.

Example 10.2

In this example, the effectiveness of the proposed outlier-resistant recursive filtering scheme is further verified on the communication-based train control (CBTC) system. In the simulation, the data of the train is obtained from a real railway system.

Consider a CBTC system described by the following model [107]:

$$\begin{cases} \dfrac{dp(t)}{dt} = v(t) \\ \dfrac{dv(t)}{dt} = \dfrac{1}{M}u(t) - \dfrac{1}{M}f_r(t) - \dfrac{1}{M}f_g(t) \\ f_r(t) = c_0 + c_1 v(t) + \omega(t) \\ f_g(t) = Mg\sin(\theta(t)) \end{cases}$$

where $p(t)$ is the position, $v(t)$ is the velocity, $u(t)$ is the known traction force, $f_r(t)$ is the resistive force, $f_g(t)$ is the force due to the gradient, and M is the mass. c_0 and c_1 are known coefficients and $\omega(t)$ is the stochastic wind force with zero mean and covariance 10. The parameters of the CBTC system are $M = 288 \times 10^3$kg, $c_0 = 3.4818036 \times 10^3$, and $c_1 = 144.9154$. The gradient angle $\theta(t)$ is dependent on the location of the train and is given in Table 10.1.

By setting $x(t) = \text{col}\{p(t), v(t)\}$, the state-space model of the CBTC system is obtained as

$$\frac{dx(t)}{dt} = Ax(t) + B\omega(t) - B(u(t) - \bar{g}(t))$$

TABLE 10.1: Gradient angle of the track.

Gradient angle	Location
-3°	[0, 1706)
3°	[1706, 2372)
-3°	[2372, 2619)

where

$$A \triangleq \begin{bmatrix} 0 & 1 \\ 0 & -\frac{c_1}{M} \end{bmatrix}, \; B \triangleq \begin{bmatrix} 0 \\ -\frac{1}{M} \end{bmatrix}, \; \bar{g}(t) = c_0 + Mg\sin(\theta(t)).$$

Discretizing the state-space model of the CBTC system with period $T =$ 1s, we have

$$x(s_{k+1}) = \bar{A}x(s_k) + \bar{B}\omega(s_k) - \bar{B}(u(s_k) - \bar{g}(s_k))$$

where

$$\bar{A} \triangleq e^{AT}, \; \bar{B} \triangleq \int_0^T e^{AT} dtB.$$

It is assumed that the CBTC system is measured by two sensors with a sampling period $2T$. The measurement models of the sensors are described as follows:

$$y_1(q_k) = C_1 x(q_k) + \nu_1(q_k),$$
$$y_2(q_k) = C_2 x(q_k) + \nu_2(q_k)$$

with

$$C_1 = \begin{bmatrix} 0 & 1 \end{bmatrix}, \; C_2 = \begin{bmatrix} 1 & 0 \end{bmatrix}$$

where $\nu_1(q_k)$ and $\nu_2(q_k)$ are zero-mean measurement noises with covariances 3 and 50, respectively.

In simulation, we set $\Omega_1 = 9$, $\Omega_2 = 0.01$, and $l_{\max} = \begin{bmatrix} 32 & 3200 \end{bmatrix}^T$. The measurement outliers are characterized by a zero-mean Gaussian random variable with covariance 2×10^4 and the occurrence of the measurement outliers is also assumed to be periodic with a period $4T$.

Simulation results are given in Figs. 10.7–10.8. Figure 10.7 shows the position of the train and its estimate. Figure 10.8 depicts the velocity of the train and its estimate. It can be seen from the simulation results that the developed outlier-resistant recursive filtering scheme is indeed effective in practical systems with measurement outliers.

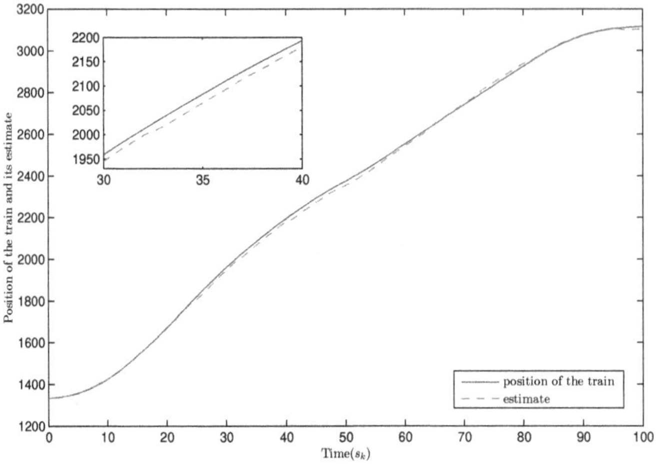

FIGURE 10.7: Position of the train and its estimate.

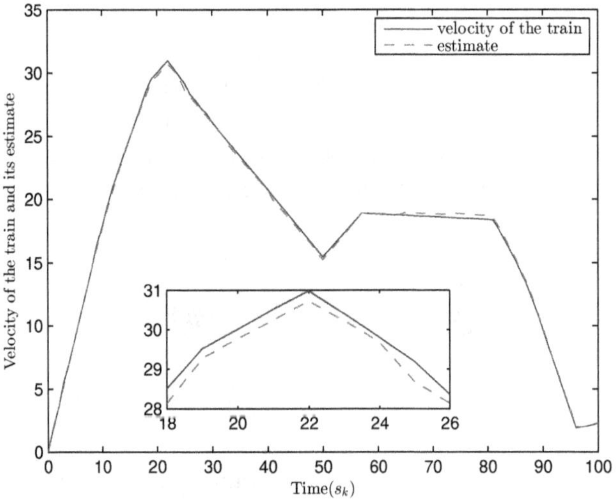

FIGURE 10.8: Velocity of the train and its estimate.

To show the scheduling effect of the WTOD protocol, the transmission instants of the sensors under the WTOD protocol are presented in Fig. 10.9. It can be seen from Fig. 10.9 that the data transmission rates of the sensor 1 and sensor 2 are 58% and 42%, respectively. Obviously, the data transmissions are reduced and therefore the communication burden is lightened. It is worth

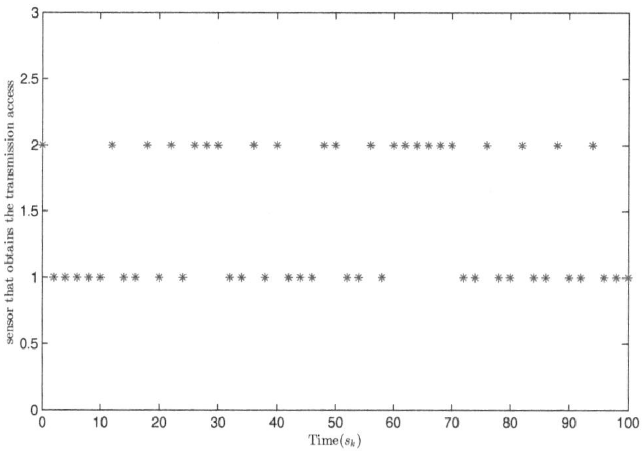

FIGURE 10.9: Transmission instants of sensors under WTOD protocol.

noting that, with the reduced data transmissions, the filtering performance of the proposed recursive filtering scheme is still satisfactory.

10.5 Conclusion

In this chapter, the recursive filtering problem has been investigated for a class of discrete time-varying multi-sensor MRSs subject to measurement outliers under the WTOD protocol. The state update period of the system and the sampling period of the sensors are allowed to be different to cater for the engineering needs. The WTOD protocol has been employed to orchestrate the transmission order of the sensors, under which only one sensor can get the access to transmit its measurement at each sampling instant. A novel filter structure has been proposed where a saturation function has been exploited to restrain the influence from the measurement outliers on the innovation, thereby maintaining a satisfactory filtering performance. With the proposed outlier-resistant filter, an upper bound has been guaranteed for the filtering error covariance, and then the minimal upper bound has been obtained by designing a proper filter gain. Moreover, the exponential boundedness (in the mean square sense) of the filtering error dynamics has been analyzed. Finally, illustrate simulations have been conducted to verify the robustness of the proposed recursive filtering scheme against the measurement outliers and also the corresponding filtering performance.

11

Dynamic Event-Based Recursive Filtering for Multi-Rate Systems with Integral Measurements over Sensor Networks

In the past few decades, motivated by military applications like battlefield surveillance, the study of sensor networks has been attracting attentions of many researchers. Nowadays, other than the applications in the military field, sensor networks have also been widely used in many industrial and civilian applications such as machine health monitoring, structural health monitoring, and natural disaster prevention. In the area of signal processing, a fundamental problem in sensor networks is the distributed filtering problem where each sensor node generates an estimate of the state of interest based on the information obtained from the node itself and the neighboring sensor nodes according to a predefined topology. Up to now, plenty of results have been obtained on the topic of distributed filtering problem. It is worth noting that, most of the available works are based on an assumption that the plant and the sensors are sampled with a uniform rate. Recently, the distributed filtering problem with a multi-rate sampling strategy has only received initial research attention despite the practical significance and the theoretical importance. The research on distributed filtering problem with multi-rate sampling strategy is still inadequate and many research problems remain open.

On the other hand, in the distributed filtering over sensor networks, each sensor node communicates with the neighboring sensor nodes to obtain their information to collaboratively estimate the state of interest. As is well known, the communication among sensor nodes consumes the most of the energy in the energy-limited sensor networks. Therefore, various communication strategies, such as communication protocols [80, 111] and event-triggered mechanism [33, 167], have been introduced to reduce the communications among sensor nodes, hence saving the limited energy. Under the event-triggered mechanism, the data transmission among sensor nodes is only triggered when a certain triggering condition is satisfied, and therefore unnecessary transmissions are avoided.

Generally speaking, the triggering condition of the event-triggered mechanism is related to the information of interest (state, measurement, or innovation) and a triggering threshold. In literature, according to the difference in the triggering threshold, there are mainly two classes of event-triggered mechanisms, i.e., the static event-triggered mechanism (SEM) [208] and the

DOI: 10.1201/9781032619507-11

dynamic event-triggered mechanism (DEM) [105]. In the SEM, the threshold in the triggering condition is fixed all the time. Unlike the SEM, the threshold in the triggering condition of the DEM has an additional internal dynamic variable which is dynamically adjusted. Compared with the SEM, the DEM is able to guarantee less triggering times, and therefore further reduces the energy consumption. Naturally, the filtering problem under the DEM has stirred interest from researchers, and several results have been obtained. Nevertheless, for MRSs over sensor networks, the filtering problem under the DEM has not been considered yet. Therefore, the main motivation of this chapter is to shorten such a gap.

In real practice such as chemical processes and nuclear reaction processes [154], the measurement output of the sensor may not only contain the information of the current system state but also contain the information of the states in a certain past period. The reason for such a kind of phenomenon, known as integral measurements, is that the time taken to collect and process the data is longer than the sampling period of the sensor, and hence can not be ignored in the measurement model. Until now, the phenomenon of integral measurements has only received minor research attention in the area of signal processing, and only a few initial research results have been available, see for examples [40, 55]. As for the distributed filtering problem over sensor networks, the phenomenon of integral measurements has not been taken into consideration yet, not to mention the case that the system is complicated by the multi-rate sampling and the DEM.

Motivated by the above discussions, two interesting yet challenging questions arise: 1) how to deal with the DEM in the framework of distributed filtering under the multi-rate sampling? and 2) how to design a recursive filtering scheme such that, with the multi-rate sampling, the integral measurements and the DEM, a minimal upper bound exists on the filtering error covariance? To find answers for the above questions, in this chapter, we aim to study the dynamic event-based recursive filtering problem for MRSs with integral measurements over sensor networks. The main contributions of this chapter can be highlighted as follows: 1) the distributed recursive filtering problem is, for the first time, investigated for MRSs over sensor networks where the DEM is implemented among sensor nodes; 2) a recursive filtering scheme is designed such that the influence from the multi-rate sampling, the integral measurements, and the DEM is taken into account; and 3) under the developed filtering scheme, an upper bound on the filtering error covariance is derived in terms of the solution to a matrix Riccati equation and then minimized at each sampling instant.

11.1 Problem Formulation

Consider a sensor network whose topology is described by a directed graph $\mathscr{G} = (\mathcal{V}, \mathscr{E}, \mathcal{R})$ of order N with the set of nodes $\mathcal{V} = \{1, 2, \ldots, N\}$, set of edges $\mathscr{E} \subseteq \mathcal{V} \times \mathcal{V}$ and weighted adjacency matrix $\mathcal{R} = [a_{ij}]$ with non-negative adjacency elements a_{ij}. An edge of $\mathscr{G}$ is denoted by (i, j). The adjacency element associated with the edge of the graph is defined as $a_{ij} > 0 \Leftrightarrow (i, j) \in \mathscr{E}$, which means that the i-th sensor node can receive information from the j-th sensor node. Throughout this chapter, we assume that $a_{ii} = 0$ for $i \in \mathcal{V}$. For sensor node $i \in \mathcal{V}$, the neighboring sensor nodes are denoted by the set $\mathcal{V}_i \triangleq \{j \in \mathcal{V} : (i, j) \in \mathscr{E}\}$.

Consider the following discrete-time linear system

$$x(s_{k+1}) = A(s_k)x(s_k) + B(s_k)w(s_k) \tag{11.1}$$

with the i-th ($i = 1, 2, \ldots, N$) sensor modeled by

$$y_i(q_k) = C_i(q_k)\sum_{a=0}^{l} x(q_k - ah) + D_i(q_k)v_i(q_k) \tag{11.2}$$

where $x(s_k) \in \mathbb{R}^{n_x}$ is the system state to be estimated, $y_i(q_k) \in \mathbb{R}^{n_y}$ is the measurement output of the sensor i, $w(s_k) \in \mathbb{R}^{n_w}$ is the zero-mean process noise with the covariance $W(s_k) > 0$, $v_i(q_k) \in \mathbb{R}^{n_v}$ is the zero-mean measurement noise on sensor i with the covariance $R_i(q_k) > 0$, and $A(s_k)$, $B(s_k)$, $C_i(q_k)$, and $D_i(q_k)$ are known matrices with appropriate dimensions. l is a positive integer representing the time taken to complete the data collection. The initial value $x(s_0)$ is a random variable with the mean $\psi(0)$. Moreover, in this chapter, the state update period of the system is $h \triangleq s_{k+1} - s_k$ where h is a known positive constant. The sampling period of the sensors is $bh \triangleq q_{k+1} - q_k$ with $b \in \mathbb{N}^+$.

Remark 11.1 *In measurement model* (11.2), *the phenomenon of integral measurements is taken into account. It can be seen from* (11.2) *that the measurement output* $y_i(q_k)$ *not only depends on the current system state* $x(q_k)$ *but also depends on the integral of the system states at the time instants* $x(q_k - lh), \ldots, x(q_k - 2h), x(q_k - h)$. *Moreover, when* $l = 0$, *the measurement model* (11.2) *becomes the general one* $y_i(q_k) = C_i(q_k)x(q_k) + D_i(q_k)v_i(q_k)$.

Note that the system under consideration is a MRS. In the following, we are going to convert the MRS to a single-rate one by using the lifting technique. By denoting $\bar{x}(q_k) \triangleq \text{col}\{x(q_{k-1} + h), \cdots, x(q_k - h), x(q_k)\}$, we have the following state equation with a state update period bh:

$$\bar{x}(q_{k+1}) = \bar{A}(q_k)\bar{x}(q_k) + \bar{B}(q_k)\bar{w}(q_k) \tag{11.3}$$

where

$$\bar{w}(q_k) \triangleq \mathrm{col}\{w(q_k), w(q_k+h), \cdots, w(q_{k+1}-h)\},$$

$$\bar{A}(q_k) \triangleq \Big[\underbrace{0_{bn_x\times n_x} \quad \cdots \quad 0_{bn_x\times n_x}}_{b-1} \quad \mathcal{A}(q_k)\Big],$$

$$\mathcal{A}(q_k) \triangleq \mathrm{col}\{\mathcal{F}_b^b(q_k), \mathcal{F}_{b-1}^b(q_k), \cdots, \mathcal{F}_1^b(q_k)\},$$

$$\bar{B}(q_k) \triangleq \bar{\mathcal{A}}(q_k)\mathcal{B}(q_k), \ \mathcal{F}_t^s(q_k) \triangleq \prod_{i=t}^{s} A(q_{k+1}-ih),$$

$$\bar{\mathcal{A}}(q_k) \triangleq \begin{bmatrix} I & 0 & \cdots & 0 \\ \mathcal{F}_{b-1}^{b-1}(q_k) & I & \cdots & 0 \\ \mathcal{F}_{b-2}^{b-1}(q_k) & \mathcal{F}_{b-2}^{b-2}(q_k) & \cdots & 0 \\ \vdots & \vdots & \cdots & 0 \\ \mathcal{F}_1^{b-1}(q_k) & \mathcal{F}_1^{b-2}(q_k) & \cdots & I \end{bmatrix},$$

$$\mathcal{B}(q_k) \triangleq \mathrm{diag}\{B(q_k), B(q_k+h), \cdots, B(q_{k+1}-h)\}.$$

Moreover, the measurement $y_i(q_k)$ can be rewritten as

$$y_i(q_k) = C_i(q_k)\sum_{a=0}^{c-1} \mathbf{I}_b \bar{x}(q_{k-a}) + C_i(q_k)E\bar{x}(q_{k-c}) + D_i(q_k)v_i(q_k) \tag{11.4}$$

where

$$\mathbf{I}_b \triangleq \Big[\underbrace{I_{n_x} \quad I_{n_x} \quad \cdots \quad I_{n_x}}_{b}\Big], \quad E \triangleq \Big[0 \quad \cdots \quad 0 \quad \underbrace{I_{n_x} \quad \cdots \quad I_{n_x}}_{l-cb}\Big]$$

and $c \triangleq \lfloor \frac{l}{b} \rfloor$ represents the largest integer not larger than $\frac{l}{b}$.

In the sensor network, the filter on the sensor node i ($i = 1, 2, \ldots, N$) collects the following innovations from the neighboring sensor nodes:

$$r_j(q_k) \triangleq y_j(q_k) - C_j(q_k)\sum_{a=0}^{c-1} \mathbf{I}_b \hat{x}_j(q_{k-a}) - C_j(q_k)E\hat{x}_j(q_{k-c}), \quad j \in \mathcal{V}_i$$

where $\hat{x}_j(q_k)$ is the estimate of $\bar{x}(q_k)$ from the filter on sensor node j.

For the purpose of reducing unnecessary data transmissions and alleviating the communication burden, in this chapter, the DEM is adopted to decide whether the sensor nodes communicate with each other at a certain time instant. The event triggering condition on sensor node i is defined as follows:

$$\Phi_i(q_k) \triangleq \frac{1}{\theta_i}\zeta_i(q_k) + \lambda_i - \varepsilon_i^T(q_k)\varepsilon_i(q_k) < 0 \tag{11.5}$$

where

$$\varepsilon_i(q_k) \triangleq r_i(q_k) - r_i^t(q_k)$$

with $r_i^t(q_k)$ being the innovation transmitted at the latest triggering instant. $\zeta_i(q_k)$ is an internal dynamic variable that satisfies

$$\zeta_i(q_{k+1}) \triangleq \mu_i \zeta_i(q_k) + \lambda_i - \varepsilon_i^T(q_k)\varepsilon_i(q_k) \tag{11.6}$$

with initial condition $\zeta_i(q_0) = \zeta_{i,0} \geq 0$. λ_i is a given positive scalar. θ_i and μ_i are given scalars satisfying

$$0 < \mu_i < 1, \ \theta_i \geq \frac{1}{\mu_i}. \tag{11.7}$$

According to the event triggering condition, the sequence of the triggering instants $0 \leq t_0^i < t_1^i < \cdots < t_l^i < \cdots$ are determined iteratively by

$$t_{l+1}^i \triangleq \min_{k \in \mathbb{N}} \{q_k > t_l^i | \Phi_i(q_k) < 0\}. \tag{11.8}$$

Remark 11.2 *From* (11.5), *we can see that the triggering threshold of the DEM contains an additional variable $\zeta_i(q_k)$ which is dynamically adjusted according to $\varepsilon_i(q_k)$. Compared with the traditional SEM, the DEM is more flexible in deciding the triggering instant and can further reduce the data transmissions among the sensor nodes. On the other hand, the DEM considered in this chapter is more general since the event triggering condition* (11.5) *will reduce to the one in SEM if θ_i approaches infinity.*

In this chapter, the filter on the sensor node i is constructed of the following form:

$$\hat{x}_i(q_{k+1}) = \bar{A}(q_k)\hat{x}_i(q_k) + L_i(q_k)r_i(q_k) + \sum_{j \in \mathcal{V}_i} a_{ij} K_{ij}(q_k) r_j^t(q_k) \tag{11.9}$$

where $L_i(q_k)$ and $K_{ij}(q_k)$ are the filter gains to be designed.

By setting the filtering error as $e_i(q_k) \triangleq \bar{x}(q_k) - \hat{x}_i(q_k)$, we have

$$\begin{aligned} e_i(q_{k+1}) =& \bar{A}(q_k)e_i(q_k) + \bar{B}(q_k)\bar{w}(q_k) - L_i(q_k)r_i(q_k) \\ &+ \sum_{j \in \mathcal{V}_i} a_{ij}K_{ij}(q_k)\varepsilon_j(q_k) - \sum_{j \in \mathcal{V}_i} a_{ij}K_{ij}(q_k)r_j(q_k). \end{aligned}$$

Denoting $e(q_k) \triangleq \text{col}\{e_1(q_k), e_2(q_k), \cdots, e_N(q_k)\}$, we have the following augmented system with a compact form:

$$\begin{aligned} e(q_{k+1}) =& (\tilde{A}(q_k) - G(q_k)C(q_k)\bar{\mathbf{I}}_b)e(q_k) - G(q_k)C(q_k)\bar{\mathbf{I}}_b\bar{e}_c(q_k) \\ &- G(q_k)C(q_k)\bar{E}e(q_{k-c}) - G(q_k)D(q_k)v(q_k) \\ &+ \tilde{B}(q_k)\bar{w}(q_k) + \bar{K}(q_k)\varepsilon(q_k) \end{aligned} \tag{11.10}$$

where

$$\begin{aligned}
G(q_k) &\triangleq L(q_k) + \bar{K}(q_k), \ \ \bar{K}(q_k) \triangleq \sum_{i=1}^{N} F_i K(q_k) H_i, \\
\bar{e}_c(q_k) &\triangleq \sum_{a=1}^{c-1} e(q_{k-a}), \ \ K(q_k) \triangleq \Big[K_{ij}(q_k)\Big]_{N\times N}, \\
\tilde{A}(q_k) &\triangleq \mathrm{diag}_N\{\bar{A}(q_k)\}, \ \ \tilde{B}(q_k) \triangleq \mathrm{col}_N\{\bar{B}(q_k)\}, \\
\bar{\mathbf{I}}_b &\triangleq \mathrm{diag}_N\{\mathbf{I}_b\}, \ \ \bar{E} \triangleq \mathrm{diag}_N\{E\}, \\
L(q_k) &\triangleq \mathrm{diag}\{L_1(q_k), L_2(q_k), \cdots, L_N(q_k)\}, \\
C(q_k) &\triangleq \mathrm{diag}\{C_1(q_k), C_2(q_k), \cdots, C_N(q_k)\}, \\
D(q_k) &\triangleq \mathrm{diag}\{D_1(q_k), D_2(q_k), \cdots, D_N(q_k)\}, \\
v(q_k) &\triangleq \mathrm{col}\{v_1(q_k), v_2(q_k), \cdots, v_N(q_k)\}, \\
H_i &\triangleq \mathrm{diag}\{a_{i1}I, a_{i2}I, \cdots, a_{iN}I\}, \\
F_i &\triangleq \mathrm{diag}\{0, \cdots, 0, I, 0, \cdots, 0\}, \\
\varepsilon(q_k) &\triangleq \mathrm{col}\{\varepsilon_1(q_k), \varepsilon_2(q_k), \cdots, \varepsilon_N(q_k)\}.
\end{aligned}$$

The purpose of this chapter is to design a set of filters in the form of (11.9) such that an upper bound exists on the filtering error covariance $P(q_k) \triangleq \mathbb{E}\{e(q_k)e^T(q_k)\}$ and such an upper bound is minimized at each sampling instant by properly designing the filter gains.

11.2 Dynamic Event-Based Filter Design

In this section, an upper bound on the filtering error covariance is first derived and then the filter gains are obtained to minimize such an upper bound at each sampling instant. First, the following lemmas are introduced to help the subsequent analysis.

Lemma 11.1 *The internal dynamic variable $\zeta_i(q_k)$ has an upper bound $\vartheta_i(q_k)$ that satisfies*

$$\vartheta_i(q_k) \triangleq \mu_i^k \zeta_{i,0} + \lambda_i \frac{1-\mu_i^k}{1-\mu_i}.$$

Proof *For $k = 0$, it is easily known that*

$$\vartheta_i(q_0) = \zeta_{i,0} = \zeta_i(q_0).$$

From (11.6) *and noting that $\varepsilon_i^T(q_{k-1})\varepsilon_i(q_{k-1}) \geq 0$, we have*

$$\zeta_i(q_k) \leq \mu_i \zeta_i(q_{k-1}) + \lambda_i.$$

Repeating the above inequality, it is derived that

$$\zeta_i(q_k) \leq \mu_i^k \zeta_{i,0} + \lambda_i \frac{1-\mu_i^k}{1-\mu_i}.$$

The proof is complete.

Lemma 11.2 *Define N event indicator variables $\alpha_i(q_k)$ $(i = 1, 2, \cdots, N)$ which satisfy $\alpha_i(q_k) = 0$ if the event triggering condition is satisfied at time instant q_k on sensor node i and $\alpha_i(q_k) = 1$ otherwise. Then, we have*

$$\mathbb{E}\{v(q_k)\varepsilon^T(q_k)\} = \bar{R}(q_k)\tilde{\alpha}(q_k)D^T(q_k)$$

where

$$\begin{aligned}\bar{R}(q_k) &\triangleq \text{diag}\{R_1(q_k), R_2(q_k), \cdots, R_N(q_k)\},\\ \tilde{\alpha}(q_k) &\triangleq \text{diag}\{\alpha_1(q_k), \alpha_2(q_k), \cdots, \alpha_N(q_k)\}.\end{aligned}$$

Proof *Let us calculate $\mathbb{E}\{v_i(q_k)\varepsilon_i^T(q_k)\}$ first. For the case that the event is triggered at time instant q_k on sensor node i, it is obvious that $\varepsilon_i(q_k) = 0$ and therefore $\mathbb{E}\{v_i(q_k)\varepsilon_i^T(q_k)\} = 0$. Otherwise, we have*

$$\begin{aligned}\mathbb{E}\{v_i(q_k)\varepsilon_i^T(q_k)\} =& \mathbb{E}\{v_i(q_k)r_i^T(q_k)\} - \mathbb{E}\{v_i(q_k)(r_i^t(q_k))^T\}\\ =& \mathbb{E}\Big\{v_i(q_k)(C_i(q_k)\sum_{a=0}^{l} e_i(q_k - ah) + D_i(q_k)v_i(q_k))^T\Big\}\\ =& R_i(q_k)D_i^T(q_k).\end{aligned}$$

Noting the definition of the variable $\alpha_i(q_k)$, it can be concluded that $\mathbb{E}\{v_i(q_k)\varepsilon_i^T(q_k)\} = R_i(q_k)\alpha_i(q_k)D_i^T(q_k)$.

Next, let us calculate $\mathbb{E}\{v_i(q_k)\varepsilon_j^T(q_k)\}$ $(i \neq j)$. Similarly, for the case that the event is triggered at time instant q_k on sensor node j, it is known that $\mathbb{E}\{v_i(q_k)\varepsilon_j^T(q_k)\} = 0$. Otherwise, we have

$$\begin{aligned}\mathbb{E}\{v_i(q_k)\varepsilon_j^T(q_k)\} =& \mathbb{E}\{v_i(q_k)r_j^T(q_k)\} - \mathbb{E}\{v_i(q_k)(r_j^t(q_k))^T\}\\ =& \mathbb{E}\Big\{v_i(q_k)(C_j(q_k)\sum_{a=0}^{l} e_j(q_k - ah) + D_j(q_k)v_j(q_k))^T\Big\}\\ =& 0.\end{aligned}$$

Summarizing the above discussions, we have

$$\mathbb{E}\{v(q_k)\varepsilon^T(q_k)\} = \bar{R}(q_k)\tilde{\alpha}(q_k)D^T(q_k).$$

The proof is complete.

With the help of Lemmas 11.1 and 11.2, in the following theorem, we are going to derive an upper bound on the filtering error covariance $P(q_k)$.

Theorem 11.1 *Let $\beta_i(q_k)$ $(i = 1, 2, \ldots, 6)$ be positive scalars. The solution $\Xi(q_{k+1})$ to the following recursion*

$$\begin{aligned}\Xi(q_{k+1}) =&\phi_k(\Xi(q_k))\\ \triangleq&\gamma_1(q_k)\tilde{A}(q_k)\Xi(q_k)\tilde{A}^T(q_k) + G(q_k)\Psi(q_k)G^T(q_k)\\ &+ \tilde{B}(q_k)\bar{W}(q_k)\tilde{B}^T(q_k) + \gamma_4(q_k)\varrho(q_k)\bar{K}(q_k)\bar{K}^T(q_k)\\ &- \Theta(q_k)G^T(q_k) - G(q_k)\Theta^T(q_k) - G(q_k)\Pi(q_k)\bar{K}^T(q_k)\\ &- \bar{K}(q_k)\Pi^T(q_k)G^T(q_k)\end{aligned} \tag{11.11}$$

where

$$\begin{aligned}\Psi(q_k) \triangleq&\gamma_1(q_k)C(q_k)\bar{\boldsymbol{I}}_b\Xi(q_k)\bar{\boldsymbol{I}}_b^T C^T(q_k)\\ &+ D(q_k)\bar{R}(q_k)D^T(q_k)\\ &+ \gamma_2(q_k)(c-1)\sum_{i=1}^{c-1} C(q_k)\bar{\boldsymbol{I}}_b\Xi(q_{k-i})\bar{\boldsymbol{I}}_b^T C^T(q_k)\\ &+ \gamma_3(q_k)C(q_k)\bar{E}\Xi(q_{k-c})\bar{E}^T C^T(q_k),\\ \Theta(q_k) \triangleq&\gamma_1(q_k)\tilde{A}(q_k)\Xi(q_k)\bar{\boldsymbol{I}}_b^T C^T(q_k),\\ \Pi(q_k) \triangleq&D(q_k)\bar{R}(q_k)\tilde{\alpha}(q_k)D^T(q_k),\\ \gamma_1(q_k) \triangleq&1 + \beta_1(q_k) + \beta_2(q_k) + \beta_3(q_k),\\ \gamma_2(q_k) \triangleq&1 + \beta_1^{-1}(q_k) + \beta_4(q_k) + \beta_5(q_k),\\ \gamma_3(q_k) \triangleq&1 + \beta_2^{-1}(q_k) + \beta_4^{-1}(q_k) + \beta_6(q_k),\\ \gamma_4(q_k) \triangleq&1 + \beta_3^{-1}(q_k) + \beta_5^{-1}(q_k) + \beta_6^{-1}(q_k),\\ \bar{W}(q_k) \triangleq&\text{diag}\{W(q_k), W(q_k+h), \cdots, W(q_{k+1}-h)\},\\ \varrho(q_k) \triangleq&\sum_{i=1}^{N}(\frac{1}{\theta_i}\vartheta_i(q_k) + \lambda_i)\end{aligned}$$

with the initial condition $\Xi(q_0) = P(q_0)$ is an upper bound on the filtering error covariance $P(q_k)$.

Proof *It is obtained from* (11.10) *that*

$$\begin{aligned}P(q_{k+1}) =&\Big(\tilde{A}(q_k) - G(q_k)C(q_k)\bar{\boldsymbol{I}}_b\Big)\mathbb{E}\{e(q_k)e^T(q_k)\}\\ &\times\Big(\tilde{A}(q_k) - G(q_k)C(q_k)\bar{\boldsymbol{I}}_b\Big)^T\\ &+ G(q_k)C(q_k)\bar{\boldsymbol{I}}_b\mathbb{E}\{\bar{e}_c(q_k)\bar{e}_c^T(q_k)\}\bar{\boldsymbol{I}}_b^T C^T(q_k)G^T(q_k)\\ &+ G(q_k)C(q_k)\bar{E}\mathbb{E}\{e(q_{k-c})e^T(q_{k-c})\}\bar{E}^T C^T(q_k)G^T(q_k)\\ &+ \tilde{B}(q_k)\mathbb{E}\{\bar{w}(q_k)\bar{w}^T(q_k)\}\tilde{B}^T(q_k)\\ &+ G(q_k)D(q_k)\mathbb{E}\{v(q_k)v^T(q_k)\}D^T(q_k)G^T(q_k)\end{aligned}$$

$$
\begin{aligned}
&+ \bar{K}(q_k)\mathbb{E}\{\varepsilon(q_k)\varepsilon^T(q_k)\}\bar{K}^T(q_k) \\
&+ \mathscr{A}(q_k) + \mathscr{A}^T(q_k) + \mathscr{B}(q_k) + \mathscr{B}^T(q_k) \\
&+ \mathscr{C}(q_k) + \mathscr{C}^T(q_k) + \mathscr{D}(q_k) + \mathscr{D}^T(q_k) \\
&+ \mathscr{E}(q_k) + \mathscr{E}^T(q_k) + \mathscr{F}(q_k) + \mathscr{F}^T(q_k) \\
&+ \mathscr{G}(q_k) + \mathscr{G}^T(q_k)
\end{aligned}
$$

where

$$
\begin{aligned}
\mathscr{A}(q_k) &\triangleq -\big(\tilde{A}(q_k) - G(q_k)C(q_k)\bar{\boldsymbol{I}}_b\big)\mathbb{E}\{e(q_k)\bar{e}_c^T(q_k)\}\bar{\boldsymbol{I}}_b^T C^T(q_k)G^T(q_k), \\
\mathscr{B}(q_k) &\triangleq -\big(\tilde{A}(q_k) - G(q_k)C(q_k)\bar{\boldsymbol{I}}_b\big)\mathbb{E}\{e(q_k)e^T(q_{k-c})\}\bar{E}^T C^T(q_k)G^T(q_k), \\
\mathscr{C}(q_k) &\triangleq \big(\tilde{A}(q_k) - G(q_k)C(q_k)\bar{\boldsymbol{I}}_b\big)\mathbb{E}\{e(q_k)\varepsilon^T(q_k)\}\bar{K}^T(q_k), \\
\mathscr{D}(q_k) &\triangleq G(q_k)C(q_k)\mathbb{E}\{\bar{\boldsymbol{I}}_b\bar{e}_c(q_k)e^T(q_{k-c})\}\bar{E}^T C^T(q_k)G^T(q_k), \\
\mathscr{E}(q_k) &\triangleq -G(q_k)C(q_k)\mathbb{E}\{\bar{\boldsymbol{I}}_b\bar{e}_c(q_k)\varepsilon^T(q_k)\}\bar{K}^T(q_k), \\
\mathscr{F}(q_k) &\triangleq -G(q_k)C(q_k)\bar{E}\mathbb{E}\{e(q_{k-c})\varepsilon^T(q_k)\}\bar{K}^T(q_k), \\
\mathscr{G}(q_k) &\triangleq -G(q_k)D(q_k)\mathbb{E}\{v(q_k)\varepsilon^T(q_k)\}\bar{K}^T(q_k).
\end{aligned}
$$

From Lemma 11.2, it is easily known that

$$
\begin{aligned}
\mathscr{G}(q_k) + \mathscr{G}^T(q_k) =& -G(q_k)D(q_k)\mathbb{E}\{v(q_k)\varepsilon^T(q_k)\}\bar{K}^T(q_k) \\
& - \bar{K}(q_k)\mathbb{E}\{\varepsilon(q_k)v^T(q_k)\}D^T(q_k)G^T(q_k) \\
=& -G(q_k)\Pi(q_k)\bar{K}^T(q_k) - \bar{K}(q_k)\Pi^T(q_k)G^T(q_k).
\end{aligned}
$$

By using the well-known inequality $ab^T + ba^T \le \beta aa^T + \beta^{-1}bb^T$ (where a and b are vectors of appropriate dimensions), we have

$$
\begin{aligned}
&\mathscr{A}(q_k) + \mathscr{A}^T(q_k) \\
\le& \beta_1(q_k)\Big(\tilde{A}(q_k) - G(q_k)C(q_k)\bar{\boldsymbol{I}}_b\Big)\mathbb{E}\{e(q_k)e^T(q_k)\}\Big(\tilde{A}(q_k) - G(q_k)C(q_k)\bar{\boldsymbol{I}}_b\Big)^T \\
&+ \beta_1^{-1}(q_k)G(q_k)C(q_k)\bar{\boldsymbol{I}}_b\mathbb{E}\{\bar{e}_c(q_k)\bar{e}_c^T(q_k)\}\bar{\boldsymbol{I}}_b^T C^T(q_k)G^T(q_k), \\
&\mathscr{B}(q_k) + \mathscr{B}^T(q_k) \\
\le& \beta_2(q_k)\Big(\tilde{A}(q_k) - G(q_k)C(q_k)\bar{\boldsymbol{I}}_b\Big)\mathbb{E}\{e(q_k)e^T(q_k)\}\Big(\tilde{A}(q_k) - G(q_k)C(q_k)\bar{\boldsymbol{I}}_b\Big)^T \\
&+ \beta_2^{-1}(q_k)G(q_k)C(q_k)\bar{E}\mathbb{E}\{e(q_{k-c})e^T(q_{k-c})\}\bar{E}^T C^T(q_k)G^T(q_k), \\
&\mathscr{C}(q_k) + \mathscr{C}^T(q_k) \\
\le& \beta_3(q_k)\Big(\tilde{A}(q_k) - G(q_k)C(q_k)\bar{\boldsymbol{I}}_b\Big)\mathbb{E}\{e(q_k)e^T(q_k)\}\Big(\tilde{A}(q_k) - G(q_k)C(q_k)\bar{\boldsymbol{I}}_b\Big)^T \\
&+ \beta_3^{-1}(q_k)\bar{K}(q_k)\mathbb{E}\{\varepsilon(q_k)\varepsilon^T(q_k)\}\bar{K}^T(q_k), \\
&\mathscr{D}(q_k) + \mathscr{D}^T(q_k) \\
\le& \beta_4(q_k)G(q_k)C(q_k)\bar{\boldsymbol{I}}_b\mathbb{E}\{\bar{e}_c(q_k)\bar{e}_c^T(q_k)\}\bar{\boldsymbol{I}}_b^T C^T(q_k)G^T(q_k)
\end{aligned}
$$

$$
\begin{aligned}
&+ \beta_4^{-1}(q_k)G(q_k)C(q_k)\bar{E}\mathbb{E}\{e(q_{k-c})e^T(q_{k-c})\}\bar{E}^TC^T(q_k)G^T(q_k),\\
&\mathscr{E}(q_k) + \mathscr{E}^T(q_k)\\
\leq&\beta_5(q_k)G(q_k)C(q_k)\bar{\boldsymbol{I}}_b\mathbb{E}\{\bar{e}_c(q_k)\bar{e}_c^T(q_k)\}\bar{\boldsymbol{I}}_b^TC^T(q_k)G^T(q_k)\\
&+ \beta_5^{-1}(q_k)\bar{K}(q_k)\mathbb{E}\{\varepsilon(q_k)\varepsilon^T(q_k)\}\bar{K}^T(q_k),\\
&\mathscr{F}(q_k) + \mathscr{F}^T(q_k)\\
\leq&\beta_6(q_k)G(q_k)C(q_k)\bar{E}\mathbb{E}\{e(q_{k-c})e^T(q_{k-c})\}\bar{E}^TC^T(q_k)G^T(q_k)\\
&+ \beta_6^{-1}(q_k)\bar{K}(q_k)\mathbb{E}\{\varepsilon(q_k)\varepsilon^T(q_k)\}\bar{K}^T(q_k).
\end{aligned}
$$

Moreover, one has

$$
\begin{aligned}
&\bar{\boldsymbol{I}}_b\mathbb{E}\{\bar{e}_c(q_k)\bar{e}_c^T(q_k)\}\bar{\boldsymbol{I}}_b^T\\
=&\sum_{i=1}^{c-1}\sum_{j=1}^{c-1}\mathbb{E}\{\bar{\boldsymbol{I}}_be(q_{k-i})e^T(q_{k-j})\bar{\boldsymbol{I}}_b^T\}\\
=&\frac{1}{2}\sum_{i=1}^{c-1}\sum_{j=1}^{c-1}\mathbb{E}\{\bar{\boldsymbol{I}}_be(q_{k-i})e^T(q_{k-j})\bar{\boldsymbol{I}}_b^T + \bar{\boldsymbol{I}}_be(q_{k-j})e^T(q_{k-i})\bar{\boldsymbol{I}}_b^T\}\\
\leq&\frac{1}{2}\sum_{i=1}^{c-1}\sum_{j=1}^{c-1}\mathbb{E}\{\bar{\boldsymbol{I}}_be(q_{k-i})e^T(q_{k-i})\bar{\boldsymbol{I}}_b^T + \bar{\boldsymbol{I}}_be(q_{k-j})e^T(q_{k-j})\bar{\boldsymbol{I}}_b^T\}\\
=&(c-1)\sum_{i=1}^{c-1}\bar{\boldsymbol{I}}_b\mathbb{E}\{e(q_{k-i})e^T(q_{k-i})\}\bar{\boldsymbol{I}}_b^T.
\end{aligned}
$$

From the event triggering condition and Lemma 11.1, it is obvious that

$$
\varepsilon_i^T(q_k)\varepsilon_i(q_k) \leq \frac{1}{\theta_i}\zeta_i(q_k) + \lambda_i \leq \frac{1}{\theta_i}\vartheta_i(q_k) + \lambda_i
$$

and, accordingly, we have

$$
\varepsilon^T(q_k)\varepsilon(q_k) \leq \varrho(q_k).
$$

Applying the properties of matrix operations, one has

$$
\begin{aligned}
\varepsilon(q_k)\varepsilon^T(q_k) &\leq \mathrm{tr}\{\varepsilon(q_k)\varepsilon^T(q_k)\}I\\
&= \varepsilon^T(q_k)\varepsilon(q_k)I \leq \varrho(q_k)I,
\end{aligned}
$$

and therefore

$$
\mathbb{E}\{\varepsilon(q_k)\varepsilon^T(q_k)\} \leq \varrho(q_k)I.
$$

Summarizing the above discussions, we have

$$\begin{aligned}
P(q_{k+1}) \leq & \gamma_1(q_k)\Big(\tilde{A}(q_k) - G(q_k)C(q_k)\bar{\boldsymbol{I}}_b\Big)P(q_k) \\
& \times \Big(\tilde{A}(q_k) - G(q_k)C(q_k)\bar{\boldsymbol{I}}_b\Big)^T \\
& + \gamma_2(q_k)(c-1)\sum_{i=1}^{c-1} G(q_k)C(q_k)\bar{\boldsymbol{I}}_b P(q_{k-i})\bar{\boldsymbol{I}}_b^T C^T(q_k)G^T(q_k) \\
& + \gamma_3(q_k)G(q_k)C(q_k)\bar{E}P(q_{k-c})\bar{E}^T C^T(q_k)G^T(q_k) \\
& + \gamma_4(q_k)\varrho(q_k)\bar{K}(q_k)\bar{K}^T(q_k) + \tilde{B}(q_k)\bar{W}(q_k)\tilde{B}^T(q_k) \\
& + G(q_k)D(q_k)\bar{R}(q_k)D^T(q_k)G^T(q_k) \\
& - G(q_k)\Pi(q_k)\bar{K}^T(q_k) - \bar{K}(q_k)\Pi^T(q_k)G^T(q_k) \\
= & \phi_k(P(q_k)).
\end{aligned}$$

Then, from Lemma 10.1, we have

$$P(q_{k+1}) \leq \Xi(q_{k+1}).$$

The proof is complete.

Remark 11.3 *In this chapter, the introduction of the internal dynamic variable $\zeta_i(q_k)$ brings additional difficulties to the filter design. In calculating upper bound on the filtering error covariance, the upper bound on $\varepsilon_i^T(q_k)\varepsilon_i(q_k)$ is needed. From the triggering condition, we know that $\varepsilon_i^T(q_k)\varepsilon_i(q_k) \leq \frac{1}{\theta_i}\zeta_i(q_k) + \lambda_i$. Nevertheless, $\zeta_i(q_k)$ is unknown on the filter side and we need to look for an alternative upper bound on $\varepsilon_i^T(q_k)\varepsilon_i(q_k)$. In Lemma 11.2, an upper bound on $\zeta_i(q_k)$ has been derived which facilitates the calculation of upper bound on $\varepsilon_i^T(q_k)\varepsilon_i(q_k)$.*

Let $K^{(i)}(q_k)$ and $\Theta^{(i)}(q_k)$ be the ith row of the matrices $K(q_k)$ and $\Theta(q_k)$, respectively. Moreover, we denote

$$\begin{aligned}
\mathcal{M}_i(q_k) &\triangleq \gamma_4(q_k)\varrho(q_k)I - \Pi(q_k)\Psi^{-1}(q_k)\Pi(q_k), \\
\mathcal{N}_i(q_k) &\triangleq \Theta^{(i)}(q_k)\Psi^{-1}(q_k)\Pi(q_k), \\
\bar{K}^{(i)}(q_k) &\triangleq \mathcal{N}_i(q_k)\bar{Y}_i\bar{\mathcal{M}}_i^{-1}(q_k)\bar{Y}_i^T \\
&\triangleq \Big[\bar{K}_{i1}(q_k) \quad \bar{K}_{i2}(q_k) \quad \cdots \quad \bar{K}_{iN}(q_k)\Big], \\
Y_i &\triangleq \mathrm{diag}\{\sqrt{a_{i1}}I, \sqrt{a_{i2}}I, \cdots, \sqrt{a_{iN}}I\}, \\
\bar{\mathcal{M}}_i(q_k) &\triangleq \bar{Y}_i^T\mathcal{M}_i(q_k)\bar{Y}_i
\end{aligned} \tag{11.12}$$

where $\bar{K}_{ij}(q_k) \in \mathbb{R}^{bn_x \times n_y}$ $(j = 1, 2, \ldots, N)$ and $\bar{Y}_i$ denotes the simplified matrix by removing the ℓth $(\ell \notin \mathcal{V}_i)$ column from the matrix Y_i. In the following theorem, we are going to minimize the upper bound obtained in Theorem 11.1 by choosing appropriate filter gains.

Theorem 11.2 *The minimal upper bound on the filtering error covariance can be obtained by choosing the filtering gains as*

$$K_{ij}(q_k) = \begin{cases} 0, & a_{ij} = 0 \\ \bar{K}_{ij}(q_k)a_{ij}^{-1}, & a_{ij} \neq 0 \end{cases}, \tag{11.13}$$

$$L(q_k) = \sum_{i=1}^{N} F_i K(q_k) H_i \Big(\Pi(q_k) - \Psi(q_k)\Big) \Psi^{-1}(q_k) + \Theta(q_k)\Psi^{-1}(q_k) \tag{11.14}$$

where $\bar{K}_{ij}(q_k)$ $(j = 1, 2, \ldots, N)$ *is defined in* (11.12).

Proof *Noting* $G(q_k) = L(q_k) + \bar{K}(q_k)$, *the upper bound* $\Xi(q_{k+1})$ *can be rewritten as*

$$\begin{aligned}
\Xi(q_{k+1}) =& L(q_k)\Psi(q_k)L^T(q_k) + \bar{K}(q_k)\Psi(q_k)\bar{K}^T(q_k) \\
&+ L(q_k)\Psi(q_k)\bar{K}^T(q_k) + \bar{K}(q_k)\Psi(q_k)L^T(q_k) \\
&- L(q_k)\Theta^T(q_k) - \bar{K}(q_k)\Theta^T(q_k) \\
&- \Theta(q_k)L^T(q_k) - \Theta(q_k)\bar{K}^T(q_k) \\
&+ \gamma_4(q_k)\varrho(q_k)\bar{K}(q_k)\bar{K}^T(q_k) - 2\bar{K}(q_k)\Pi(q_k)\bar{K}^T(q_k) \\
&+ \gamma_1(q_k)\tilde{A}(q_k)\Xi(q_k)\tilde{A}^T(q_k) + \tilde{B}(q_k)\bar{W}(q_k)\tilde{B}^T(q_k) \\
&- \bar{K}(q_k)\Pi(q_k)L^T(q_k) - L(q_k)\Pi(q_k)\bar{K}^T(q_k).
\end{aligned}$$

To obtain the minimal upper bound, we need to derive the filter gains that minimize $\mathrm{tr}\{\Xi(q_{k+1})\}$. *Taking the partial derivative of* $\mathrm{tr}\{\Xi(q_{k+1})\}$ *with respect to* $L(q_k)$, *one has*

$$\frac{\partial}{\partial L(q_k)}\mathrm{tr}\{\Xi(q_{k+1})\} = 2L(q_k)\Psi(q_k) + 2\bar{K}(q_k)\Psi(q_k) - 2\Theta(q_k) - 2\bar{K}(q_k)\Pi(q_k).$$

Noting that $\mathrm{tr}\{F_i X F_j^T\} = 0$ *where* X *is a matrix with appropriate dimensions and* $F_i^T F_i = F_i$, *the partial derivative of* $\mathrm{tr}\{\Xi(q_{k+1})\}$ *with respect to* $K(q_k)$ *is derived as*

$$\begin{aligned}
\frac{\partial}{\partial K(q_k)}\mathrm{tr}\{\Xi(q_{k+1})\} =& 2\sum_{i=1}^{N} F_i K(q_k) H_i \Psi(q_k) H_i^T + 2\sum_{i=1}^{N} F_i L(q_k)\Psi(q_k) H_i^T \\
&- 2\sum_{i=1}^{N} F_i \Theta(q_k) H_i^T - 2\sum_{i=1}^{N} F_i L(q_k)\Pi(q_k) H_i^T \\
&+ 2\gamma_4(q_k)\varrho(q_k)\sum_{i=1}^{N} F_i K(q_k) H_i H_i^T \\
&- 4\sum_{i=1}^{N} F_i K(q_k) H_i \Pi(q_k) H_i^T.
\end{aligned}$$

Letting

$$\frac{\partial}{\partial L(q_k)}\mathrm{tr}\{\Xi(q_{k+1})\} = 0, \quad \frac{\partial}{\partial K(q_k)}\mathrm{tr}\{\Xi(q_{k+1})\} = 0,$$

one has

$$L(q_k) = \bar{K}(q_k)\Big(\Pi(q_k) - \Psi(q_k)\Big)\Psi^{-1}(q_k) + \Theta(q_k)\Psi^{-1}(q_k) \tag{11.15}$$

and

$$\begin{aligned}&\sum_{i=1}^{N} F_i K(q_k) H_i\Big(\Psi(q_k) + \gamma_4(q_k)\varrho(q_k)I - 2\Pi(q_k)\Big)H_i^T \\ &+ \sum_{i=1}^{N} F_i L(q_k)\Big(\Psi(q_k) - \Pi(q_k)\Big)H_i^T - \sum_{i=1}^{N} F_i \Theta(q_k) H_i^T = 0.\end{aligned} \tag{11.16}$$

Substituting (11.13) *into* (11.14), *we have*

$$\begin{aligned}&\sum_{i=1}^{N} F_i K(q_k) H_i\Big(\Psi(q_k) + \gamma_4(q_k)\varrho(q_k)I - 2\Pi(q_k)\Big)H_i^T \\ &+ \sum_{i=1}^{N} F_i \Theta(q_k)\Psi^{-1}(q_k)\Big(\Psi(q_k) - \Pi(q_k)\Big)H_i^T \\ &+ \sum_{i=1}^{N} F_i\Big(\sum_{i=1}^{N} F_i K(q_k) H_i\Big(\Pi(q_k) - \Psi(q_k)\Big)\Psi^{-1}(q_k)\Big) \\ &\times \Big(\Psi(q_k) - \Pi(q_k)\Big)H_i^T - \sum_{i=1}^{N} F_i\Theta(q_k)H_i^T = 0\end{aligned}$$

which means

$$\sum_{i=1}^{N} F_i K(q_k) H_i \mathcal{M}_i(q_k) H_i^T = \sum_{i=1}^{N} F_i \Theta(q_k)\Psi^{-1}(q_k)\Pi(q_k)H_i^T.$$

Note that $K^{(i)}(q_k)$ *and* $\Theta^{(i)}(q_k)$ *are the* i*th row of the matrices* $K(q_k)$ *and* $\Theta(q_k)$, *respectively. Then, it is derived that*

$$K^{(i)}(q_k)H_i\mathcal{M}_i(q_k)H_i^T = \mathcal{N}_i(q_k)H_i^T.$$

Noting $H_i = \mathrm{diag}\{a_{i1}I, a_{i2}I, \cdots, a_{iN}I\}$ *with* $a_{ij} = 0$ *for* $j \notin \mathcal{V}_i$, *we know that* $H_i = \bar{Y}_i\bar{Y}_i^T$. *Then, we have*

$$K^{(i)}(q_k)\bar{Y}_i\bar{\mathcal{M}}_i(q_k)\bar{Y}_i^T = \mathcal{N}_i(q_k)\bar{Y}_i\bar{Y}_i^T.$$

Since the matrix $\bar{Y}_i$ *has full row rank and the matrix* $\bar{\mathcal{M}}_i(q_k)$ *is invertible, it is obtained that*

$$K^{(i)}(q_k)\bar{Y}_i = \mathcal{N}_i(q_k)\bar{Y}_i\bar{\mathcal{M}}_i^{-1}(q_k).$$

Accordingly, one has

$$K^{(i)}(q_k)H_i = \bar{K}^{(i)}(q_k)$$

where $\bar{K}^{(i)}(q_k)$ *is defined in* (11.12). *Therefore, the filter gain* $K(q_k) = \begin{bmatrix} K_{ij}(q_k) \end{bmatrix}_{N\times N}$ *can be calculated by* (11.13), *and the filter gain* $L(q_k)$ *can be derived by substituting the matrix* $K(q_k)$ *into* (11.14). *The proof is complete.*

Remark 11.4 *Until now, the dynamic event-based recursive filtering problem has been studied for MRSs with integral measurements over sensor networks. In Theorem 11.1, an upper bound on the filtering error covariance has been derived by solving a matrix Riccati equation* (11.11). *Then, in Theorem 11.2, such an upper bound has been minimized by selecting the filter gains as* (11.13) *and* (11.14).

11.3 Simulation Examples

In this section, comprehensive simulations are conducted to show the effectiveness and superiority of the proposed dynamic event-based recursive filtering algorithm.

Example 11.1

Consider the following discrete-time linear system:

$$x(s_{k+1}) = \begin{bmatrix} 0.92 + 0.2\cos(s_k) & 0.39 \\ -0.18 & 0.98 \end{bmatrix} x(s_k) + \begin{bmatrix} 0.08 \\ 0.06 \end{bmatrix} w(s_k)$$

where the state update period of the system is $s_{k+1} - s_k = h = 1$.

The measurements of the system are measured by three sensors with a sampling period $q_{k+1} - q_k = 2$ and the following parameters:

$$\begin{aligned}
C_1(q_k) &= \begin{bmatrix} 8 + \sin(q_k) & 8 - \cos(q_k) \end{bmatrix}, \\
C_2(q_k) &= \begin{bmatrix} 6\sin(q_k) & 6\cos(q_k) \end{bmatrix}, \\
C_3(q_k) &= \begin{bmatrix} 7\sin(q_k) & 7\cos(q_k) \end{bmatrix}, \\
D_1(q_k) &= \sin(q_k),\ D_2(q_k) = \sin(q_k), \\
D_3(q_k) &= \sin(q_k),\ l = 3.
\end{aligned}$$

The topology of the sensor network is represented by a directed graph $\mathscr{G} = (\mathcal{V}, \mathscr{E}, \mathcal{R})$ with the set of nodes $\mathcal{V} = \{1, 2, 3\}$, set of edges $\mathscr{E} = \{(1, 2), (2, 3), (3, 1)\}$, and the weighted adjacency matrix $\mathcal{R} = [a_{ij}]_{3\times 3}$ with adjacency elements $a_{ij} = 1$ for $(i, j) \in \mathscr{E}$ and $a_{ij} = 0$ otherwise. In the simulation, the covariances of the process noise and the measurement noises are set as $W(s_k) = 0.05$, $R_1(q_k) = 0.03$, $R_2(q_k) = 0.05$, and $R_3(q_k) = 0.04$, respectively.

11.3.1 Effectiveness of the Proposed Filtering Scheme

In this subsection, we aim to demonstrate the effectiveness of the proposed dynamic event-based recursive filtering scheme. For the conditions (11.5) and (11.6) of the DEM, we set the thresholds as $\lambda_1 = \lambda_2 = \lambda_3 = 1.8$, and other parameters as $\theta_1 = \theta_2 = \theta_3 = 20$, $\mu_1 = \mu_2 = \mu_3 = 0.7$, and $\zeta_{1,0} = \zeta_{2,0} = \zeta_{3,0} = 2$.

With the given parameters, the filter gains $K_{ij}(q_k)$ and $L_i(q_k)$ are derived according to (11.13) and (11.14), and the minimal upper bound on the filtering error covariance is calculated by solving (11.11). Figs. 11.1 and 11.2 depict the estimates of the state $x_1(s_k)$ and $x_2(s_k)$ ($x_i(s_k)$ $(i = 1, 2)$ is the i-th element of $x(s_k)$) from the three filters, respectively. The corresponding estimation errors are also given in Figs. 11.1 and 11.2, respectively. Figure 11.3 shows the trace of the minimal upper bound and the mean square error (MSE) with $M = 500$ times independent experiments. The MSE is defined as

$$\text{MSE}(s_k) \triangleq \frac{1}{M}\sum_{i=1}^{M}\sum_{j=1}^{3}(x(s_k) - \vec{x}_j(s_k))^T(x(s_k) - \vec{x}_j(s_k))$$

where $\vec{x}_j(s_k)$, which is obtained by applying matrix operations to $\hat{x}_j(q_k)$, is the estimate of $x(s_k)$ from the j-th filter.

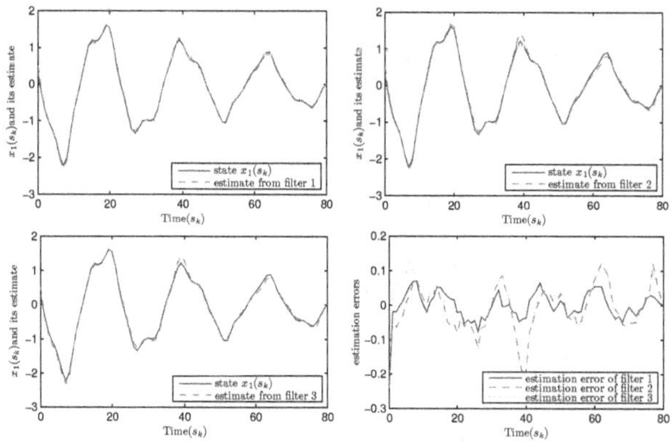

FIGURE 11.1: Estimates of the state $x_1(s_k)$ and the estimation errors.

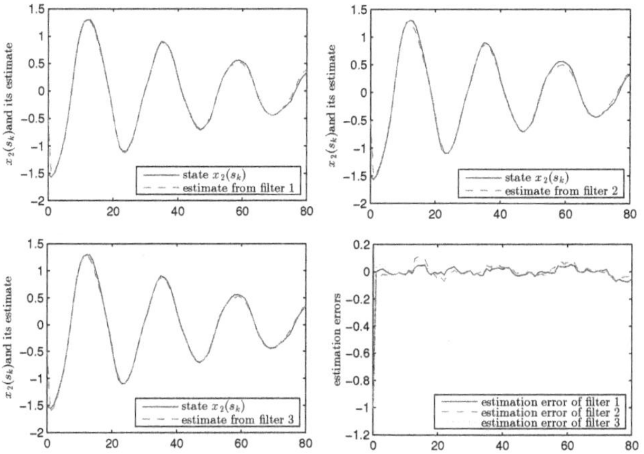

FIGURE 11.2: Estimates of the state $x_2(s_k)$ and the estimation errors.

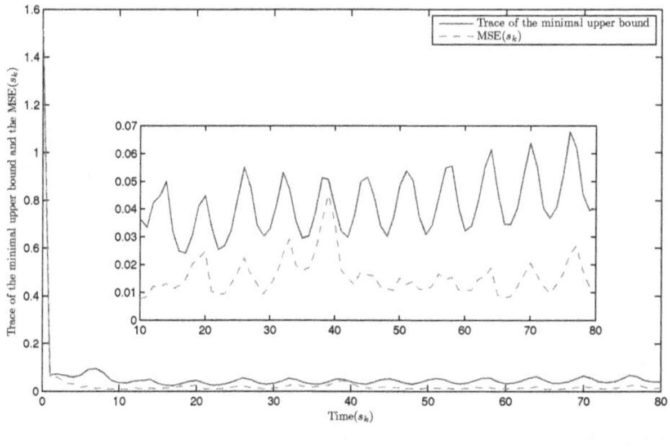

FIGURE 11.3: Trace of the minimal upper bound and the MSE.

Figs. 11.1 and 11.2 show that the estimates of the proposed filtering scheme follow the system state well. Therefore, the proposed dynamic event-based recursive filtering scheme is indeed effective.

11.3.2 Comparison of Results

In this subsection, to show the superiority of the DEM over the SEM, the filtering performance under the DEM and the SEM is compared. The parameters of the event-triggering condition are chosen as those in subsection 11.3.1, and other parameters are all the same in both DEM case and SEM case.

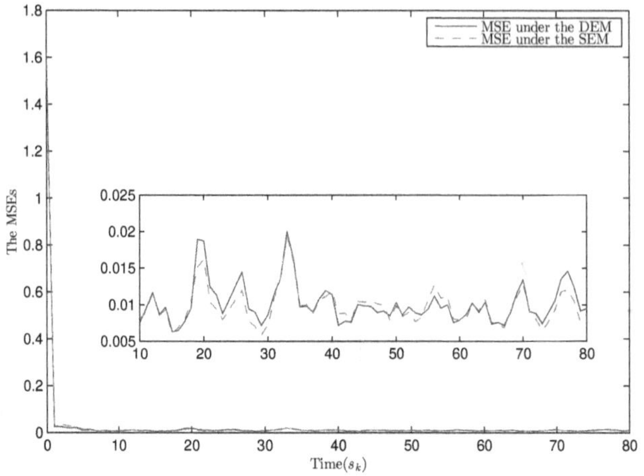

FIGURE 11.4: The MSEs under the DEM and the SEM.

Simulation results are given in Figs. 11.4–11.5 and Table 11.1. Figure 11.4 shows the MSEs under the DEM and the SEM. Figure 11.5 shows the triggering instants under the DEM and the SEM. Table 11.1 gives the triggering rates and the averages of the MSEs under the DEM and the SEM. It can be seen that, compared with the triggering times under the SEM, the triggering times under the DEM are significantly reduced. Although the filtering performance under the DEM is slightly decreased compared with the filtering performance under the SEM, the resource consumption of the overall systems is effectively reduced.

Example 11.2

In this example, we consider a practical system to further demonstrate the usefulness of the developed dynamic event-based recursive filtering scheme.

TABLE 11.1: Triggering rates and the averages of the MSEs under the DEM and the SEM.

	Triggering rate		Average of the MSE	
	DEM	SEM	DEM	SEM
Node 1	25%	58%	0.0095	0.0088
Node 2	30%	48%	0.0109	0.0099
Node 3	30%	50%	0.0104	0.0093

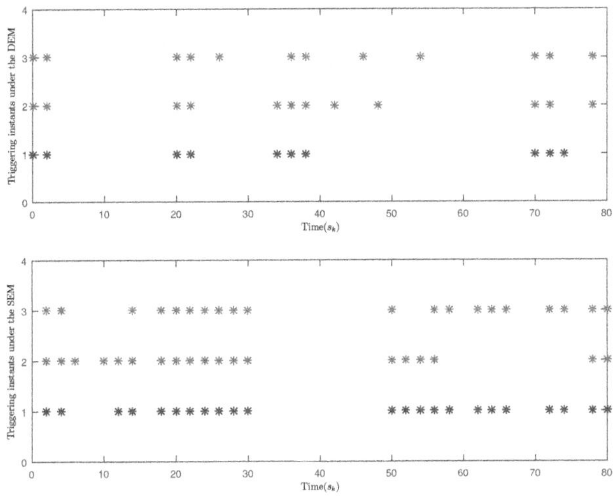

FIGURE 11.5: Triggering instants on the nodes with the DEM and the SEM.

Modified from [212], the DC servo system can be characterized by the following model:

$$x(s_{k+1}) = \begin{bmatrix} 1.12 + 0.3\sin(s_k) & 0.213 & -0.333 \\ 1 & 0 & 0 \\ 0 & 1 & 0 \end{bmatrix} x(s_k) + \begin{bmatrix} 0.8 \\ 0 \\ 0 \end{bmatrix} w(s_k)$$

where $h = 1$. Moreover, the parameters of the measurement model are $b = 2$, $l = 3$, and

$$\begin{aligned} C_1(q_k) &= \begin{bmatrix} 1 & 1 & 0 \end{bmatrix}, \quad D_1(q_k) = 0.1\sin(q_k), \\ C_2(q_k) &= \begin{bmatrix} 1 & 1 & 0 \end{bmatrix}, \quad D_2(q_k) = 0.1\cos(q_k), \\ C_3(q_k) &= \begin{bmatrix} 1 & 1 & 0 \end{bmatrix}, \quad D_3(q_k) = 0.1\sin(q_k). \end{aligned}$$

The topology of the sensor network is the same as that in Example 11.1. The parameters of the DEM are set as $\lambda_1 = \lambda_2 = \lambda_3 = 1.8$, $\theta_1 = \theta_2 = \theta_3 = 4$, $\mu_1 = \mu_2 = \mu_3 = 0.3$, and $\zeta_{1,0} = \zeta_{2,0} = \zeta_{3,0} = 1$. With the given parameters, the filter gains and the minimal upper bound on the filtering error covariance are obtained. Simulation results are shown in Figs. 11.6–11.9, from which it can be verified that the proposed dynamic event-based recursive filtering algorithm is effective in real practical system.

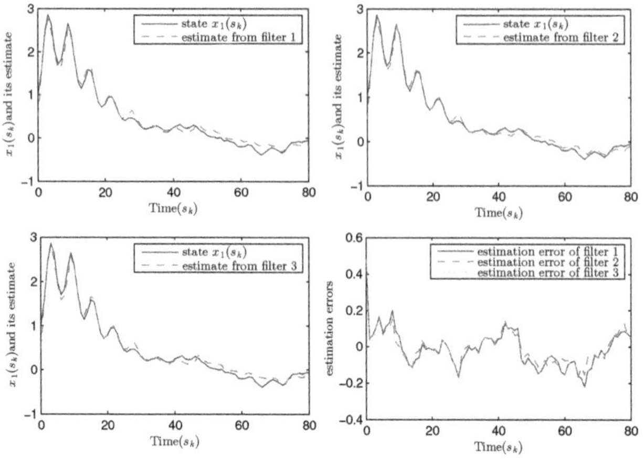

FIGURE 11.6: Estimates of the state $x_1(s_k)$ and the estimation errors.

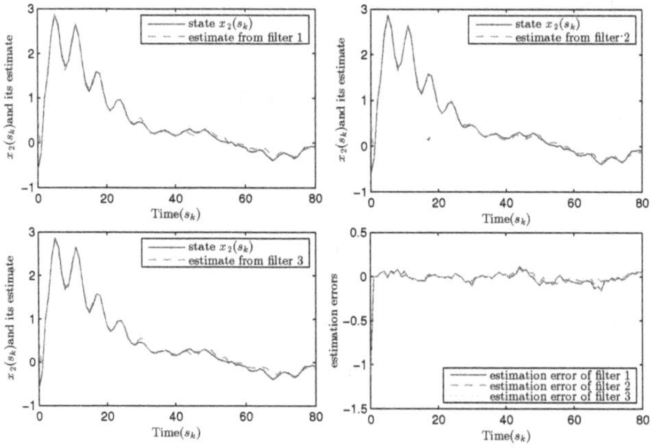

FIGURE 11.7: Estimates of the state $x_2(s_k)$ and the estimation errors.

11.4 Conclusion

In this chapter, the dynamic event-based recursive filtering problem has been studied for MRSs over sensor networks. A general situation has been considered where the state update rate of the system and the sampling rate of the sensors are allowed to be different. The phenomenon of integral measurements has been considered in the measurement model. The DEM has been used to reduce data transmissions among sensor nodes. An upper bound has been first derived on the filtering error covariance and then minimized by

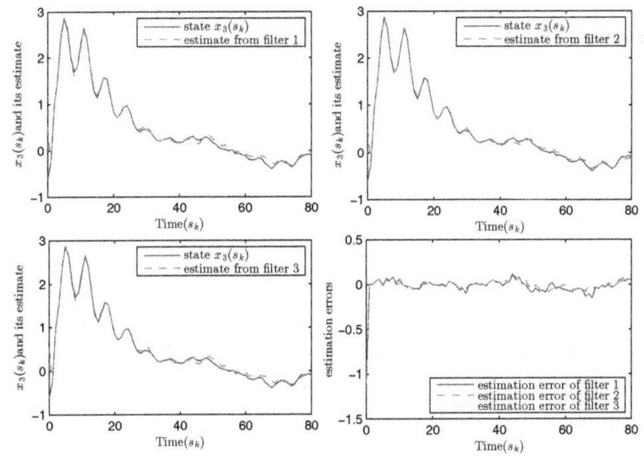

FIGURE 11.8: Estimates of the state $x_3(s_k)$ and the estimation errors.

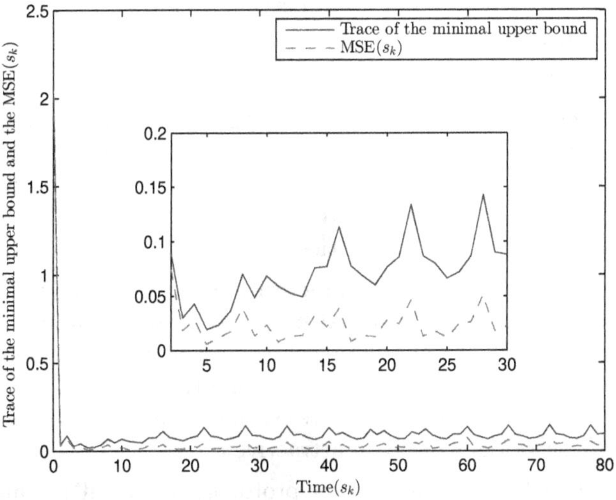

FIGURE 11.9: Trace of the minimal upper bound and the MSE.

choosing appropriate filter gains. Finally, simulation examples have been given to verify the effectiveness and superiority of the developed dynamic event-based recursive filtering scheme.

12

Conclusion and Further Work

The focus of the book has been placed on state estimation problems for MRSs. Firstly, the general state-space model of the MRSs has been introduced and the available research results on the state estimation problems for MRSs have been reviewed. Then, the state estimation problems for MRSs with various network-induced phenomena have been studied in Chapters 2–5. Subsequently, the state estimation problems for MRSs under different communication protocols have been investigated in Chapters 6–11.

Generally speaking, this book has established a unified theoretical framework for estimator synthesis for MRSs while addressing difficulties induced by different factors such as network-induced phenomena and communication protocols. Nevertheless, the results in this book are still quite limited. Some future research topics are given as follows:

1. In the existing literature, the state estimation problems are mostly investigated for linear MRSs or non-linear MRSs with strict assumption on the non-linear function. It would be an interesting and challenging topic to investigate the state estimation problems for MRSs with general nonlinearities.
2. The complex networks can characterize plenty of real-world dynamical systems and have received consistent research interest in the past decades. Unfortunately, the studies on the state estimation problems for complex networks with multi-rate sampling strategy are quite few. Therefore, the state estimation problem for multi-rate complex networks will be an attractive area.
3. Other than the communication protocols discussed in this book, there are also some effective communication protocols, e.g. the FlexRay protocol, introduced in networked systems and the corresponding state estimation algorithms have been developed. Nevertheless, how to modify the existing results to make them applicable to MRSs or how to develop novel estimation algorithms suitable for MRSs with above-mentioned protocols remain open and challenging.
4. In the estimate fusion process, due to the asynchronism of the local estimates, only the estimates available at the current time instant are fused. As such, the valuable estimate information are not fully used which would lead to deterioration of the estimation accuracy.

DOI: 10.1201/9781032619507-12

To this end, a trend for future research is to propose novel fusion strategies that make full use of the asynchronous estimates.

5. Another promising research topic is to study the fault detection, diagnosis, and isolation problems for MRSs. Although some initial results have been obtained, the corresponding theories for MRSs are far from mature. Hence, it is of great importance to develop fault detection, diagnosis, and isolation theories for MRSs.

Bibliography

[1] A. Alessandri and L. Zaccarian, Stubborn state observers for linear time-invariant systems, *Automatica*, vol. 88, pp. 1–9, 2018.

[2] D. Andrisani and C.-F. Gau, Estimation using a multirate filter, *IEEE Transactions on Automatic Control*, vol. 32, no. 7, pp. 653–656, 1987.

[3] L. Armesto, J. Tornero and M. Vincze, On multi-rate fusion for non-linear sampled-data systems: Application to a 6D tracking system, *Robotics and Autonomous Systems*, vol. 56, no. 8, pp. 706–715, 2008.

[4] J. Bae, Y. Kim and J. M. Lee, Multirate moving horizon estimation combined with parameter subset selection, *Computers & Chemical Engineering*, vol. 147, art. no. 107253, 2021.

[5] V. Barnett and T. Lewis, *Outliers in Statistical Data*, Wiley Series in Probability and Statistics, John Wiley & Sons, 1994.

[6] V. P. Bhuvana, C. Preissl, A. M. Tonello and M. Huemer, Multi-sensor information filtering with information-based sensor selection and outlier rejection, *IEEE Sensors Journal*, vol. 18, no. 6, pp. 2442–2452, 2018.

[7] T. Bonargent, T. Ménard, O. Gehan and E. Pigeon, Adaptive observer design for a class of Lipschitz nonlinear systems with multirate outputs and uncertainties: Application to attitude estimation with gyro bias, *International of Robust and Nonlinear Control*, vol. 31, no. 8, pp. 3137–3162, 2021.

[8] S. Boyd, L. Ghaoui, E. Feron and V. Balakrishnan, *Linear Matrix Inequalities in System and Control Theory*, Philadelphia, PA, USA: SIAM, 1994.

[9] R. Caballero-Águila, A. Hermoso-Carazo and J. Linares-Pérez, Optimal state estimation for networked systems with random parameter matrices, correlated noises and delayed measurements, *International Journal of General Systems*, vol. 44, no. 2, pp. 142–154, 2015.

[10] Z. Cao, J. Lu, R. Zhang and F. Gao, Iterative learning Kalman filter for repetitive processes, *Journal of Process Control*, vol. 46, pp. 92–104, 2016.

[11] Z. Cao, Y. Niu, H.-K. Lam and J. Zhao, Sliding mode control of Markovian jump fuzzy systems: A dynamic event-triggered method, *IEEE Transactions on Fuzzy Systems*, vol. 29, no. 10, pp. 2902–2915, 2021.

[12] N. A. Carlson, Federated square root filter for decentralized parallel processors, *IEEE Transactions on Aerospace and Electronic Systems*, vol. 26, no. 3, pp. 517–525, 1990.

[13] P. Casoli, N. Authier, X. Jacquet and J. Cartier, Characterization of the caliban and prospero critical assemblies neutron spectra for integral measurements experiments, *Nuclear Data Sheets*, vol. 118, pp. 554–557, 2014.

[14] G. Cena, L. Durante and A. Valenzano, Standard field bus networks for industrial applications, *Computer Standards & Interfaces*, vol. 17, no. 2, pp. 155–167, 1995.

[15] L. Chen, Y. Chen and N. Zhang, Synchronization control for chaotic neural networks with mixed delays under input saturations, *Neural Processing Letters*, vol. 53, pp. 3735–3755, 2021.

[16] Q. Chen, C. Yin, J. Zhou, Y. Wang, X. Wang and C. Chen, Hybrid consensus-based cubature Kalman filtering for distributed state estimation in sensor setworks, *IEEE Sensors Journal*, vol. 18, no. 11, pp. 4561–4569, 2018.

[17] S. Chen, L. Ma and Y. Ma, Distributed set-membership filtering for nonlinear systems subject to round-robin protocol and stochastic communication protocol over sensor networks, *Neurocomputing*, vol. 385, pp. 13–21, 2020.

[18] W. Chen, J. Hu, Z. Wu, X. Yu and D. Chen, Finite-time memory fault detection filter design for nonlinear discrete systems with deception attacks, *International Journal of Systems Science*, vol. 51, no. 8, pp. 1464–1481, 2020.

[19] Y. Chen, Z. Chen, Z. Chen and A. Xue, Observer-based passive control of non-homogeneous Markov jump systems with random communication delays, *International Journal of Systems Science*, vol. 51, no. 6, pp. 1133–1147, 2020.

[20] Y. Chen, J. Ren, X. Zhao and A. Xue, State estimation of Markov jump neural networks with random delays by redundant channels, *Neurocomputing*, vol. 453, pp. 493–501, 2021.

[21] Y. Chen, Z. Wang, B. Shen and H. Dong, Exponential synchronization for delayed dynamical networks via intermittent control: Dealing with actuator saturations, *IEEE Transactions on Neural Networks and Learning Systems*, vol. 30, no. 4, pp. 1000–1012, 2019.

[22] Y. Chen and W. X. Zheng, L_2-L_∞ filtering for stochastic Markovian jump delay systems with nonlinear perturbations, *Signal Processing*, vol. 109, pp. 154–164, 2015.

[23] X. Chi, X. Jia, F. Cheng and M. Fan, Networked H_∞ filtering for Takagi-Sugeno fuzzy systems under multi-output multi-rate sampling, *Journal of the Franklin Institute*, vol. 356, no. 6, pp. 3661–3691, 2019.

[24] H. D. Choi, C. K. Ahn, H. R. Karimi and M. T. Lim, Filtering of discrete-time switched neural networks ensuring exponential dissipative and l_2-l_∞ performances, *IEEE Transactions on Cybernetics*, vol. 47, no. 10, pp. 3195–3207, 2017.

[25] M. B. G. Cloosterman, L. Hetel, N. van de Wouw, W. P. M. H. Heemels, J. Daafouz and H. Nijmeijer, Controller synthesis for networked control systems, *Automatica*, vol. 46, no. 10, pp. 1584–1594, 2010.

[26] R. Cristi and M. Tummala, Multirate, multiresolution, recursive Kalman filter, *Signal Processing*, vol. 80, no. 9, pp. 1945–1958, 2000.

[27] Y. Cui, Y. Liu, W. Zhang and F. E. Alsaadi, Sampled-based consensus for nonlinear multiagent systems with deception attacks: The decoupled method, *IEEE Transactions on Systems, Man and Cybernetics: Systems*, vol. 51, no. 1, pp. 561–573, 2021.

[28] Y. Cui, L. Xu and Y. Shen, H_2/H_∞ filtering for networked systems with data transmission time-varying delay, data packet dropout and sequence disorder, *Journal of the Franklin Institute*, vol. 354, no. 13, pp. 5443–5462, 2017.

[29] N. Davari and A. Gholami, Variational Bayesian adaptive Kalman filter for asynchronous multirate multi-sensor integrated navigation system, *Ocean Engineering*, vol. 174, pp. 108–116, 2019.

[30] Z. Deng, P. Zhang, W. Qi, J. Liu and Y. Gao, Sequential covariance intersection fusion Kalman filter, *Information Sciences*, vol. 189, pp. 293–309, 2012.

[31] D. Ding, Q.-L. Han, X. Ge and J. Wang, Secure state estimation and control of cyber-physical systems: A survey, *IEEE Transactions on Systems, Man, and Cybernetics: Systems*, vol. 51, no. 1, pp. 176–190, 2021.

[32] D. Ding, Z. Wang, Q.-L. Han and G. Wei, Neural-network-based output-feedback control under Round-Robin scheduling protocols, *IEEE Transactions on Cybernetics*, vol. 49, no. 6, pp. 2372–2384, 2019.

[33] L. Ding, Q.-L. Han, X. Ge and X.-M. Zhang, An overview of recent advances in event-triggered consensus of multiagent systems, *IEEE Transactions on Cybernetics*, vol. 48, no. 4, pp. 1110–1123, 2018.

[34] H. Dong, Z. Wang, J. Lam and H. Gao, Fuzzy-model-based robust fault detection with stochastic mixed time delays and successive packet dropouts, *IEEE Transactions on Systems Man and Cybernetics: Cybernetics*, vol. 42, no. 2, pp. 365–376, 2012.

[35] Y. Dong, Y. Song and G. Wei, Efficient model-predictive control for networked interval type-2 T-S fuzzy system with stochastic communication protocol, *IEEE Transactions on Fuzzy Systems*, vol. 29, no. 2, pp. 286–297, 2021.

[36] E. Elliott, Estimates of error rates for codes on burst-noise channels, *Bell System Technical Journal*, vol. 42, no. 5, pp. 1977–1997, 1963.

[37] M. Elsheikh, R. Hille, A. T.-Codrean and S. Krämer, A comparative review of multi-rate moving horizon estimation schemes for bioprocess applications, *Computers & Chemical Engineering*, vol. 146, art. no. 107219, 2021.

[38] M. Enkhtur, S. Y. Cho and K.-H. Kim. Modified unscented Kalman filter for a multirate INS/GPS integrated navigation system, *ETRI Journal*, vol. 35, no. 5, pp. 943–946, 2013.

[39] A. Fatehi and B. Huang, Kalman filtering approach to multi-rate information fusion in the presence of irregular sampling rate and variable measurement delay, *Journal of Process Control*, vol. 53, pp. 15–25, 2017.

[40] A. Fatehi and B. Huang, State estimation and fusion in the presence of integrated measurement, *IEEE Transactions on Instrumentation and Measurement*, vol. 66, no. 9, pp. 2490–2499, 2017.

[41] X. Feng, C. Wen and J. H. Park, Sequential fusion H_∞ filtering for multi-rate multi-sensor time-varying systems–A Krein-space approach, *IET Control Theory & Applications*, vol. 11, no. 3, pp. 369–381, 2017.

[42] E. Fornasini and G. Marchesini, State-space realization theory of two dimensional filters, *IEEE Transactions on Automatic Control*, vol. 21, no. 4, pp. 484–492, 1976.

[43] E. Fridman, A. Seuret and J.-P. Richard, Robust sampled-data stabilization of linear systems: An input delay approach, *Automatica*, vol. 40, no. 8, pp. 1441–1446, 2004.

[44] T. Fukuda and T. Shibata, Theory and applications of neural networks for industrial control systems, *IEEE Transactions on Industrial Electronics*, vol. 39, no. 6, pp. 472–489, 1992.

[45] Z. Gao, Fault estimation and fault-tolerant control for discrete-time dynamic systems, *IEEE Transactions on Industrial Electronics*, vol. 62, no. 6, pp. 3874–3884, 2015.

[46] H. Geng, Y. Liang and Y. Cheng, Target state and Markovian jump ionospheric height bias estimation for OTHR tracking systems, *IEEE Transactions on Systems, Man, and Cybernetics: Systems*, vol. 50, no. 7, pp. 2599–2611, 2020.

[47] H. Geng, Y. Liang and X. Zhang, Linear-minimum-mean-square-error observer for multi-rate sensor fusion with missing measurements, *IET Control Theory & Applications*, vol. 8, no. 14, pp. 1375–1383, 2014.

[48] M. Ghosal and V. Rao, Fusion of multirate measurements for nonlinear dynamic state estimation of the power systems, *IEEE Transactions on Smart Grid*, vol. 10, no. 1, pp. 216–226, 2019.

[49] E. Gilbert, Capacity of a burst-noise channel, *Bell System Technical Journal*, vol. 39, no. 5, pp. 1253–1265, 1960.

[50] A. Gopalakrishnan, N. S. Kaisare and S. Narasimhan, Incorporating delayed and infrequent measurements in Extended Kalman Filter based nonlinear state estimation, *Journal of Process Control*, vol. 21, no. 1, pp. 119–129, 2011.

[51] X. Gu, T. Jia and Y. Niu, Consensus tracking for multi-agent systems subject to channel fading: A sliding mode control method, *International Journal of Systems Science*, vol. 51, no. 14, pp. 2703–2711, 2020.

[52] Y. Gu, Y. Chou, J. Liu and Y. Ji, Moving horizon estimation for multirate systems with time-varying time-delays, *Journal of the Franklin Institute*, vol. 356, no. 4, pp. 2325–2345, 2019.

[53] X. Guan and C. Chen, Delay-dependent guaranteed cost control for T-S fuzzy systems with time delays, *IEEE Transactions on Fuzzy Systems*, vol. 12, no. 2, pp. 236–249, 2004.

[54] X. Guan, J. Hu, J. Qi, D. Chen, F. Zhang and G. Yang, Observer-based H_∞ sliding mode control for networked systems subject to communication channel fading and randomly varying nonlinearities, *Neurocomputing*, vol. 437, pp. 312–324, 2021.

[55] Y. Guo and B. Huang, State estimation incorporating infrequent, delayed and integral measurements, *Automatica*, vol. 58, pp. 32–38, 2015.

[56] C. Han, H. Zhang and M. Fu, Optimal filtering for networked systems with Markovian communication delays, *Automatica*, vol. 49, no. 10, pp. 3097–3104, 2013.

[57] F. Han, Z. Wang, H. Dong and H. Liu, Partial-nodes-based scalable H_∞-consensus filtering with censored measurements over sensor networks, *IEEE Transactions on Systems, Man and Cybernetics: Systems*, vol. 51, no. 3, pp. 1892–1903, 2021.

[58] D. Haßkerl, M. Arshad, R. Hashemi, S. Subramanian and S. Engell, Simulation study of the particle filter and the EKF for state estimation of a large-scale DAE-system with multi-rate sampling, *IFAC-PapersOnLine*, vol. 49, no. 7, pp. 490–495, 2016.

[59] N. Hou, H. Dong, Z. Wang, W. Ren and F. E. Alsaadi, Non-fragile state estimation for discrete Markovian jumping neural networks, *Neurocomputing*, vol. 179, pp. 238–245, 2016.

[60] G. Hu, B. Gao, Y. Zhong and C. Gu, Unscented Kalman filter with process noise covariance estimation for vehicular ins/gps integration system, *Information Fusion*, vol. 64, pp. 194–204, 2020.

[61] J. Hu, G.-P. Liu, H. Zhang and H. Liu, On state estimation for nonlinear dynamical networks with random sensor delays and coupling strength under event-based communication mechanism, *Information Sciences*, vol. 511, pp. 265–283, 2020.

[62] J. Hu, Z. Wang and H. Gao, Joint state and fault estimation for time-varying nonlinear systems with randomly occurring faults and sensor saturations, *Automatica*, vol. 97, pp. 150–160, 2018.

[63] J. Hu, Z. Wang, H. Gao and L. K. Stergioulas, Extended Kalman filtering with stochastic nonlinearities and multiple missing measurements, *Automatica*, vol. 48, no. 9, pp. 2007–2015, 2012.

[64] J. Hu, Z. Wang, B. Shen and H. Gao, Quantised recursive filtering for a class of nonlinear systems with multiplicative noises and missing measurements, *International Journal of Control*, vol. 86, no. 4, pp. 650–663, 2013.

[65] J. Hu, H. Zhang, H. Liu and X. Yu, A survey on sliding mode control for networked control systems, *International Journal of Systems Science*, vol. 52, no. 6, pp. 1129–1147, 2021.

[66] J. Hu, H. Zhang, X. Yu, H. Liu and D. Chen, Design of sliding-mode-based control for nonlinear systems with mixed-delays and packet losses under uncertain missing probability, *IEEE Transactions on Systems, Man, and Cybernetics: Systems*, vol. 51, no. 5, pp. 3217–3228, 2021.

[67] Y. Hu, Z. Duan and D. H. Zhou, Estimation fusion with general asynchronous multi-rate sensors, *IEEE Transactions on Aerospace and Electronic Systems*, vol. 46, no. 4, pp. 2090–2102, 2010.

[68] H. Huang, T. Huang and X. Chen, Guaranteed H_∞ performance state estimation of delayed static neural networks, *IEEE Transactions on Circuits and Systems II: Express Briefs*, vol. 60, no. 6, pp. 371–375, 2013.

[69] H. Huang, T. Huang and X. Chen, Further result on guaranteed H_∞ performance state estimation of delayed static neural networks, *IEEE Transactions on Neural Networks and Learning Systems*, vol. 26, no. 6, pp. 1335–1341, 2014.

[70] K. Huang, S. Wu, F. Li, C. Yang and W. Gui, Fault diagnosis of hydraulic systems based on deep learning model with multirate data samples, *IEEE Transactions on Neural Networks and Learning Systems*, vol. 33, no. 11, pp. 6789–6801, 2022.

[71] M. B. Ignagni, Separate bias Kalman estimator with bias state noise, *IEEE Transactions on Automatic Control*, vol. 35, no. 3, pp. 338–341, 1990.

[72] L. Jiang, L. P. Yan, Y. Q. Xia and M. Y. Fu, Optimal distributed fusion for state estimation in multirate wireless sensor networks, in *processdings of the 34th Chinese Control Conference*, Hangzhou, China, Jul. 28-30, 2015.

[73] Y. Ju, Y. Liu, X. He and B. Zhang, Finite-horizon H_∞ filtering and fault isolation for a class of time-varying systems with sensor saturation, *International Journal of Systems Science*, vol. 52, no. 2, pp. 321–333, 2021.

[74] Y. Ju, G. Wei, D. Ding and S. Liu, A novel fault detection method under weighted try-once-discard scheduling over sensor networks, *IEEE Transactions on Control of Network Systems*, vol. 7, no. 3, pp. 1489–1499, 2020.

[75] T. Kaczorek, *Two-Dimensional Linear Systems*, Springer-Verlag: Berlin, 1985.

[76] L. Ke, Y. Zhang, B. Yang, Z. Luo and Z. Liu, Fault diagnosis with synchrosqueezing transform and optimized deep convolutional neural network: An application in modular multilevel converters, *Neurocomputing*, vol. 430, pp. 24–33, 2021.

[77] B. Khaleghi, A. Khamis, F. O. Karray and S. N. Razavi, Multisensor data fusion: A review of the state-of-the-art, *Information Fusion*, vol. 14, no. 1, pp. 28–44, 2013.

[78] M. Kordestani, M. Dehghani, B. Moshiri and M. Saif, A new fusion estimation method for multi-rate multi-sensor systems with missing measurements, *IEEE Access*, vol. 8, pp. 47522–47532, 2020.

[79] M. V. Kulikova and G. Y. Kulikov, SVD-based factored-form cubature Kalman filtering for continuous-time stochastic systems with discrete measurements, *Automatica*, vol. 120, art. no. 109110, 2020.

[80] D. Li, J. Liang, F. Wang and X. Ren, Observer-based H_∞ control of two-dimensional delayed networks under the random access protocol, *Neurocomputing*, vol. 401, pp. 353–363, 2020.

[81] H. Li, M. Lyu and B. Du, Event-based multi-objective filtering for multi-rate time-varying systems with random sensor saturation, *Kybernetika*, vol. 56, no. 1, pp. 81–106, 2020.

[82] J. Li, H. Dong, Z. Wang and X. Bu, Partial-neurons-based passivity-guaranteed state estimation for neural networks with randomly occurring time-delays, *IEEE Transactions on Neural Networks and Learning Systems*, vol. 31, no. 9, pp. 3747–3753, 2020.

[83] J. Li, G. Wei, D. Ding and E. Tian, Protocol-based H_∞ filtering for piecewise linear systems: A measurement-dependent equivalent reduction approach, *International Journal of Robust and Nonlinear Control*, vol. 31, no. 8, pp. 3163–3178, 2021.

[84] J.-Y. Li, B. Zhang, R. Lu and Y. Xu, Robust distributed H_∞ state estimation for stochastic periodic systems over constraint sensor networks, *IEEE Transactions on Systems, Man, and Cybernetics: Systems*, vol. 50, no. 11, pp. 4396–4407, 2020.

[85] J.-Y. Li, B. Zhang, R. Lu, Y. Xu and T. Huang, Distributed H_∞ state estimator design for time-delay periodic systems over scheduling sensor networks, *IEEE Transactions on Cybernetics*, vol. 51, no. 1, pp. 462–472, 2021.

[86] L. Li, T. Wang, Y. Xia and N. Zhou, Trajectory tracking control for wheeled mobile robots based on nonlinear disturbance observer with extended Kalman filter, *Journal of the Franklin Institute*, vol. 357, no. 13, pp. 8491–8507, 2020.

[87] Q. Li, J. Liang and H. Qu, H_∞ estimation for stochastic semi-Markovian switching CVNNs with missing measurements and mode-dependent delays, *Neural Networks*, vol. 141, pp. 281–293, 2021.

[88] Q. Li, B. Shen, Y. Liu and T. Huang, Event-triggered H_∞ state estimation for discrete-time neural networks with mixed time delays and sensor saturations, *Neural Computing and Applications*, vol. 28, no. 12, pp. 3815–3825, 2017.

[89] Q. Li, B. Shen, Z. Wang and F. E. Alsaadi, An event-triggered approach to distributed H_∞ state estimation for state-saturated systems with randomly occurring mixed delays, *Journal of the Franklin Institute*, vol. 355, no. 6, pp. 3104–3121, 2018.

[90] W. Li, S. L. Shah and D. Xiao, Kalman filters in non-uniformly sampled multirate systems: For FDI and beyond, *Automatica*, vol. 44, no. 1, pp. 199–208, 2008.

[91] W. Li, G. Wei, F. Han and Y. Liu, Weighted average consensus-based unscented Kalman filtering, *IEEE Transactions on Cybernetics*, vol. 46, no. 2, pp. 558–567, 2015.

[92] X. Li, F. Han, N. Hou, H. Dong and H. Liu, Set-membership filtering for piecewise linear systems with censored measurements under Round-Robin protocol, *International Journal of Systems Science*, vol. 51, no. 9, pp. 1578–1588, 2020.

[93] Y. Liang, T. Chen and Q. Pan, Multi-rate stochastic H_∞ filtering for networked multi-sensor fusion, *Automatica*, vol. 46, no. 2, pp. 437–444, 2010.

[94] C.-H. Lien, W.-C. Cheng, C.-H. Tsai and K.-W. Yu, Non-fragile observer-based controls of linear system via LMI approach, *Chaos Solitons and Fractals*, vol. 32, no. 4, pp. 1530–1537, 2007.

[95] H. Lin and S. Sun, An overview of multirate multisensor systems: Modelling and estimation, *Information Fusion*, vol. 52, pp. 335–343, 2019.

[96] A. Liu, W.-A. Zhang, M. Z. Q. Chen and L. Yu, Moving horizon estimation for mobile robots with multirate sampling, *IEEE Transactions on Industrial Electronics*, vol. 64, no. 2, pp. 1457–1467, 2017.

[97] D. Liu, Z. Wang, Y. Liu, F. E. Alsaadi and F. E. Alsaadi, Recursive state estimation for stochastic complex networks under Round-Robin communication protocol: Handling packet disorders, *IEEE Transactions on Network Science and Engineering*, vol. 8, no. 3, pp. 2455–2468, 2021.

[98] K. Liu, E. Fridman, K. H. Johansson and Y. Xia, Quantized control under round-robin communication protocol, *IEEE Transactions on Industrial Electronics*, vol. 63, no. 7, pp. 4461–4471, 2016.

[99] L. Liu, L. Ma, J. Guo, J. Zhang and Y. Bo, Distributed set-membership filtering for time-varying systems: A coding-decoding-based approach, *Automatica*, vol. 129, art. no. 109684, 2021.

[100] L. Liu, L. Ma, J. Zhang and Y. Bo, Distributed non-fragile set-membership filtering for nonlinear systems under fading channels and bias injection attacks, *International Journal of Systems Science*, vol. 52, no. 6, pp. 1192–1205, 2021.

[101] Q. Liu, Z. Wang, Q.-L. Han and C. Jiang, Quadratic estimation for discrete time-varying non-Gaussian systems with multiplicative noises and quantization effects, *Automatica*, vol. 113, art. no. 108714, 2020.

[102] Q. Liu, Z. Wang, J. Zhang and D. H. Zhou, Necessary and sufficient conditions for fault diagnosability of linear open- and closed-loop stochastic systems under sensor and actuator faults, *IEEE Transactions on Automatic Control*, vol. 67, no. 8, pp. 4178–4185, 2022.

[103] S. Liu, Z. Wang, L. Wang and G. Wei, On quantized H_∞ filtering for multi-rate systems under stochastic communication protocols: The finite-horizon case, *Information Sciences*, vol. 459, pp. 211–223, 2018.

[104] S. Liu, Z. Wang, G. Wei and M. Li, Distributed set-membership filtering for multirate systems under the Round-Robin scheduling over sensor networks, *IEEE Transactions on Cybernetics*, vol. 50, no. 5, pp. 1910–1920, 2020.

[105] Y. Liu, B. Shen and H. Shu, Finite-time resilient H_∞ state estimation for discrete-time delayed neural networks under dynamic event-triggered mechanism, *Neural Networks*, vol. 121, pp. 356–365, 2020.

[106] Y. Liu, Z. Wang, L. Ma and F. E. Alsaadi, Robust H_∞ control for a class of uncertain nonlinear systems with mixed time-delays, *Journal of the Franklin Institute*, vol. 355, no. 14, pp. 6339–6352, 2018.

[107] G. V. Lomonossoff, *Introduction to railway mechanism*, Oxford University Press, London, 1993.

[108] R. López-Negrete and L. T. Biegler, A moving horizon estimator for processes with multi-rate measurements: A nonlinear programming sensitivity approach, *Journal of Process Control*, vol. 22, no. 4, pp. 677–688, 2012.

[109] M. S. Mahmoud and M. F. Emzir, State estimation with asynchronous multi-rate multi-smart sensors, *Information Sciences*, vol. 196, pp. 15–27, 2012.

[110] J. Mao, D. Ding, Y. Song, Y. Liu and F. E. Alsaadi, Event-based recursive filtering for time-delayed stochastic nonlinear systems with missing measurements, *Signal Process*, vol. 134, pp. 158–165, 2017.

[111] J. Mao, Y. Sun, X. Yi, H. Liu and D. Ding, Recursive filtering of networked nonlinear systems: A survey, *International Journal of Systems Science*, vol. 52, no. 6, pp. 1110–1128, 2021.

[112] W. Marszalek, Two-dimensional state-space discrete models for hyperbolic partial differential equations, *Applied Mathematical Modelling*, vol. 8, no. 1, pp. 11–14, 1984.

[113] D. G. Meyer, A new class of shift-varying operators, their shift-invariant equivalents, and multirate digital systems, *IEEE Transactions on Automatic Control*, vol. 35, no. 4, pp. 429–433, 1990.

[114] F. Milano and M. Anghel, Impact of time delays on power system stability, *IEEE Transactions on Circuits and Systems I: Regular Papers*, vol. 59, no. 4, pp. 889–900, 2012.

[115] J. Misra and I. Saha, Artificial neural networks in hardware: A survey of two decades of progress, *Neurocomputing*, vol. 74, pp. 239–255, 2010.

[116] M. Moarref and L. Rodrigues, Observer design for linear models of multi-rate asynchronous aerospace systems, *Aerospace Systems*, vol. 3, pp. 127–137, 2020.

[117] I. Moreira, C. Pimentel, F. P. Barros and D. P. B. Chaves, Modeling fading channels with binary erasure finite-state Markov channels, *IEEE Transactions on Vehicular Technology*, vol. 66, pp. 4429–4434, 2017.

[118] M. Mosalanejad and M. M. Arefi, UKF-based soft sensor design for joint estimation of chemical processes with multi-sensor information fusion and infrequent measurements, *IET Science Measurement & Technology*, vol. 12, no. 6, pp. 755–763, 2018.

[119] Y. Niu, L. Sheng, M. Gao and D. Zhou, Dynamic event-triggered state estimation for continuous-time polynomial nonlinear systems with external disturbances, *IEEE Transactions on Industrial Informatics*, vol. 17, no. 6, pp. 3962–3970, 2020.

[120] L. Orihuela, S. Roshany-Yamchi, R. A. García and P. Millán, Distributed set-membership observers for interconnected multi-rate systems, *Automatica*, vol. 85, pp. 221–226, 2017.

[121] H. Park, Outlier-resistant high-dimensional regression modelling based on distribution-free outlier detection and tuning parameter selection, *Journal of Statistical Computation and Simulation*, vol. 87, no. 9, pp. 1799–1812, 2017.

[122] F. Peng and S. Sun, Distributed fusion estimation for multisensor multirate systems with stochastic observation multiplicative noises, *Mathematical Problems in Engineering*, vol. 2014, art. no. 373270, 2014.

[123] W. Qian, Y. Li, Y. Chen and W. Liu, L_2-L_∞ filtering for stochastic delayed systems with randomly occurring nonlinearities and sensor saturation, *International Journal of Systems Science*, vol. 51, no. 13, pp. 2360–2377, 2020.

[124] W. Qian, Y. Li, Y. Zhao and Y. Chen, New optimal method for L_2-L_∞ state estimation of delayed neural networks, *Neurocomputing*, vol. 415, pp. 258–265, 2020.

[125] B. Qu, N. Li, Y. Liu and F. E. Alsaadi, Estimation for power quality disturbances with multiplicative noises and correlated noises: A recursive estimation approach, *International Journal of Systems Science*, vol. 51, no. 7, pp. 1200–1217, 2020.

[126] B. Qu, B. Shen, Y. Shen and Q. Li, Dynamic state estimation for islanded microgrids with multiple fading measurements, *Neurocomputing*, vol. 406, pp. 196–203, 2020.

[127] B. Qu, Z. Wang and B. Shen, Fusion estimation for a class of multi-rate power systems with randomly occurring SCADA measurement delays, *Automatica*, vol. 125, art. no. 109408, 2021.

[128] R. Roesser, A discrete state-space model for linear image processing, *IEEE Transactions on Automatic Control*, vol. 20, no. 1, pp. 1–10, 1975.

[129] E. Rogers, K. Galkowski and D. H. Owens, Two decades of research on linear repetitive processes Part I: Theory, In *Proceedings of the 8th Internaitonal Workshop on Multidimensional Systems*, pp. 27–32, Erlangen, Germany, 9–11 Sept. 2013.

[130] S. Safari, F. Shabani and D. Simon, Multirate multisensor data fusion for linear systems using Kalman filters and a neural network, *Aerospace Science and Technology*, vol. 39, pp. 465–471, 2014.

[131] Y. Salehi and B. Huang, Offline and online parameter learning for switching multirate processes with varying delays and integrated measurements, *IEEE Transactions on Industrial Electronics*, vol. 69, no. 7, pp. 7213–7222, 2021.

[132] J. Salt, J. Alcaina, Á. Cuenca and A. Baños, Multirate control strategies for avoiding sample losses. Application to UGV path tracking, *ISA Transactions*, vol. 101, pp. 130–146, 2020.

[133] M. Sathishkumar, R. Sakthivel, C. Wang, B. Kaviarasan and S. M. Anthoni, Non-fragile filtering for singular Markovian jump systems with missing measurements, *Signal Processing*, vol. 142, pp. 125–136, 2018.

[134] B. Shen, Z. Wang and Y. S. Hung, Distributed H_∞-consensus filtering in sensor networks with multiple missing measurements: The finite-horizon case, *Automatica*, vol. 46, no. 10, pp. 1682–1688, 2010.

[135] B. Shen, Z. Wang and X. Liu, Sampled-data synchronization control of complex dynamical networks with stochastic sampling, *IEEE Transactions on Automatic Control*, vol. 57, no. 10, pp. 2644–2650, 2012.

[136] B. Shen, Z. Wang, D. Wang, J. Luo, H. Pu and Y. Peng, Finite-horizon filtering for a class of nonlinear time-delayed systems with an energy harvesting sensor, *Automatica*, vol. 100, pp. 144–152, 2019.

[137] J. Sheng, T. Chen and S. L. Shah, Optimal filtering for multirate systems, *IEEE Transactions on Circuits and Systems II: Express Briefs*, vol. 52, no. 4, pp. 228–232, 2005.

[138] R. G. Shenoy, D. Burnside and T. W. Parks, Linear periodic systems and multirate filter design, *IEEE Transactions on Signal Processing*, vol. 42, no. 9, pp. 2242–2256, 1994.

[139] Y. Shi, E. Tian, S. Shen and X. Zhao, Adaptive memory-event-triggered H_∞ control for network-based T-S fuzzy systems with asynchronous premise constraints, *IET Control Theory & Applications*, vol. 15, no. 4, pp. 534–544, 2021.

[140] A. Smyth and M. Wu, Multi-rate Kalman filtering for the data fusion of displacement and acceleration response measurements in dynamic system monitoring, *Mechanical Systems and Signal Processing*, vol. 21, no. 2, pp. 706–723, 2007.

[141] H. Song, L. Yu and W.-A. Zhang, Multi-sensor-based H_∞ estimation in heterogeneous sensor networks with stochastic competitive transmission and random sensor failures, *IET Control Theory & Applications*, vol. 8, no. 3, pp. 202–210, 2014.

[142] J. Song, D. Ding, H. Liu and X. Wang, Non-fragile distributed state estimation over sensor networks subject to DoS attacks: The almost sure stability, *International Journal of Systems Science*, vol. 51, no. 6, pp. 1119–1132, 2020.

[143] Q. Song, Y. Chen, Z. Zhao, Y. Liu and F. E. Alsaadi, Robust stability of fractional-order quaternion-valued neural networks with neutral delays and parameter uncertainties, *Neurocomputing*, vol. 420, pp. 70–81, 2021.

[144] W. Song, J. Wang, S. Zhao and J. Shan, Event-triggered cooperative unscented Kalman filtering and its application in multi-UAV systems, *Automatica*, vol. 105, pp. 264–273, 2019.

[145] W. Song, Z. Wang, J. Wang, F. E. Alsaadi and J. Shan, Distributed auxiliary particle filtering with diffusion strategy for target tracking: A dynamic event-triggered approach, *IEEE Transactions on Signal Processing*, vol. 69, pp. 328–340, 2021.

[146] D. Spinello, Asymptotic agreement in a class of networked Kalman filters with intermittent stochastic communications, *IEEE Transactions on Automatic Control*, vol. 61, no. 4, pp. 1093–1098, 2016.

[147] X. Su, F. Xia, Y.-D. Song, M. V. Basin and L. Zhao, L_2-L_∞ output feedback controller design for fuzzy systems over switching parameters, *IEEE Transactions on Fuzzy Systems*, vol. 26, no. 6, pp. 3755–3769, 2018.

[148] J. Suo, N. Li and Q. Li, Event-triggered H_∞ state estimation for discrete-time delayed switched stochastic neural networks with persistent dwell-time switching regularities and sensor saturations, *Neurocomputing*, vol. 455, pp. 297–307, 2021.

[149] M. Tabbara and D. Nesic, Input-output stability of networked control systems with stochastic protocols and channels, *IEEE Transactions on Automatic Control*, vol. 53, no. 5, pp. 1160–1175, 2008.

[150] H. Tan, B. Shen, Y. Liu, A. Alsaedi and B. Ahmad, Event-triggered multi-rate fusion estimation for uncertain system with stochastic nonlinearities and colored measurement noises, *Information Fusion*, vol. 36, pp. 313–320, 2017.

[151] H. Tan, B. Shen, K. Peng and H. Liu, Robust recursive filtering for uncertain stochastic systems with amplify-and-forward relays, *International Journal of Systems Science*, vol. 51, no. 7, pp. 1188–1199, 2020.

[152] X. Tian, G. Wei, L. Wang and J. Zhou, Wireless-sensor-network-based target localization: A semidefinite relaxation approach with adaptive threshold correction, *Neurocomputing*, vol. 405, pp. 229–238, 2020.

[153] J.-A. Ting, E. Theodorou and S. Schaal, A Kalman filter for robust outlier detection, in: *2007 IEEE/RSJ International Conference on Intelligent Robots and Systems*, pp. 1520–1525, San Diego, USA, 29 Oct-02 Nov. 2007.

[154] S. Tripathi, M. Gehlot, J. K. Hussain, G. Mishra, V. Kumar and S. Chouksey, Field integral measurement of a six period undulator in a pulsed wire set up, *Optics Communications*, vol. 284, no. 1, pp. 350–357, 2011.

[155] D. Wang, B. Shen, Z. Wang, F. E. Alsaadi and A. M. Dobaie, Reliable H_∞ filtering for stochastic spatial-temporal systems with sensor saturations and failures, *IET Control Theory and Applications*, vol. 9, no. 17, pp. 2590–2597, 2015.

[156] F. Wang, Z. Wang, J. Liang and X. Liu, Resilient filtering for linear time-varying repetitive processes under uniform quantizations and Round-Robin protocols, *IEEE Transactions on Circuits and Systems I-Regular Papers*, vol. 65, no. 9, pp. 2992–3004, 2018.

[157] F. Wang, Z. Wang, J. Liang and X. Liu, Recursive state estimation for two-dimensional shift-varying systems with random parameter perturbation and dynamical bias, *Automatica*, vol. 112, art. no. 108658, 2020.

[158] H. Wang, H. Li, J. Fang and H. Wang, Robust Gaussian Kalman filter with outlier detection, *IEEE Signal Processing Letters*, vol. 25, no. 8, pp. 1236–1240, 2018.

[159] J. Wang, Y. Alipouri and B. Huang, Dual neural extended Kalman filtering approach for multirate sensor data fusion, *IEEE Transactions on Instrumentation and Measurement*, vol. 70, art. no. 6502109, 2021.

[160] L. Wang, Z. Wang, G. Wei and F. E. Alsaadi, Finite-time state estimation for recurrent delayed neural networks with component-based event-triggering protocol, *IEEE Transactions on Neural Networks and Learning Systems*, vol. 29, no. 4, pp. 1046–1057, 2018.

[161] L. Wang, Z. Wang, G. Wei and F. E. Alsaadi, Variance-constrained H_∞ state estimation for time-varying multi-rate systems with redundant channels: The finite-horizon case, *Information Sciences*, vol. 501, pp. 222–235, 2019.

[162] Z. Wang and H. Shang, Kalman filter based fault detection for two-dimensional systems, *Journal of Process Control*, vol. 28, pp. 83–94, 2015.

[163] Z. Wang, B. Shen and X. Liu, H_∞ filtering with randomly occurring sensor saturations and missing measurements, *Automatica*, vol. 48, no. 3, pp. 556–562, 2012.

[164] Z. Wang, G. Wei and G. Feng, Reliable H_∞ control for discrete-time piecewise linear systems with infinite distributed delays, *Automatica*, vol. 45, no. 12, pp. 2991–2994, 2009.

[165] Z. Wang, F. Yang, D. W. C. Ho and X. Liu, Robust H_∞ filtering for stochastic time-delay systems with missing measurements, *IEEE Transactions on Signal Processing*, vol. 54, no. 7, pp. 2579–2587, 2006.

[166] G. Wei, W. Li, D. Ding and Y. Liu, Stability analysis of covariance intersection-based Kalman consensus filtering for time-varying systems, *IEEE Transactions on Systems, Man, and Cybernetics: Systems*, vol. 50, no. 11, pp. 4611–4622, 2020.

[167] G. Wei, L. Liu, L. Wang and D. Ding, Event-triggered control for discrete-time systems with unknown nonlinearities: An interval observer-based approach, *International Journal of Systems Science*, vol. 51, no. 6, pp. 1019–1031, 2020.

[168] S. Wen, T. Huang, X. Yu, M. Z. Q. Chen and Z. Zeng, Aperiodic sampled-data sliding-mode control of fuzzy systems with communication delays via the event-triggered method, *IEEE Transactions on Fuzzy Systems*, vol. 24, no. 5, pp. 1048–1057, 2016.

[169] T. Wen, L. Zou, J. Liang and C. Roberts, Recursive filtering for communication-based train control systems with packet dropouts, *Neurocomputing*, vol. 275, pp. 948–957, 2018.

[170] A. Willig, Polling-based MAC protocols for improving real-time performance in a wireless PROFIBUS, *IEEE Transactions on Industrial Electronics*, vol. 50, no. 4, pp. 806–817, 2003.

[171] A. Willig, M. Kubisch, C. Hoene and A. Wolisz, Measurements of a wireless link in an industrial environment using an IEEE 802.11-compliant physical layer, *IEEE Transactions on Industrial Electronics*, vol. 49, no. 6, pp. 1265–1282, 2002.

[172] L. Wu, D. W. C. Ho and C. W. Li, Sliding mode control of switched hybrid systems with stochastic perturbation, *Systems & Control Letters*, vol. 60, no. 8, pp. 531–539, 2011.

[173] Q. Wu, Q. Song, Z. Zhao, Y. Liu and F. E. Alsaadi, Stabilization of T-S fuzzy fractional rectangular descriptor time-delay system, *International Journal of Systems Science*, vol. 52, no. 11, pp. 2268–2282, 2021.

[174] N. Xu, Y. Chen, A. Xue and Q. Yu, A new approach to stability analysis of switched systems composed of two subsystems with average dwell time switching, *International Journal of Systems Science*, vol. 51, no. 10, pp. 1862–1872, 2020.

[175] W. Xu, D. W. C. Ho, J. Zhong and B. Chen, Event/self-triggered control for leader-following consensus over unreliable network with DoS attacks, *IEEE Transactions on Neural Networks and Learning Systems*, vol. 30, no. 10, pp. 3137–3149, 2019.

[176] W. Xu, S. Yang and J. Cao, Fully distributed self-triggered control for second-order consensus of multiagent systems, *IEEE Transactions on Systems, Man and Cybernetics: Systems*, vol. 51, no. 6, pp. 3541–3551, 2021.

[177] Y. Xu, R. Lu, P. Shi, J. Tao and S. Xie, Robust estimation for neural networks with randomly occurring distributed delays and Markovian jump coupling, *IEEE Transactions on Neural Networks and Learning Systems*, vol. 29, no. 4, pp. 845–855, 2018.

[178] L. P. Yan, B. S. Liu and D. H. Zhou, The modeling and estimation of asynchronous multirate multisensor dynamic systems, *Aerospace Science and Technology*, vol. 10, no. 1, pp. 63–71, 2006.

[179] L. P. Yan, B. S. Liu and D. H. Zhou, Asynchronous multirate multisensor information fusion algorithm, *IEEE Transactions on Aerospace and Electronic Systems*, vol. 43, no. 3, pp. 1135–1146, 2007.

[180] L. P. Yan, B. Xiao, Y. Q. Xia and M. Y. Fu, State estimation for asynchronous multirate multisensor nonlinear dynamic systems with missing measurements, *International Journal of Adaptive Control and Signal Processing*, vol. 26, no. 6, pp. 516–529, 2012.

[181] L. P. Yan, D. H. Zhou, M. Y. Fu and Y. Q. Xia, State estimation for asynchronous multirate multisensor dynamic systems with missing measurements, *IET Signal Processing*, vol. 4, no. 6, pp. 728–739, 2010.

[182] C. Yang, Z. Yang and Z. Deng, Robust weighted state fusion Kalman estimators for networked systems with mixed uncertainties, *Information Fusion*, vol. 45, pp. 246–265, 2019.

[183] X. Yin and J. Liu, Event-triggered state estimation of linear systems using moving horizon estimation, *IEEE Transactions on Control Systems Technology*, vol. 29, no. 2, pp. 901–909, 2020.

[184] A. Yogarathinam, J. Kaur and N. R. Chaudhuri, Modeling adequacy for studying power oscillation damping in grids with wind farms and networked control systems (NCS), in *2016 IEEE Power and Energy Society General Meeting*, Boston, USA, 17–21 July 2016.

[185] B. Yu, Y. Shi and H. N. Huang, l_2-l_∞ filtering for multirate systems based on lifted models, *Circuits Systems and Signal Processing*, vol. 27, no. 5, pp. 699–711, 2008.

[186] Y. Yuan, Y. Wang and L. Guo, Sliding-mode-observer-based time-varying formation tracking for multi-spacecrafts subjected to switching topologies and time-delays, *IEEE Transactions on Automatic Control*, vol. 66, no. 8, pp. 3848–3855, 2021.

[187] J. Zhang, X. He and D. Zhou, Filtering for stochastic uncertain systems with non-logarithmic sensor resolution, *Automatica*, vol. 89, pp. 194–200, 2018.

[188] J. Zhang, Q. Zhang, X. He, G. Sun and D. Zhou, Compound-fault diagnosis of rotating machinery: A fused imbalance learning method, *IEEE Transactions on Control Systems Technology*, vol. 29, no. 4, pp. 1462–1474, 2021.

[189] L. Zhang, Y. Zhu and W. X. Zheng, Energy-to-peak state estimation for Markov jump RNNs with time-varying delays via nonsynchronous filter with nonstationary mode transitions, *IEEE Transactions on Neural Networks and Learning Systems*, vol. 26, no. 10, pp. 2346–2356, 2015.

[190] P. Zhang, Y. Yuan and L. Guo, Fault-tolerant optimal control for discrete-time nonlinear system subjected to input saturation: A dynamic event-triggered approach, *IEEE Transactions on Cybernetics*, vol. 51, no. 6, pp. 2956–2968, 2021.

[191] W.-A. Zhang, B. Chen and M. Z. Q. Chen, Hierarchical fusion estimation for clustered asynchronous sensor networks, *IEEE Transactions on Automatic Control*, vol. 61, no. 10, pp. 3064–3069, 2016.

[192] W-A. Zhang, H. Dong, G. Guo and L. Yu, Distributed sampled-data H_∞ filtering for sensor networks with nonuniform sampling periods, *IEEE Transactions on Industrial Informatics*, vol. 10, no. 2, pp. 871–881, 2014.

[193] W.-A. Zhang, G. Feng and L. Yu, Multi-rate distributed fusion estimation for sensor networks with packet losses, *Automatica*, vol. 48, no. 9, pp. 2016–2028, 2012.

[194] X.-M. Zhang, Q.-L. Han and X. Ge, An overview of neuronal state estimation of neural networks with time-varying delays, *Information Sciences*, vol. 478, pp. 83–99, 2019.

[195] X.-M. Zhang, Q.-L. Han, X. Ge, D. Ding, L. Ding, D. Yue and C. Peng, Networked control systems: A survey of trends and techniques, *IEEE/CAA Journal of Automatica Sinica*, vol. 7, no. 1, pp. 1–17, 2020.

[196] X.-M. Zhang, Q.-L. Han and X. Yu, Survey on recent advances in networked control systems, *IEEE Transactions on Industrial Informatics*, vol. 12, no. 5, pp. 1740–1752, 2016.

[197] Y. Zhang, Z. Wang and F. E. Alsaadi, Detection of intermittent faults for nonuniformly sampled multi-rate systems with dynamic quantisation and missing measurements, *International Journal of Control*, vol. 93, no. 4, pp. 898–909, 2020.

[198] Y. Zhang, Z. Wang and L. Ma, Variance-constrained state estimation for networked multi-rate systems with measurement quantization and probabilistic sensor failures, *International Journal of Robust and Nonlinear Control*, vol. 26, no. 16, pp. 3507–3523, 2016.

[199] Y. Zhang, Z. Wang, L. Zou and H. Fang, Event-based finite-time filtering for multirate systems with fading measurements, *IEEE Transactions on Aerospace and Electronic Systems*, vol. 53, no. 3, pp. 1431–1441, 2017.

[200] Y. Zhang, Z. Wang, L. Zou and Z. Liu, Fault detection filter design for networked multi-rate systems with fading measurements and randomly occurring faults, *IET Control Theory & Applications*, vol. 10, no. 5, pp. 573–581, 2016.

[201] Z. Zhang, Y. Niu, Z. Cao and J. Song, Security sliding mode control of interval type-2 fuzzy systems subject to cyber attacks: The stochastic communication protocol case, *IEEE Transactions on Fuzzy Systems*, vol. 29, no. 2, pp. 240–251, 2020.

[202] Z. Zhang, Y. Niu and H.-K. Lam, Sliding-mode control of T-S fuzzy systems under weighted Try-Once-Discard protocol, *IEEE Transactions on Cybernetics*, vol. 50, no. 12, pp. 4972–4982, 2020.

[203] Z. Zhang, S.-F. Su and Y. Niu, Dynamic event-triggered control for interval type-2 fuzzy systems under fading channel, *IEEE Transactions on Cybernetics*, vol. 51, no. 11, pp. 5342–5351, 2021.

[204] Y. Zhao, X. He, J. Zhang, H. Ji, D. Zhou and M. G. Pecht, Detection of intermittent faults based on an optimally weighted moving average T^2 control chart with stationary observations, *Automatica*, vol. 123, art. no. 109298, 2021.

[205] Z. Zhao, Z. Wang, L. Zou and J. Guo, Set-membership filtering for time-varying complex networks with uniform quantisations over randomly delayed redundant channels, *International Journal of Systems Science*, vol. 51, no. 16, pp. 3364–3377, 2020.

[206] X. Zheng, H. Zhang, Z. Wang and H. Yan, Almost surely state estimation for multi-rate networked systems under random and malicious packet losses, *Journal of the Franklin Institute*, vol. 356, no. 17, pp. 10593–10607, 2019.

[207] Y. Zheng, W. Zhou, W. Yang, L. Liu, Y. Liu and Y. Zhang, Multivariate/minor fault diagnosis with severity level based on Bayesian decision theory and multidimensional RBC, *Journal of Process Control*, vol. 101, pp. 68–77, 2021.

[208] X. Zhou, Y. Chen, Q. Wang, K. Zhang and A. Xue, Event-triggered finite-time H_∞ control of networked state-saturated switched systems, *International Journal of Systems Science*, vol. 51, no. 10, pp. 1744–1758, 2020.

[209] K. Zhu, J. Hu, Y. Liu, N. D. Alotaibi and F. E. Alsaadi, On ℓ_2-ℓ_∞ output-feedback control scheduled by stochastic communication protocol for Two-dimensional switched systems, *International Journal of Systems Science*, vol. 52, no. 14, pp. 2961–2976, 2021.

[210] M. Zhu, Y. Chen, Y. Kong, C. Chen and J. Bai, Distributed filtering for Markov jump systems with randomly occurring one-sided lipschitz nonlinearities under Round-Robin scheduling, *Neurocomputing*, vol. 417, pp. 396–405, 2020.

[211] L. Zou, Z. Wang, H. Dong and Q.-L. Han, Moving horizon estimation with multirate measurements and correlated noises, *International Journal of Robust and Nonlinear Control*, vol. 30, no. 17, pp. 7429–7445, 2020.

[212] L. Zou, Z. Wang and H. Gao, Observer-based H_∞ control of networked systems with stochastic communication protocol: The finite-horizon case, *Automatica*, vol. 63, no. 1, pp. 366–373, 2016.

[213] L. Zou, Z. Wang and H. Gao, Set-membership filtering for time-varying systems with mixed time-delays under Round-Robin and Weighted Try-Once-Discard protocols, *Automatica*, vol. 74, pp. 341–348, 2016.

[214] L. Zou, Z. Wang, H. Gao and F. E. Alsaadi, Finite-horizon H_∞ consensus control of time-varying multiagent systems with stochastic communication protocol, *IEEE Transactions on Cybernetics*, vol. 47, no. 8, pp. 1830–1840, 2017.

[215] L. Zou, Z. Wang, H. Gao and X. Liu, State estimation for discrete-time dynamical networks with time-varying delays and stochastic disturbances under the Round-Robin protocol, *IEEE Transactions on Neural Networks and Learning Systems*, vol. 28, no. 5, pp. 1139–1151, 2017.

[216] L. Zou, T. Wen, Z. Wang, L. Chen and C. Roberts, State estimation for communication-based train control systems with CSMA protocol, *IEEE Transactions on Intelligent Transportation Systems*, vol. 20, no. 3, pp. 843–854, 2019.

Index

www.ingramcontent.com/pod-product-compliance
Lightning Source LLC
Chambersburg PA
CBHW060355220326
41598CB00023B/2931
9781032619491